ORGANIC HORTICULTURE
Principles, Practices and Technologies

H.P. Singh
George V. Thomas

WESTVILLE

WESTVILLE PUBLISHING HOUSE
NEW DELHI

H.P. Singh and George V. Thomas (2010) Organic Horticulture — Principles, Practices and Technologies pp 440, Westville Publishing House, New Delhi

First Edition 2010, Reprint 2014, 2017, 2025

ISBN 978-81-85873-61-9

Published by Mrinal Goel, Westville Publishing House
47, B-5, Paschim Vihar, New Delhi – 110 063
Tel: 011-25284742, Fax: 011-25267469
Email: westville_2002@yahoo.co.in
Printed in New Delhi

Foreword

Indian horticulture has made rapid strides during the last decades and the growth achieved has been highly appreciable. Plan investment made for the development of horticulture has been rewarding in terms of increased production, productivity, availability and export of horticultural produce. Resultantly, India has emerged as 2nd largest producer of fruits and vegetables, and continues to be the largest exporter of spices and cashewnut. India has also provided leadership role in many of the commodities like banana, cashew, spices and many vegetable crops. However, due to increase in health consciousness, there has been increasing demand for produce grown without the use of chemicals, referred as "organic produce".

The organic production system largely relies on farming system of growing crops without the use of chemical and cares for soil health. Indian Council of Agricultural Research (ICAR) has initiated several programmes for the organic production using organics and microbes. Soil amendments based on microbes, vermicompost and bio-intensive pest management have been developed. Solarization alone, takes care of many soil borne pathogens. Thus, it is becoming apparent that organic farming would be feasible but, it cannot be adopted across the country and crops. However, we may have to proceed in phased manner so that production output is not affected adversely. Since adoption of organic farming is assuming significance, and area is increasing national dialogue is warranted to assess the needs and impact. The major challenges in the organic farming are to develop the package which do not sacrifice the yield gains and utilizes natural resources fruitfully. In this context, organization of National Conference on Organic Farming in Horticultural Crops, with special emphasis to Plantation Crops at Central Plantation Crops Research Institute, Kasaragod, in October, 2008 was timely, which provided opportunity for all the stakeholders to deliberate upon various issues. The presentations and discussions have been documented in a form of a book entitled, "Organic Horticulture — Principles, Practices and Technologies". I compliment Dr H.P. Singh, DDG (Horticulture) and Dr George V. Thomas, Director, CPCRI, Kasaragod for their efforts in compiling and editing this book.

The book contains valuable information on the status of organic horticulture — national and international scenario, principles, practices, prospects of organic farming, technological advancements made in organic farming of fruits, plantation crops, tuber crops, vegetables and spices, input management and biomass recycling comparison of organic and conventional agriculture, soil health management, plant health management, issues, quality control and certification. Additional information provided on the certification process, accreditation agancies, certification agencies and sample certification form will help to familiarize the farming community with the certification process and the agencies involved. I am sure, that this book will be of immense use to all those interested in organic horticulture.

5th March, 2010
New Delhi

S. Ayyappan
Secretary, Department of Agricultural Research & Education and
Director General, Indian Council of Agricultural Research
Ministry of Agriculture, Krishi Bhawan, New Delhi

Preface

India is endowed with a wide range of climatic conditions, which provide opportunities for growing large number of crops. Cultivation of crops without chemicals and fertilizer has been practiced for long, adopting the technique relevant to the regions like crop rotation, use of fallow and green manuring and management of pests with indigenous system. However, to enhance the production, and to meet the needs of growing population, intensive production systems using chemicals and fertilizers were adopted. This had dramatic impact on enhancement of production and productivity, but this gain was not without cost. Land degradation, decline in factor productivity and above all health hazards were observed, which attracted the attention calling for integrated management for sustainable production. Current production systems are more dependent on non-renewable energy based fertilizers and pesticides. With changing scenario, understanding the impact of this system of cultivation, which is detrimental to the sustainability, has become inevitable. Thus, there is a growing interest to have system of cultivation which ensures soil health, human health, animal health and environmental health which is becoming essential with changing preference for nutritious and safe food. Consequently, organically grown foods are becoming popular among the consumers. Resultantly, market for organic food is expanding at the rate of 12–15% annually. In the past two decades, standards have been set for the production, which is a process certification and not the product certification, and needs supervision for the adoption of organic farming. Conversion period is 2-3 years and various regulations have to be followed. The certification has to ensure maintenance of soil health and a chemical free produce for better human health.

Horticulture, which includes fruits, vegetables, flowers, tuber crops including potato, spices, medicinal and aromatic plants and plantation crops contributes 30.7 per cent to GDP of agriculture and the past efforts have succeeded in higher growth rate with increased production and productivity. Since, many of horticultural crops especially fruits, vegetables, plantation crops are eaten afresh and organic production to meet consumer's needs is becoming imperative. In order to adopt the organic production, it becomes essential that technologies are made available for growing of crops without the use of the chemicals and fertilizers. Many technologies have been generated but, there continue to be a considerable gap between the needs and availability. In this context, it is essential that the available information is brought together and emerging issues are deliberated to make organic farming a reality. Accordingly, a National Conference on "Organic Farming in Horticultural Crops with Special Reference to Plantation Crops" was held at Central Plantation Crops Research Institute, Kasaragod with the theme of promoting organic horticulture for safe food and health environment.

This book based on the papers presented in the conference has eight sections, viz. current scenario and principles, technological advances, input management, organic *vs.* inorganic, soil health management, plant health management, issues, quality control and certification. Advancements have been made in developing technologies designed to increase soil organic matter, biological activity and nutrient availability through the use of renewable resources, conservation of energy, soil, water and environmental maintenance and enhancement, while producing optimum quantities of produce without the use of artificial fertilizers or synthetic chemicals. Besides, a chapter on standards and certification is added to provide information on principles, basic standards of production, documentation, inspection and certification procedures as approved by National Standards Committee. Information on certification procedure, contact details of approved accreditation and certification agencies and model application forms are included in the annexure for the benefit of the farmers taking up organic farming.

The editors are grateful to all the contributors to the book for providing latest information on diverse aspects of organic horticulture and the papers based on the experimental data generated from the research programmes undertaken in different institutions. We are grateful to Dr S. Ayyappan, Director General, ICAR for his encouragement and inspiration. Editorial assistance provided by Dr C. Palaniswami and Dr V. Krishnakumar, Senior Scientists, CPCRI is gratefully acknowledged. We are sure that this book will prove to be a valuable source of information to all those involved in organic horticulture including scientists, developmental personnel, policy makers, NGOs and farmers. It is hoped that this comprehensive treatise will stimulate and motivate more intensified research, accelerated developmental efforts, favourable policy initiatives and spread of organic farming at the grass roots for the production of safe food under healthy environmental conditions. Once again we thank all those directly or indirectly associated with compilation of the information.

H.P. Singh
George V. Thomas

About the Editors

Born on 2nd July, 1950, Dr H.P. Singh has served horticulture in various capacities namely, Project Coordinator (Fruits); Director, NRC for Banana; Horticulture Commissioner, Govt. of India; Chairman, Coconut Development Board; Dean (Agriculture), GBPUA&T, Pantnagar; Vice Chancellor, Rajendra Agricultural University, Pusa and now Deputy Director General (Horticulture), ICAR.

He has also served international organisations like the Chairman of APCC, Jakarta, National Director for internationally-aided projects and Chairperson of Steering Committee of BAPNET of Bioversity International and implemented various international programmes. Besides, excellent academic record, Dr Singh has contributed to research and agricultural education which has earned him three international, twenty one national awards and five fellowships including the Fellow of NAAS. He has developed many cultivars and technologies which has shaped the horticulture. Recognising his immense contribution, OUAT, Bhubaneshwar, has conferred D.Sc. (*Honoris causa*) last year.

Dr Singh is the architect of Mission mode approach for horticultural development, starting from planning to execution which has been due to his insight. He has provided leadership to horticulture and is credited to have authored and edited 39 books and 30 bulletins, besides 175 research papers. He has widely travelled in India and abroad. His efforts have provided a strong foundation for horticultural development, leading to ***Golden Revolution*** in the country.

Born on 17th August, 1952, Dr. George V. Thomas is currently Director, Central Plantation Crops Research Institute (CPCRI), Kasaragod, Kerala. He joined Agricultural Research Service of ICAR as Scientist S-1 at CPCRI in 1978 and has contributed immensely to plantation crops research during the last three decades. He served as President, Indian Society for Plantation Crops (2007-2009) and Editor, Journal of Plantation Crops (2001-2006) and

Fellow of India Society for Plantation Crops. He is recipient of ICAR Outstanding Team is Research Award for Horticulture (2005-06) and Dr C.S. Venkataram Memorial Award. His main areas of specilisation are bioresource management in production system of plantation crops, organic farming, recycling of crop residues, microbial activity and functions in cropping systems and biofertilizers. He has to his credit more than 250 publications including research papers, reviews, book chapters and technical bulletins.

Contents

Contributors

Abraham Amith
Rubber Research Institute of India
Kottayam – 686 009
Kerala

Ajitha Mohan T.V.
Rubber Research Institute of India
Kottayam – 686 009
Kerala

Ambily K.K.
Rubber Research Institute of India
Kottayam – 686 009
Kerala

Anithakumari P.
Krishi Vigyan Kendra
Central Plantation Crops Research Institute
Regional Station, Krishnapuram P.O.
Kayamkulam – 690533
Kerala

Annamma George
College of Agriculture
Vellayani – 695522
Kerala

Annie Philip
Rubber Research Institute of India
Kottayam
Kerala – 686 009

Anoop K.
Indian Institute of Spices Research
Marikunnu (P.O.)
Kozhikkode – 673 012
Kerala

Arulraj S.
Project Co-ordinator (Palms)
Central Plantation Crops Research Institute
Kasaragod – 671 124, Kerala

Arunkumar V.K.
Rubber Research Institute of India
Kottayam – 686 009
Kerala

Basavaraju T.B.
AICRP on Palms
ARSUAS
Arsikere – 573103
Karnataka

Bavappa K.V. Ahamed
Formerly Chief Technical Adviser
UNDP/FAO Projects
Kappur (P.O) – 679 552
Kerala

Bhat M.G.
National Research Centre for Cashew
Puttur – 574 202
DK, Karnataka

Boraiah B.
AICRP on Palms, Agricultural Research Station
University of Agricultural Sciences
Arsikere – 573103
Karnataka

Chandramohanan R.
Central Plantation Crops Research Institute
Kasaragod – 671 124
Kerala

Chandrika Mohan
CPCRI
Regional Station, Kayangulam
P.O. Krishnapuram – 690 533
Kerala

Dash D.K.
Department of Horticulture
Orissa University of Agriculture and Technology
Bhubaneswar
Orissa

Desai V.S.
Regional Coconut Research Station, Bhatye
Ratnagiri Dr.B.S.Konkan Krishi Vidyapeeth
Dapoli
Maharashtra

Devarajan L.
Professor (Retd.)
P.N. Puder
Coimbatore – 641 003
Tamil Nadu

Dhanapal R.
Central Plantation Crops Research Institute
Kasaragod – 671124
Kerala

Dinesh R.
Indian Institute of Spices Research
PB No. 1701, Marikunnu PO
Calicut – 673 012
Kerala

Ganeshamurthy A.N.
Indian Institute of horticultural Research
Hessaraghatta Lake – PO
Bangalore – 560 089
Karnataka

Geetha Jose
Rubber Research Institute of India
Kottayam – 686 009
Kerala

Gupta Alka
Central Plantation Crops Research Institute
Kasaragod – 671124
Kerala

Haldandar P.M.
Regional Coconut Research Station
Bhatye, Ratnagiri
Dr. B.S. Konkan Krishi Vidyapeeth
Dapoli
Maharashtra

Hamza S.
Indian Institute of Spices Research
Calicut – 673 012
Kerala

Hanumanthappa M.
Institute of Horticultural Research
Hessaraghatta Lake P.O.
Bangalore – 560 089
Karnataka

Hebbar S.S.
Indian Institute of Horticultural Research
Hessaraghatta Lake P.O.
Bangalore – 560 089
Karnataka

Hegde Hemant G.
Programme Co-ordinator
Krishi Vigyan Kendra
Sirsi – 581 401
Uttara Kannada
Karnataka

Jacob J.
Cropping Systems Research Centre
Kerala Agricultural University
Karamana
Thiruvananthapuram – 695 522
Kerala

Jacob Mathew
Rubber Research Institute of India
Kottayam – 686 009
Kerala

Jadhav B.B.
Regional Coconut Research Station
Bhatye
Ratnagiri
Dr.B.S. Konkan Krishi Vidyapeeth
Dapoli
Maharashtra

Jeeva M.L.
Central Tuber Crops Research Institute
Srrekaryam
Thiruvananthapuram – 695 107
Kerala

Jessy M.D.
Rubber Research Institute of India
Kottayam – 686 009
Kerala

Jiji T.
Department of Agricultural Entomology
College of Agriculture, Vellayani
Thiruvanthapuram – 695 552
Kerala

Joycy Rodrigues
Department of Biotechnology
St. Aloysius College
Mangalore – 575 003
Karnataka

Kochu Babu M.
Directorate of Oil Palm Research
Pedavegi
Andhra Pradesh

Kochuthresiamma Joseph
Rubber Research Institute of India
Kottayam – 686 009
Kerala

Krishnakumar V.
Central Plantation Crops Research Institute
Kasaragod – 671 124, Kerala

Kumar A.
Indian Institute of Spices Research
Calicut – 673 012, Kerala

Kuruvilla Jacob C.
Rubber Research Institute of India
Kottayam – 686 009, Kerala

Lokesh M.S.
Agricultural Research Station (Pepper)
Sirsi – 581 401
Uttara Kannada
Karnataka

Mahendran P.
UPASI Tea Research Foundation Regional Centre
Coonoor – 643 101
The Nilgiris

Maheswarappa H.P.
Central Plantation Crops Research Institute
Kasaragod – 671 124
Kerala

Malhotra S.K.
KAB–II, Indian Council of Agricultural Research
New Delhi – 110 012

Mathew M.
Rubber Research Institute of India
Kottayam – 686 009
Kerala

Mathew Sebastian
INDOCERT
Thottumugham P.O.
Aluva – 683 105
Kerala

Meena B.
Coconut Research Station
Tamil Nadu Agricultural University
Aliyarnagar – 642 101
Tamil Nadu

Misra R.S.
Central Tuber Crops Research Institute
Sreekariyam
Thiruvananthapuram – 17, Kerala

More T.A.
Central Institute for Arid Horticulture
Bikaner – 334 006
Rajasthan

Murali Gopal
Central Plantation Crops Research Institute
Kasaragod – 671 124
Kerala

Muralidharan K
Central Plantation Crops Research Institute
Kasaragod – 671 124
Kerala

Nagwekar D.D.
Regional Coconut Research Station
Bhatye
Ratnagiri
Maharashtra

Nair A.N.S.
Rubber Research Institute of India
Kottayam – 686 009
Kerala

Nair C.P.R.
Central Plantation Crops Researh Institute
Regional Station
Kayangulam P.O.
Krishnapuram – 690 533
Kerala

Nandakumar C.
College of Agriculture
Vėllayani – 695522
Kerala

Naseema A.
Department of Agricultural Entomology
College of Agriculture
Vellayani
Thiruvanthapuram – 695 552
Kerala

Nayak M.G.
National Research Centre for Cashew
Puttur – 574 202
DK, Karnataka

Palaniswami C.
Central Plantation Crops Research Institute
Kasaragod – 671124
Kerala

Palanna K.B.
ARSUAS
Arsikere – 573103
Karnataka

Panneerselvam P.
Indian Institute of Horticultural Research
Hessaraghatta Lake – PO
Bangalore – 560 089
Karnataka

Parashurama Chandravanshi
Agricultural Research Station
Kankanady – 575 002
Mangalore, Karnataka

Parthasarathy V.A.
Indian Institute of Spices Research
PB No. 1701, Marikunnu P.O.
Calicut – 673 012, Kerala

Poduval Mini
Regional Research Station
Bidhan Chandra Krishi Viswavidyalaya Jhargram
Paschim Medinipur
West Bengal

Poornima R.
Central Tuber Crops Research Institute
Srrekaryam
Thiruvananthapuram – 695 107
Kerala

Prabhakar B.S.
Division of Vegetable Science
Indian Institute of Horticultural Research
Hessaaraghatta Lake P.O.
Bangalore – 560 089
Karnataka

Prabhakar M.
Indian Institute of Horticultural Research
Hessaraghatta Lake P.O.
Bangalore – 560 089
Karnataka

Prasannakumari P.
Rubber Research Institute of India
Kottayam – 686 009
Kerala

Prathapan K.
College of Agriculture
Vellayani
Thiruvananthapuram – 695 522
Kerala

Praveena R.
College of Agriculture
Vellayani
Thiruvananthapuram – 695 522, Kerala

Priya Nair
Central Tuber Crops Research Institute
Sreekariyam
Thiruvanthapuram – 695 552
Kerala

Pushpakumari R.
College of Agriculture
Vellayani – 695522, Kerala

Radhakrishnan B.
UPASI Tea Research Foundation
Regional Centre
Glenview
Coonoor – 643 101
Tamil Nadu

Rajamanickam K.
Coconut Research Station
Tamil Nadu Agricultural University
Aliyarnagar – 642 101
Tamil Nadu

Rajan P.
Central Plantation Crops Researh Institute
Regional Station
Kayangulam P.O.
Krishnapuram – 690 533
Kerala

Ram R.A.
Central Institute for Sub-tropical Horticulture
Lucknow – 227 017
Uttar Pradesh

Ramesh P.R.
Division of Soil Science and Agricultural Chemistry
Indian Institute of Horticultural Research
Hessaraghatta Lake P.O.
Bangalore – 560 089, Karnataka

Rangwala A.D.
Regional Coconut Research Station, Bhatye
Ratnagiri, Dr. B.S. Konkan Krishi Vidyapeeth
Dapoli, Maharashtra

Raviprasad T.N.
National Research Centre for Cashew
Puttur (D.K.) – 574 202
Karnataka

Ray D.P.
Department of Horticulture
Orissa University of Agriculture and Technology
Bhubaneswar
Orissa

Rayudu B.T.
Krishi Vigyan Kendra
Central Plantation Crops Research Institute
Kasaragod – 671124
Kerala

Reddy B.M.C.
Central Institute for Subtropical Horticulture
Lucknow – 227 017
Uttar Pradesh

Reddy D.V. Srinivasa
ICAR Zonal Project Directorate
Zone 8, Hebbal
MRS HA-Farm Post
Bangalore – 560 024
Karnataka

Roshini Susan Elias
Rubber Research Institute of India
Kottayam – 686 009
Kerala

Sahu S.C.
Department of Horticulture
Orissa University of Agriculture and Technology
Bhubaneswar
Orissa

Samiyappan R.
Centre for Plant Protection Studies
Tamil Nadu Agricultural University
Coimbatore – 641 003
Tamil Nadu

Sangeeth K.P.
Indian Institute of Spices Research
Calicut – 673 012
Kerala

Sankar Meti
Rubber Research Institute of India
Kottayam – 686 009
Kerala

Sankaranarayanan P.
UPASI Tea Research Foundation, Regional Centre
Glenview
Coonoor – 643 101
Tamil Nadu

Saritha N.S.
Cropping Systems Research Centre
Kerala Agricultural University
Karamana
Thiruvanthapuram – 695 522
Kerala

Sasidhar R.
UPASI Tea Research Foundation
Regional Centre, Glenview
Coonoor – 643 101
Tamil Nadu

Sasidharan Nair A.N.
Rubber Research Institute of India
Kottayam – 686 009
Kerala

Satisha G.C.
Directorate of Oil Palm Research
Pedavegi – 534 450
Andhra Pradesh

Sawant V.S.
Regional Coconut Research Station
Bhatye, Ratnagiri
Dr.B.S.Konkan Krishi Vidyapeeth
Dapoli
Maharashtra

Selvamani V.
Central Plantation Crops Research Institute
Kasaragod – 671 124
Kerala

Shaji Philip
Rubber Research Institute of India
Kottayam – 686 009
Kerala

Shamsudeen M.
Indian Institute of Spices Research
Calicut – 673 012
Kerala

Shareef M.V.M.
National Research Centre for Oil Palm
Regional Station
Palode
Trivandrum
Kerala

Sharma B.D.
Central Institute for Arid Horticulture
Bikaner – 334 006
Rajasthan

Sheeba U.
Cropping Systems Research Centre
Kerala Agricultural University
Karamana
Thiruvanthapuram – 695 522
Kerala

Sheela K.R.
College of Agriculture
Vellayani – 695 522
Kerala

Sherin George
Rubber Research Institute of India
Kottayam – 686 009
Kerala

Shirmila J.
Cropping Systems Research Centre
Kerala Agricultural University
Karamana
Thiruvanthapuram – 695 522, Kerala

Shivarama Bhat P.
National Research Centre for Cashew
Puttur, D.K.
Karnataka – 574 202

Singh H.P.
Deputy Director General (Horticulture)
ICAR, KAB – II, Pusa Campus
New Delhi – 110 012

Sinoby V.
Cropping Systems Research Centre
Kerala Agricultural University
Karamana
Thiruvanthapuram – 695 522, Kerala

Sivakumar T.
Krishi Vigyan Kendra
Central Plantation Crops Research Institute
Regional Station
Krishnapuram. P. O. Kayamkulam – 690533
Kerala

Sosamma V.K.
Nematology Laboratory
Central Plantation Crops Research Institute
Regional Station
Kayangulam – 690 533
Kerala

Srinivasan V.
Indian Institute of Spices Research
PB No. 1701
Marikunnu P.O.
Calicut – 673 012
Kerala

Subramanian P.
Central Plantation Crops Research Institute
Kasaragod – 671 124
Kerala

Suja G.
Central Tuber Crops Research Institute
Sreekariyam
Thiruvantahapuram – 695 017
Kerala

Sundararaju D.
National Research Centre for Cashew
Puttur (D.K.) – 574 202
Karnataka

Sundaresan S.
Central Tuber Crops Research Institute
Sreekariyam
Thiruvantahapuram – 695 017
Kerala

Sunitha S.
National Research Centre for Oil Palm
Regional Station
Palode
Trivandrum
Kerala

Susan John K.
Central Tuber Crops Research Institute
Sreekariyam
Thiruvanthapuram – 17
Kerala

Suseela Bhai R.
Indian Institute of Spices Research
Marikkunnu
Calicut
Kerala

Syamala V.K.
Rubber Research Institute of India
Kottayam – 686 009
Kerala

Talashikar S.C.
Regional Coconut Research Station, Bhatye
Ratnagiri, Dr.B.S.Konkan Krishi Vidyapeeth
Dapoli
Maharashtra

Tapas Biswas
Regional Research Station
Bidhan Chandra Krishi Viswavidyalaya
Jhargram, Paschim Medinipur
West Bengal

Thamban C.
Central Plantation Crops Research Institute
Kasaragod – 671 124
Kerala

Thankamani C.K.
Indian Institute of Spices Research
Marikkunnu
Calicut – 673 012
Kerala

Thomas George V.
Central Plantation Crops Research Institute
Kasaragod – 671 124
Kerala

Thomas Mathew
Coconut Development Board
Kochi
Kerala

Thyagaraj N.E.
AICRP on Palms
ARSUAS
Arsikere – 573103
Karnataka

Ulaganathan A.
Rubber Research Institute of India
Kottayam – 686 009
Kerala

Usha Nair N.
Rubber Research Institute of India
Kottayam – 686 009
Kerala

Venugopal S.
P.G. Department of Botany
N.S.S. College
Pandalam
Pathanamthitta
Kerala

Vidhan Singh T.
Central Plantation Crops Research Institute
Kasaragod – 671 124
Kerala

Yadukumar N.
National Research Centre for Cashew
Puttur DK – 574 202
Karnataka

Zachariah T.J.
Indian Institute of Spices Research
Marikkunnu
Calicut – 673 012
Kerala

SECTION – I
CURRENT SCENARIO AND PRINCIPLES

1

Organic Horticulture – Current Trend, Prospects and Strategies

H.P. Singh

Organic horticulture is the oldest form of agriculture on the earth. Farming without the use of petroleum-based chemicals (fertilizers and pesticides) was the sole option for farmers until post Second World War. The war brought with it technologies that were useful for agricultural production. For example, ammonium nitrate used for ammunitions during Second World War evolved into ammonium nitrate fertilizer; organophosphate nerve gas production led to the development of powerful insecticides. Organic farming or biodynamic farming is a specialized form of diversified agriculture that not only aims to produce nutritious food of high quality but also leads to the generation of more income and employment opportunities for rural population.

Organic horticulture seeks to utilize the advances that consistently yield benefits (new varieties of crops; precision technologies; more efficient machinery). Instead of using synthetic fertilizers and pesticides, organic farmers utilize crop rotations, cover crops, and natural–based products to maintain or enhance soil fertility. Use of synthetic fertilizers, pesticides, growth hormones or regulators, etc. are prohibited under this system.

Organic horticulture is defined as practices that emphasize the use of renewable resources, conservation of energy, soil and water and environmental maintenance and enhancement, while producing optimum quantities of produce without the use of artificial fertilizer or synthetic chemicals. In the organic horticulture, utilization of solar radiation for soil sterilization, etc. are used and soil fertility is maintained through organic manure and symbiotic soil micro flora. Recognizing the impact of excessive use of chemical fertilizer on soil health and pesticides on human health, there has been a realization for integrated management system.

India has converted over 2.5 million ha including 1.1 million ha of cultivable land into organic (the world have over 31 million ha under certified organic production). As per recent report of International Federation of Organic Agriculture Movement (IFOAM), the total organically managed area is over 24 million ha worldwide and further growing. Organic farming is practiced in approximately 130 countries of the world. The distribution of Global organic land and farms by continents are given in Fig. 1.1 and organic area in the world in Table 1.1.

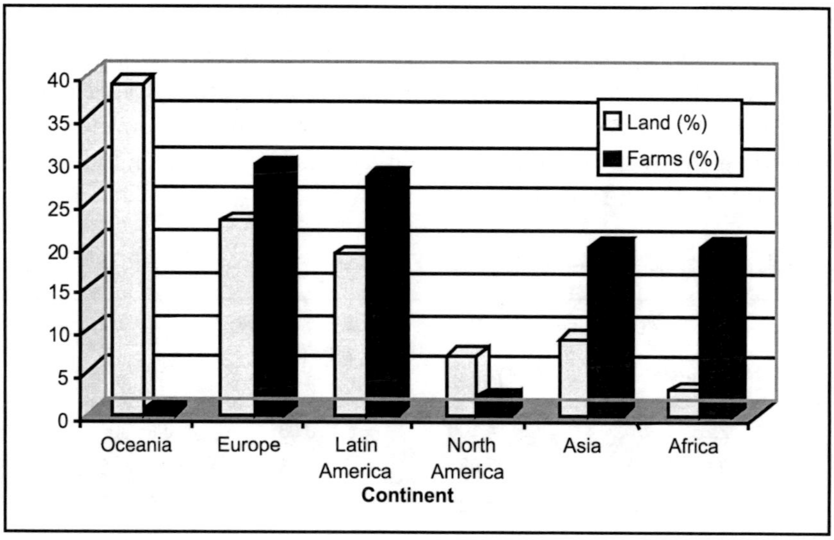

Fig. 1.1. Distribution of global organic land and farms by continent

Table 1.1. Organic area in the world (Continent–wise)

Continent	*Organic land (ha)*	*Share to total agricultural area (%)*	*Organic farms*
Africa	8,90,504	0.11	1,24,805
Asia	2,89,3572	0.21	1,29,927
Europe	69,20,462	1.38	1,87,697
Latin America	58,09,320	0.93	1,76,710
North America	21,99,225	0.56	12,063
Oceania	1,18,45,100	2.59	2,689
Total	2,05,58,183	0.74	6,33,891

Source: SOEL-FiBL Survey 2007

Table 1.2. Global market of organic products (2006-07)

Market	38.6 Billion $
Major markets	USA – 16 Billion $, Europe – 18 Billion $ Japan and others – 4-5 Billion $
Annual growth rate	20–30%
Market share	1% of total food sales, Indian share is only 0.02%

The trade by the Indian organic farming industry, which was estimated to be US$ 20 million in 2002, has grown nearly five times to about US$ 100 million by 2005 and it is expected to touch US$ 29 to 31 billions by 2007-08. The market for organic products is growing, not only in Europe and North America but also in many other countries. The demand for organic food is steadily increasing both in developed and developing countries, with annual average growth rate of 20–25%. Organic farming status in India in relation to the world status is given in Table 1.3.

Table 1.3. Organic status in India *vis-a-vis* world status

Year	*Organic area (ha)*		*% share to agricultural area (India)*	*No.of organic farms*	
	World	*India*		*World*	*India*
2003	2,28,11,267	4,100	0.03	3,98,804	5,661
2004	2,40,70,010	37,050	0.03	4,62,475	5,147
2005	2,64,58,270	76,326	0.05	5,58,449	5,147
2006	3,15,02,786	1,14,037	0.06	6,22,782	5,147

Source: SOEL-FiBL Survey, 2003–06

India ranks 33rd in terms of total land under organic and occupies 88th position for agricultural land under organic crops to total farming area. The total cultivated area under certification is 2.8 million ha (2007-08), which includes 1 million ha under cultivation and rest under forest area (wild cultivation).

Organic spices exported from India are black pepper, small cardamom, ginger, turmeric, cloves, etc; However, the quantity of export is negligible. There is also a scope for encouraging the export of organic fruits and flowers. Changing dietary habits among many segments of the population coupled with health consciousness has resulted in growing demand for organic food. India also encouraged the production and export of organic food by developing capacity and capability which exists more in horticultural crops. The details of Export of organic products from India from 2002-03–2006-07 are furnished in Fig. 1.2.

To take the benefit of organic farming in terms of higher value of produce, efforts have been made by Non-Government Organizations (NGOs) and individuals to get the

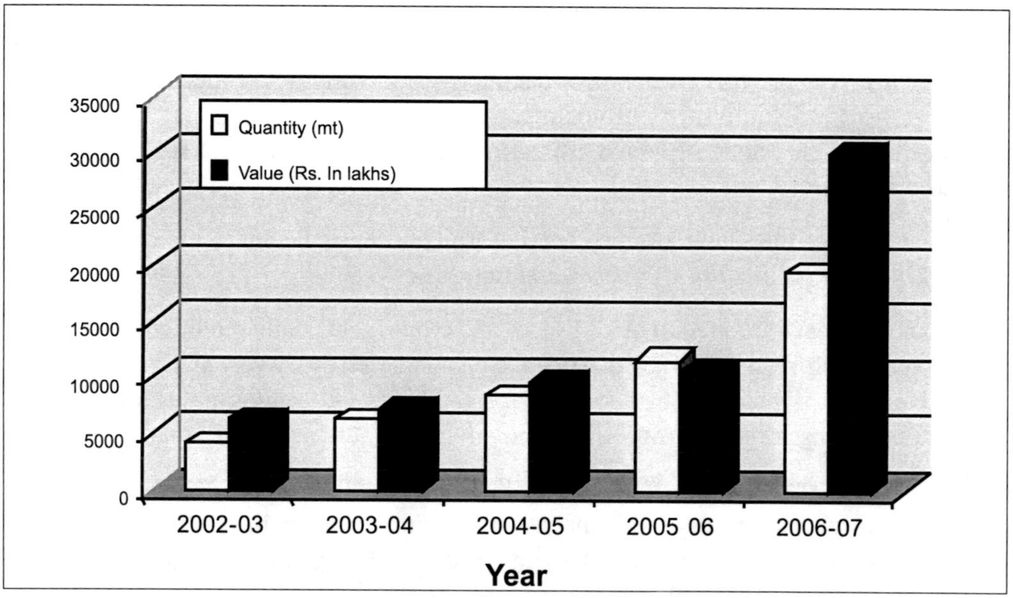

Fig. 1.2. Export of organic products from India

certification done through accredited agencies of International Federation of Organic Agriculture Movement (IFOAM) or other accredited agencies, operating in India. System developed in India for export of organically grown produce, is now recognized as a system for production for domestic market and export. The Government of India has established a 'National Centre of Organic Farming' at Ghaziabad to integrate all the efforts together to give a boost in organic production.

ORGANIC SEED

The term 'organic seed' means 'seed produced under an organic production system, ideally one that is certified'. This definition was proposed as it was considered that there was potential for 'organic seed' to also mean a cultivar that has been bred to perform optimally in organic systems. This differentiation is important because the long term desires of the organic movement are not just to have organic seed but to have organic cultivars that have been produced in a breeding system that is also organic, which could be called 'organic breeding'. Such systems would not only outlaw such practices as transgenic but also other current breeding and multiplication procedures that are incompatible with organic philosophy.

PROSPECTS FOR ORGANIC HORTICULTURE IN INDIA

There are more and more evidences that chemical based fertilizers, herbicides and pesticides are extremely harmful to our health. Conventional agriculture relies heavily on chemical fertilizers and toxic pesticides, which are harmful to human health, and which enter the food supply, penetrate water sources, harm the livestock, deplete the soil, devastate natural eco-systems, and contaminate the foods grown for the general public. In addition to which, there are over 7000 artificial and/or chemical preservatives and additives permitted in non-organic food including colourings, stabilizers, fillers, residual antibiotics, hydrogenated fats, etc., none of which are permitted in organic foods. Besides this other dangerous effects are:

- In many places trees have stopped bearing fruits because the heavy use of pesticides has killed the pollinators – the bees and butterflies.
- The most recent introduction of GM seeds and crops (genetically modified, also called GE, genetically engineered), which need to be purchased every year at a high cost, and need to be heavily sprayed with the killer pesticides that the same companies produce, evokes a strong response from health and environment conscious people all over the world.
- Through organic agriculture, farmers learn healthy, sustainable farming practices, and are regaining the wisdom of their agricultural heritage. Organic agriculture enriches the soil season by season. It increases the long-term yield, nutrient value and potency of their crops. It allows for a naturally clean water supply, and provides overall richness, health and well-being of their families, livestock, farmlands and communities.
- Many of India's farmers have lost much of their ancestral farming knowledge and ancient seed varieties since the introduction of chemical based commercial agriculture and hybrid seeds.

The genetic base of crops is very important and a reduction of genetic diversity, as happens in conventional agriculture, leads to the emergence of pests on a large scale. Farmers, in older times, apart from using the crop rotation methods to maintain the soil fertility also relied on the genetic means to increase crop production. The fertilizers also destroy soil organisms. They damage the rhizobia that fix nitrogen and other microorganisms that make phosphates available to plants.

The long-term effect of modern or conventional farming was damage of soil and damaged soil was easily eroded by wind and water. The eroding soil needed use of continuously increasing quantities of fertilizers, much of which was washed/leached into surface and underground water sources. Sustainable agriculture based on technologies that combine increased production with improved environmental protection has been accepted as absolutely essential for the maximization of the consumer welfare. The consumers are increasingly concerned about the quality of the products they consume and food safety has become a crucial requirement.

Organically cultivated soils are relatively better attuned to withstand water stress and nutrient loss. Their potential to counter soil degradation is high and several experiments in arid areas reveal that organic farming may help to combat desertification. It is reported that about 70 hectares of desert in Egypt could be converted into fertile soil supporting livestock through organic and biodynamic practices. India, which has some areas of semi-arid and arid nature, can benefit from the experiment. The organic agriculture movement in India received inspiration and assistance from IFOAM which has about 600 organizational members from 120 countries. All India Federation of Organic Farming (AIFOF) is a member of IFOAM and consists of a number of NGOs, farmers' organizations, promotional bodies and institutions.

Soil organic matter content is the direct measure of soil fertility. The enrichment of soil organic matter will improve the soil fertility along with physical and chemical properties in the long-run. Sustainability of soil health depends to a large extent on the maintenance of soil organic carbon, the oxidation of which is very rapid in the tropics, especially in the arid and semi-arid regions. Minimum tillage, cover cropping and mulching reduce oxidation of soil organic matter. Perennial horticultural crops have a definite advantage in this respect as they are amenable to these practices and also produce lot of biomass that can be recycled to replenish the soil organic carbon. For example, herbaceous perennials like banana is known to give 13.94 and 8.17 t dry matter/ ha. for cultivars 'Robusta' and 'Ney Poovan', respectively, whereas, 'Coorg Honey Dew' cv. of papaya yields 2.5 t dry matter/ha as recyclable crop residue. Woody perennials like mango, sapota, guava, etc., yield reasonable amounts of leaf litter even under rain fed conditions in the arid and semi-arid regions.

Since the soil under cultivation and perennial horticultural crops like plantation crops and fruit crops would have accumulated lot of organic matter, especially in the humid tropics, through leaf litter and cultivation methods followed earlier, these could be ideal for switching over to organic farming without any drastic yield reduction, in the initial phase. Also, the organically rich soils of north-eastern and south-western parts of India, where traditional farming without the use of chemicals is being largely followed, can be identified for growing organic fruits like pineapple, avocado, passion fruit, etc. and plantation crops like tea and cashew for export.

Covercropping, mulching and other practices of organic farming help to reduce soil moisture losses. Soil organic matter enhances the water holding capacity. This is of particular relevance to rain fed horticulture, particularly in the arid and semi-arid tracts. Reduction in energy costs for the manufacture of fertilizers and pesticides and in foreign exchange outgo for the import of base materials for their manufacture are added advantages. This aspect of organic horticulture is of great importance to Indian economy.

In India, 25.1 million tones of nutrients are removed from the soil annually, whereas, only 15.0 million tones are supplied annually from all sources including organic. The fertilizer production is less than the required amount with estimated gap of 3.2 and 6.4 million tones for 1966 and 2031, respectively. If these gaps were to be bridged through inorganic sources, lot of energy expenditure and foreign exchange out go would result.

Chemical fertilizer is applied only in 30% of the cultivated area, which is irrigated, and remaining land is under rain fed agriculture with almost no fertilizer application. Also the rain fed area under cultivation accounts for only 40% of the food-grain production of the country. The introduction of organic farming in this area will allay the fears of a sudden sharp decline of food production. The demand for biomass for the production of organic manures can also be controlled in a phased manner.

Horticultural produce with less chemical residues will be a boon for the public health. This is of special importance for fruits and vegetables, which are consumed either raw or mildly cooked. Enhanced shelf life of organically produced fruits and vegetables will go a long way in reducing the post-harvest losses of these commodities which at present is about 25–40%. Perennial fruit trees are amenable to inter-cropping with green manure crops. This would also facilitate optimum use of diminishing land resources especially in small and marginal holdings.

Promotion of organic recycling and farming systems through organic horticulture can improve rural employment generation. Since organic farming operates on 'on-farm system' requiring integration of farm and animal management and recycling of farm waste, there is scope for increased employment generation in the rural areas. Organic farming is considered to be the best among all because of its scientific approach and wider acceptance all over the world.

ORGANIC FARMING PRACTICES

***Crop and variety selection*:** The crops chosen to be grown for sale is an economic and/ or personal decision. The potential market should be determined even before deciding what crops to grow and the acreage of each. Market availability will determine to a great extent the crops grown, acreage, and the time of year best to be in production. Next comes the matching of soil type to crop and which variety to select. The ideal soils for growing vegetables are well-drained, fairly deep, and relatively high in organic matter. Deep sands may be too well-drained and very low in organic matter for vegetable production. Muck soils, when the water can be controlled, are ideally suited for growing celery, leafy vegetables, bulb and root crops. The better the job in matching crop and soil, the greater chance of producing a successful crop.

Which variety to grow? Organic vegetable growers have fewer choices. Disease and, if possible, insect resistance must be considered when selecting the variety. Variety trial results will provide information on adaptability of varieties under a range of environmental conditions. Pests and disease resistance is the most economical and effective means of their management. Yield, horticultural quality, and market acceptability also have to be considered in making your variety selections.

Cover crops can be beneficial for intensive organic vegetable production in a number of ways. Water penetration and infiltration can be improved by root growth of a cover crop and by returning organic matter to soils. Increased organic matter may improve the ability of soil to retain moisture. If leguminous cover crops are grown, soil nitrogen can be increased through nitrogen fixation. Grasses are particularly helpful in promoting soil structure and soil aggregate stability because of their fibrous root systems. Microbial activity, often stimulated by cover crop root exudates and organic matter additions to soils, has also been shown to promote aggregate stability. As microbes decompose organic matter, nutrients are released. Weed suppression for subsequent crops may be another benefit. Furthermore, cover crops can provide a favourable environment to attract and sustain beneficial arthropods.

***Plant nutrition requirements*:** Recent papers contrast the characteristics of old and modern varieties, or of varieties selected under low or high input conditions. A major difference lies in the ability of older varieties to thrive under low fertility conditions relative to modern varieties. Possible reasons for such differences are based on whether the older varieties are more efficient at scavenging for nutrients, or whether they are more efficient at using limited nutrients.

***Disease and pest resistance*:** Organic farmers have observed consistently that the potential for disease under organic conditions appears to be lower than under conventional conditions (though seed-borne disease is still a crucial issue). This may be due to the higher levels of nitrogen available in plant tissues grown conventionally, but the reasons may be more complex. It has been observed that under conventional conditions with high synthetic nitrogen inputs, the quantity of phenolic substances produced as secondary metabolites in barley leaves was less than that produced under low input conditions.

***Weed competition*:** Because of the ready availability of herbicides, the resistance of crop varieties to competition from weeds has been increasingly overlooked. However, it seems that this can be rectified by selection of lines that produce vigorous, fast-growing seedlings with prostrate habit. Under organic conditions, without fungicide protection, weeds frequently become diseased and this can serve first, to reduce their competition potential and second to generate spores and semi-chemicals that can induce resistance in neighbouring crop plants against potential threats.

NUTRIENT MANAGEMENT

Organic farming systems rely on the management of soil organic matter to enhance the chemical, biological and physical properties of the soil. One of the basic principles of soil fertility management in organic systems is that plant nutrition depends on 'biologically-derived nutrients' instead of using readily soluble forms of nutrients; less available forms of nutrients such as those in bulky organic materials are used. This requires release of nutrients to the plant via the activity of soil microbes and soil animals. In organic horticulture, nutrients are supplied through organic sources, i.e., FYM, compost, vermicompost, poultry manure, green manures, oil cakes, etc. In mango it has been found that supply of 40–80% N through organic sources increases the yield. Banana and papaya respond very well to organic culture.

A large number of fast growing nitrogen fixing crops like dhaincha (*Sesbania* sp.), sunhemp and cowpea may be used as green manure that can fix N to the extent of 60–100 kg/ha. Biofertilizers are very important components in organic agriculture which convert and fix unavailable sources of elemental N, bound phosphates and decomposed plant residues into available forms and at the same time produce growth promoting substances. The legume– *Rhizobium* association can fix N ranging from 40–120 kg/ha under optimum conditions. Vegetable crops in general give good response to *Azotobacter*, a free-living heterotrophic N fixing bacterium. P-solubilizers like *Aspergillus*, *P. striata*, *B. polymixa* and use of VAM are also found to be very effective in vegetable crops. In general, the application of 10 tonnes/ha of any locally available organic sources like compost, FYM, biogas slurry, press mud, carpet waste, vermicompost coupled with the inoculation of bio-fertilizers like *Rhizobium*, *Azotobacter/Azospirillum*, PSM and green manuring in alternate years can take care of the nutrient requirement for most of the vegetable crops.

Perennial horticultural crops produce lot of biomass that can be recycled to replenish the soil organic carbon. For example, herbaceous perennials like banana is known to produce 13.94 and 8.17 t dry matter/ha for cvs.Robusta and Ney Poovan respectively whereas papaya cv.Coorg Honey Dew yield 2.5 t dry matter/ha as a recyclable crop residue. Plantation crops produce large quantities of biomass viz. husk, dried leaves, coir dust, coffee husk, tea wastes, shedding , oil palm waste which can be effectively utilized as a source of organic manure to improve the productivity of the crops.

Livestock keeping at farms is an age old practice. Livestock play major role in organic agriculture as the intermediary between the utilization of crop residues or fodder produced at the farm and the return of nutrients as manure. Dairying in particular has helped number of small and marginal farmers to improve their income. Field survey revealed that marginal and small farmers, even in progressive states like Punjab, have helped to raise farm profitability as well as availability of cattle dung in sufficient amounts. Storage and application of their resources seldom attract proper attention of the farmers resulting in 40–60% losses in nutrients, especially N. Leaching of NO_3-N polluting the ground and surface water resources is usually observed from cattle dung pits. Organic farmers and farming methods take adequate care in minimizing these losses through adoption of technologies on composting, vermicomposting, etc. This not only improves the nutrients availability from organic sources but also prevents potential hazard of ground water pollution. The power of microorganisms in ameliorating such adversaries in soil should be harnessed.

Based on the principles of organic farming, certain inputs have been identified by IFOAM as permitted and some as restricted for the use in fertilization and soil conditioning. A clear-cut policy decision about the permitted and restricted inputs for fertilization and soil conditioning based on the availability and use under Indian perspective for organic agriculture is to be framed. Wealth of information is available about the effect of different organic sources on specific crop and feasibility of nutrient supplementation in soil system under organic farming. The important bottleneck to the success of organic agriculture lies more to the limited option for crop protection measures rather than soil fertilization.

IRRIGATION MANAGEMENT

The irrigation delivery method, number of applications, and total irrigation amount varies depending on specific crop requirements, soil type, the amount of water stored in the soil profile, and water availability. A crop's rooting depth and growing season greatly affect the water needs. For example, shallow rooted crops and crops grown over the summer months generally require more frequent irrigations than deeper rooted crops or crops grown in the winter months. Soils with high levels of organic matter may be better able to retain moisture and, therefore, need less amount of applied water. Furrow or sprinkler irrigation is most often used by organic vegetable growers in western counties. Surface and subsurface drip irrigation systems are also used by some growers; however, these systems require a larger capital investment initially.

PEST MANAGEMENT

Biological control is use of organisms to regulate a pest or pathogen below its economic threshold level. It assumes importance in sustainable agriculture and organic farming and production with reduced cost without chemical residues. However, it has several inherent disadvantages like the availability of natural enemies in sufficient numbers to utilize on a large-scale which can be addressed by encouraging commercial insectaries, which can supply quality natural enemies to farmers at a very short notice. The use of commercial nuclear polyhedrosis virus (NPV) is gaining importance all over the world. In India too, private industries are bringing out field compatible formulations. It has been found that using NPVs at early stage brings in excellent control of *Helicoverpa armigera* on tomato. NPV is further, compatible with *Trichogramma* egg parasitoids and pheromone traps. These, in turn would constitute an ideal Integrated Pest Management (IPM). One of the advantages of NPV is its specificity. It is necessary, therefore, to test specificity using restriction endo-nuclease analysis of viral DNA. Safe and sound technologies for IPM in several crop pest situations like tomato fruit-borer and mealy bugs of several fruit crops are available.

The use of botanicals such as neem seed kernel extract, neem and *Pongamia* powder and cakes are very effective in keeping ever such notorious pests, diamond back moth and *Helicoverpa armigera,* etc., well below the threshold. Pest management measures such as use of insect pathogen (Bt, NPV), pheromones, trap crops, natural enemies such as predators and parasitoids if integrated scientifically into the pest management can contribute very significantly towards organic (pesticide free) cultivation in horticultural crops. A number of insect/ disease/viruses resistant varieties have evolved based on intensive research, which laid the foundation for effective organic production of horticultural crops.

PROBLEMS IN ADOPTING ORGANIC FARMING

It is quite natural that a change in the system of agriculture in a country of more than a billion people should be a well thought out process, which requires utmost care and caution.

There may be several impediments on the way. The following are found to be the major problem areas for the growth of organic farming in the country.

***Lack of awareness*:** It is a fact that many farmers in the country have only vague ideas about organic farming. Use of bio-fertilizers and bio-pesticides requires awareness and willingness on the part of the farming community. Lack of awareness among farmers about scientific procedures of compost making and application leads to dispersal of pernicious weeds (*Cynadon, Cyprus, Oxalis* and *Parthenium*), insect pests (*Rhinocerous beetle* - a serious pest of coconut), and diseases (Soil borne pathogens- *Phytophthora, Fusarium, Rhizoctonia*) etc. The maximum they do is making a pit and filling it with small quantities of wastes. Often the pit is flooded with rainwater and result is that the top of the compost remains under composted, the bottom becomes like a hard cake. Proper training to the farmers will be necessary to make vermi-compost on the modern lines. In general, the organic matter is spread in the field during the months when the right moisture level is absent on the soil. The whole manure turns into wastes in the process. The required operation is of course, labour intensive and costly, but it is necessary to obtain the desired results.

***Output marketing problems*:** It is found that before the beginning of the cultivation of organic crops, their marketability and that too at a premium over the conventional produce has to be assured. Inability to obtain a premium price, at least during the period required to achieve the productivity levels of the conventional crop will be a setback. The cost of marketing of both types of products was also same and the buyers of wheat were not prepared to pay higher prices to the organic.

***Shortage of biomass*:** Many experts and well-informed farmers are not sure whether all the nutrients with the required quantities can be made available by the organic materials. Even if this problem can be surmounted, they are of the view that the available organic matter is not simply enough to meet the requirements. The crop residues useful to prepare vermicompost are removed after harvest from the farms and they are used as fodder and fuel. Even if some are left out on the farms termites, etc., destroy them. Experiments have shown that the crop residues ploughed back into soil will increase productivity and a better alternative is conversion into compost. The small and marginal cultivators have difficulties in getting the organic manures compared to the chemical fertilizers, which can be bought easily.

***Inadequate supporting infrastructure*:** In spite of the adoption of the National Programme of Organic Production (NPOP) during 2000, the state governments are yet to formulate policies and a credible mechanism to implement them. There are only four agencies for accreditation and their expertise is limited to fruits and vegetables, tea, coffee and spices. The certifying agencies are inadequate, the recognized green markets are non-existent, the trade channels are yet to be formed and the infrastructure facilities for verification leading to certification of the farms are inadequate.

***High input costs*:** The small and marginal farmers in India have been practicing a sort of organic farming in the form of the traditional farming system. They use local or own farm renewable resources and carry on the agricultural practices in an ecologically friendly environment. The groundnut cake, neem seed and cake, vermicompost, silt, cow dung, other manures, etc., applied as organic manure are increasingly becoming costly making them unaffordable to the small cultivators. Bio-fertilizers and bio-pesticides are yet to become popular in the country. There is a lack of marketing and distribution network for them because the retailers are not interested to deal in these products, as the demand is low. The erratic supplies and the low level of awareness of the cultivators also add to the problem.

***Lack of financial support*:** The developing countries like India have to design a plethora of national and regional standards in attune with those of the developed countries. The adoption and maintenance of such a regulatory framework and its implementation will be costly. The cost of certification, a major component of which is the periodical inspections carried out by the certifying agencies, which have freedom to fix the timings, type and number of such inspections appears to be burdensome for the small and marginal farmers. Of course, the fees charged by the international agencies working in India before the NPOP was prohibitive and that was a reason for the weak response to organic agriculture even among the large farms in the country. No financial support as being provided in advanced countries like Germany is available in India.

***Low yield*:** In many cases the farmers experience some loss in yields on discarding synthetic inputs on conversion of their farming method from conventional to organic. Restoration of full biological activity in terms of growth of beneficial insect populations, nitrogen fixation from legumes, pest suppression and fertility problems will take some time and the reduction in the yield rates is the result in the interregnum. It may also be possible that it will take years to make organic production possible on the farm. Small and marginal farmers cannot take the risk of low yields for the initial two-three years on the conversion to organic farming.

***Lack of quality standards for biomanures*:** There are a very large number of brands of organic manures, claiming the high levels of natural nutrients and essential elements. But most farmers are not aware of the pitfalls of using the commercially available bio-manure products. While the concept of organic farming itself lays great stress on the manures produced on the farm and the farmers' household, many of the branded products available in the market may not be really organic. Elements of chemicals slipping into the manures through faulty production methods could make the product not certifiable as organic. Most farmers are still unaware of the difference between bio-manure and bio-fertilizer, it is pointed out. While bio-manure contains organic matter, which improves the soil quality, bio-fertilizers are nutritional additives separated from the organic material, which could be added to the soil, much like taking vitamin pills. Bio-fertilizers do nothing to enhance soil quality while the loss of soil quality has been the major problem faced by farmers these days.

***Limited domestic market of horticultural produce*:** It is well-established that organic farming can lead to higher consumer prices. This would be of particular disadvantage for the high value horticultural produce, which is already beyond the reach of common Indian consumers. The farmers resorting to organic farming may be handicapped to restrict their domestic marketing to the elite consumers of major cities at least during the initial years. The nutritional security of the rural population in the country may face a threat as such if large-scale organic production of fruits is followed. A sudden, widespread shift from intensive horticulture to organic horticulture can result in drastic reduction of horticultural output during the initial years. This may have adverse bearing on the nutritional security of the nation.

ORGANIC HORTICULTURE PRODUCTION

Horticulture and plantation crops grown organically and exported from the country are tea, coffee, spices, fruits and vegetables and their value added products. India is best known as an exporter of organic tea and also has great potential for many other products. The major products produced in India by organic farming are given in Table 1.4.

Table 1.4. Major products produced in India by organic farming

Type	*Products*
Commodity	Tea, Coffee, Rice, Wheat
Spices	Cardamom, Black Pepper, White Pepper, Ginger, Turmeric, Vanilla, Mustard, Tamarind, Clove, Cinnamon, Nutmeg, Mace, Chilli
Pulses	Red gram, Black gram
Fruits	Mango, Banana, Pineapple, Grapes, Passion Fruit, Orange, Cashew nut, Walnut
Vegetables	Okra, Brinjal, Garlic, Onion, Tomato, Potato
Oil seeds	Sesame, Castor, Sunflower
Others	Cotton, Herbal extract

***Organic tea*:** According to Exim Bank Agro sources, total world production of organic tea is estimated at around 9 million kg which represents a negligible (less than 1%) share of global tea production. India produced around 3.15 million kg of organic tea (around one-third of the world production). India is the pioneer and one of the leading producers of organic tea in the world.

***Organic coffee*:** The total world production of organic coffee is estimated at 48,000 tonnes, valued at approximately U$ 150 million. This forms a very negligible (though rapidly growing) part of the world conventional coffee production (around 1% share quantity-wise and 3% share value-wise). India currently accounts for nearly 1% (around 370 tonnes valued at around US $ 1 million) of the world organic coffee market. Organic coffee in India represents only 0.12% of total Indian coffee production and was nearly 370 tonnes on a certified, cultivated area of 382 hectares. The main geographical regions cultivating certified organic coffee are Karnataka (60%), Kerala (27%) and Tamil Nadu (13%).

Organic spices: The country exported around 700 tonnes of different varieties of organic spices in 2002-03 and Spices Board expects to export organic spices worth Rs. 240–260 crore by 2012. The current market size is small but demand is expected to grow at a rate of 15–20% and may grow to 57,000 tonnes in the coming decade.

***Organic fruits and vegetables*:** The organic fruits and vegetables market is growing at annual rates ranging between 20 and 30% during the last few years. The size of the Indian organic fruits and vegetables farming is valued at Rs. 5-6 crore producing approximately 1500–2000 tonnes. India is in a very nascent stage when it comes to exports of organic fruits and vegetables. So far, only a select few corporates in India have exported organic fruits and vegetables. However, some more firms are in the process of setting up their infrastructure and may be ready to export in a year or two. The Indian organic fruits and vegetables market is a very small and niche segment due to various reasons. Some private corporates and governmental agencies such as Khadi and Village Industries Commission (KVIC) are actively trying to promote Indian organic foods in the domestic market through setting up of a distribution network or making use of their existing networks.

***Organic herbs*:** The US $ 14 billion global market for organic herbal plants and medicines is growing at 15–25 % a year and is likely to be worth US $ 5 trillion in 2050, according to the World Health Organisation. The global trade in processed herbal medicines and food supplements is estimated at US $ 60 billion. Unfortunately, India has not yet benefited much from exporting organic foods or herbal medicines, despite its potential.

The Organic production scenario of various horticultural produces in the World and India is given in Table 1.5.

Table 1.5. Organic horticulture — Indian *vis-à-vis* world production scenario

Organic tea	Global	9 million kg
	Indian	3.15 million kg
Organic coffee	Global	48,000 tonnes
	Indian	370 tonnes
	Karnataka	60 %
	Kerala	27 %
	Tamil Nadu	13 %
Organic spices	Global	875 tonnes
	Indian	115 tonnes
Organic fruits and vegetables	Indian	1,500–2,000 tonnes valued at 5-6 crores showing 20–30 % increase

GOVERNMENT SUPPORT

Institutional support for organic export was created by launching a National Programme for Organic Production (NPOP) by Agriculture and Processed Food Export Development Authority (APEDA), Ministry of Commerce. The NPOP supports promotional initiatives, accreditation of inspection and certification agencies, and offers support to agribusiness enterprises to facilitate export. APEDA has been interacting with European Union (EU), United States Department of Agriculture (USDA), Japan and IFOAM for recognition of equivalence of Indian quality assurance system. While the country follows IFOAM basic standards, it has been recently provided equivalence recognition by EU.

National Programme aims at declaring standards for organic production, recognizing specific bodies for preparing approval package of practices for specified products and approving certification programmes, recognized bodies such as accrediting agencies, inspect and certify the products, seeking recognition and accord reciprocal recognition to standards of the nation and trading blocks and instituting a logo and prescribe its award by accrediting bodies on products qualifying for bearing Indian organic label. Scope of Standards include laying down policies for development and certification of organic products, facilitating certification of organic products confirming to the National Programme containing the standards for organic production and encouraging the development of organic farming and organic processing.

The standards deal with conversion requirement, maintenance of organic management, general principles, crop production, diversity, fertilizers, nutritional management, pest and disease management, contamination control, soil and water conservation, animal husbandry, food processing and handling, labeling, list of products and list of fertilizers and chemicals which could be used. For the implementation of the programme a Steering Committee is functioning under the Ministry of Commerce supported by Technical Committee, headed by representative from Ministry of Agriculture, Government of India.

In the year 2004-05 a 'National Centre of Organic Farming' was established under the Ministry of Agriculture at Ghaziabad to provide institutional support and to facilitate farmers to move into organic crop production, by providing suitable logistics of knowledge, material inputs such as bio-fertilizers. Initiatives have also been taken to include organic farming as curriculum in degree programme, and training modules are developed, which shall help in capacity building for organic farming.

National Horticulture Mission has several programmes to promote organic horticulture in the country. During the last three years, approximately 76,000 hectare areas have been covered under organic farming. It includes wide range of crops, i.e., fruits, vegetables, spices, flowers, medicinal and aromatic crops. Similarly 41,685 vermiunits have also been established particularly in areas identified for organic adoption.

Spices Board has been implementing the programmes for encouraging the growers to adopt organic farming by laying demonstrations. Organic spices exported from India are black pepper, small cardamom, ginger, turmeric, cloves, etc.

Among different spices, organic pepper has established its market share better than other products. Most of the North Eastern states have virgin land very much suitable for organic cultivation of spices and major spices like black pepper, ginger, turmeric and chillies have larger area in this tract, enabling it to become the "Organic Spices Hub of India". As the pressure on land under cropping is increasing, area expansion for spice cultivation is questioned. Hence, the best alternative would be cultivation of spices as a component crop in the existing coffee, tea, coconut and areca nut plantations. Vines like black pepper and vanilla and rhizomatous crops like ginger, turmeric, cardamom and galangal are ideally suited for mixed/inter-cropping under tree plantations with partial light and are found to be highly remunerative. In a cropping system, the addition of organics due to litter fall would also result in organic matter build up and enhanced nutrient availability. It is reported that there is great scope of bringing tea and coffee estates under black pepper production by covering all the shade trees with pepper which can result in additional pepper yield up to 20,500 tonnes per year.

QUALITY CERTIFICATION

For gaining consumer's confidence, valid certification is an essential pre-requisite for marketing, especially in the export market, certification is based on a range of inspection activities covering the production and processing system, including auditing of quality assurance systems and examination of finished products. A large number of certifying agencies are functioning world-wide in this aspect and the number of such agencies is given in Table. 1.6.

Table 1.6. Number of organic certifying agencies (Continent-wise)

Name of continent	*No. of certifying agencies*
Europe	143
North America	97
Asia	91
Latin America	34
Oceania Australia	11
Africa	9

To improve the availability of certification services, a group of organizations and corporate bodies promoting organic agriculture took the initiative in 2001 to set up Indian Organic Certification Agency, named INDOCERT, which will offer the following services.

- It offers a positive distinction for organic products and avoids confusion.
- It helps to identify weaknesses and strengths of production systems.
- It helps to collect accurate data to describe the organic movement.
- It helps in market planning and lobbying.
- It adds another layer of value added service to the cost of organic products.

ACCREDITATION AGENCIES

The National Steering Committee has designated the following statutory Boards/ Authority as Accreditation Agencies:

- Agricultural and Processed Food Products Export Development Authority (APEDA)
- Coffee Board
- Spices Board
- Tea Board
- Coconut Development Board
- Directorate of Cashew and Cocoa Development

APEDA has been assisting the Ministry of Commerce in preparing various documents under the National Programme for Organic Production and has also developed the National Organic Logo. There are about 100 certified organic producers in the country for the products covered under APEDA. The export of organic products covered under APEDA is estimated as Rs. 10 crore, but in view of the number of projects coming up in the country, the exports are expected to go up substantially.

APEDA has recognized 11 inspection certification bodies, some of which being branches of foreign certification bodies, others being local certification bodies, all of these are able to certify based on the NPOP. Indocert is the only purely indigenously controlled certification body that is internationally accredited. These agencies are listed below with their branch location in brackets.

- BVQI (India) Pvt. Ltd. (Mumbai)*
- Ecocert SA (India Branch Office) (Aurangabad)*
- IMO Control Private Limited (Bangalore)*
- Indian Organic Certification Agency (Indocert) (Cochin, Kerala)
- International Resources for Fairer Trade (Mumbai)
- Lacon Quality Certification Pvt. Ltd. (Theepany, Kerala)*
- Natural Organic Certification Association (Pune)
- OneCert Asia Agri Certification Pvt. Ltd. (Jaipur, Rajasthan)
- SGS India Pvt. Ltd. (Gurgaon, Haryana)*
- Skal International (Bangalore)*
- Uttaranchal State Organic Certification Agency (Uttaranchal)

*based internationally

Tea Board, Coffee Board, Spices Board and Coconut Development Board have developed the guidelines for production and certification and encourage the production and export of organic produce and products. Recognizing the need for organic farming, the Ministry of Agriculture has also taken a major initiative and production and certification for domestic market shall be the responsibility of Ministry of Agriculture. It will promote the organic farming through capacity building, creation of awareness and promotion of related activities. District Agricultural Officers are designated to register the interested farmers. The registered farmers shall receive guidance from service provider regarding rules and regulations for certification and would be certified by accredited certifying agencies based on norms and rules.

REGULATORY MECHANISMS

There are several statutory provisions which regulate the quality of marketed produce in the country. Some of these sttatutory provisions lay standards which have to be met and relate to quality and hygienic requirement of the product. These statutory provisions have been put in position to ensure that the marketed products should not adversely affect the health of the consumers. On the other hand, some of these statutory provisions lay down standards which are the optional but conform to the value of the product. Existing statutory standards are Essential Commodity Act, 1955, the Edible Oils Packaging (Regulation) Order, 1998, Ministry of Food and Consumer Affairs, Department of Sugar and Edible Oils, Notification, New Delhi, 17th September, 1998; Fruit Product Order (FPO) 1955 and the Fruit Products (Amendment) Order, 1997 Ministry of Food Processing Industries, New Delhi, 30th May, 1997, the Meat Food Products Order, 1973, SO 176(E) dated 28th March, 1973. Milk and Milk Product Order, 1992, (Notification no. SO 405{E, dated June 9, 1992) for maintaining and increasing the supply of liquid milk of the desired quality in the interest of the general public to provide for regulating the production, supply and distribution of milk and milk product and the Prevention of Food Adulteration Act, 1955.

PRIORITIES AND FUTURE THRUST

- Since organic horticulture technology is location specific, with a wide variation in agro-climatic conditions, R&D work in India has to be taken up on a national level through AICRP or similar network projects involving on site research under various agro-ecological zones and multi-disciplinary approach.
- Focussed efforts needed to identify crops and commodities with high potential for organic production and export and delineation of specific zones having ideal conditions for development of sustainable organic production systems.
- Research efforts are to be undertaken to evaluate specific advantages of organic farming in terms of qualitative differences of produce, long-term soil fertility and economic viability and higher economic returns.
- There is a need to develop on-farm organic manure production technology especially suited for small and marginal holdings through approaches like vermi-compost, vermi-wash, use of microbial technology for rapid decomposition of organic biomass, use of bio-fertilizes, use of botanicals for pest and disease management, inter-cropping with green manure crops, etc., cost effective technology for large-scale production

of compost from city and horticultural farm wastes and evaluation of such products for various horticultural crops, long range field evaluation of organic horticulture technology to generate adequate scientific information on the yield, quality and post-harvest aspects, survey and critical study for evaluation of indigenous technological knowledge of organic horticultural practices prevalent in different parts of the country.

- Training of scientific manpower in horticulture on advanced organic farming technologies available in developed countries like Japan, USA, Netherlands, Germany, France, Israel, Canada, etc. and their fine tuning to suit Indian conditions and to understand strategies adopted in quality standards and certification will be needed.
- Critical study to evaluate indigenous technological knowledge of organic horticultural practices prevalent in different parts of the country.
- Government initiatives to provide requisite financial and institutional support for a strong R & D backup, certification at affordable rates, market infrastructure and export promotion.
- Creating awareness among the farmers to encourage the adoption of organic concept of farming.

CONCLUSION

Organic production of crop is not new, which has been practiced in the past, and the quest to achieve higher production resulted in use of chemicals and pesticides. With growing health and environmental concern and to address the issue of sustainability, organic production is receiving acceptance. To achieve larger goal of sustainable production and minimize the use of chemicals and pesticides, attention is being paid to create awareness and facilitate the availability of required inputs. With the initiatives, it is expected that demand for organic food would grow to the tune of 20 % annually. However, frequently it is asked, how to control the pests and diseases, do we have complete package of practices for growing of crops in different environment. We do have information but it is isolated, there are information but is not validated scientifically. Therefore, to harness the potential there is an urgent need to have horizontal and vertical integration of the efforts to address the issue in the production to consumption chain to make organic farming a people movement. Concerted efforts are needed to address various issues in production, processing, marketing and export with efforts of all stakeholders for accelerated growth of organic horticulture in the country.

SECTION – II
TECHNOLOGICAL ADVANCES

2

Emerging Technologies on Organic Production of Fruit Crops

B. M. C. Reddy and R. A. Ram

Present days' conventional farming in fruit crops is not sustainable in consonance with economics, ecology, equity, energy and socio-cultural dimensions. Indiscriminate use of chemical fertilizers, weedicides and pesticides has resulted in various environmental and health hazards along with socio-economic problems. In general, 6-8 sprayings are required for mango, citrus and apple production. The situation is even much serious in production of large number of vegetables and spices. The degenerative effects of intensive farming practices have forced for alternative system of farming. Though fruit production has continued to increase, productivity rate per unit area has started to decline. The entire orchardist community is trying to find out an alternative sustainable farming system, which is ecologically sound, economically and socially acceptable. Sustainable farming is unifying concept, which considers ecological, environmental, philosophical, ethical and social impacts, balanced with cost effectiveness. The answer to the problem probably lies in returning to our own roots. Traditional farming practices, which are, based on natural and organic methods of farming offer several effective, feasible and cost effective solutions to most of the basic problems being faced in conventional farming. There is also need to conserve our traditional seed, some of which have drought resistant properties and resistant to different pests and diseases.

In case to initiate organic farming in fruit crops, the following issues will be required to be addressed.

ISSUES

- on-farm input generation,

- quantum production equal to or higher than what is expected from optimum combination of agro-chemicals,
- continuous improvement in physico-chemical and biological properties of soil,
- par excellence produce quality with respect to nutrition, essential constituents, therapeutic value and storability and
- eco-friendly and cost effective technology.

CONCEPT OF ORGANIC FARMING IN FRUIT CROPS

Organic farming is one among the broad spectrum of production methods that is supportive of the environment. Organic production systems in fruits are based on specific standards precisely formulated for fruit production and aim at achieving agro-ecosystems, which are socially and ecologically sustainable. It is based on minimizing the use of external inputs through use of on-farm resources efficiently compared to industrial farming. Thus the use of synthetic fertilizers and pesticides is avoided.

Crop rotations and varieties are selected to suit local conditions having the potential to sufficiently balance the nitrogen demand of crops. Requirements for phosphorus, sulphur and micronutrients are met with local, preferably renewable resources. Organic farming is, therefore, often termed as knowledge-based rather than input-based farming. Furthermore, organic farms aim to optimize the crop productivity under a given set of farm conditions. There are ample evidences to show that agro-chemical-based, high input farming is not sustainable for long periods due to gradual decline in factor productivity, with adverse impact on soil health and quality.

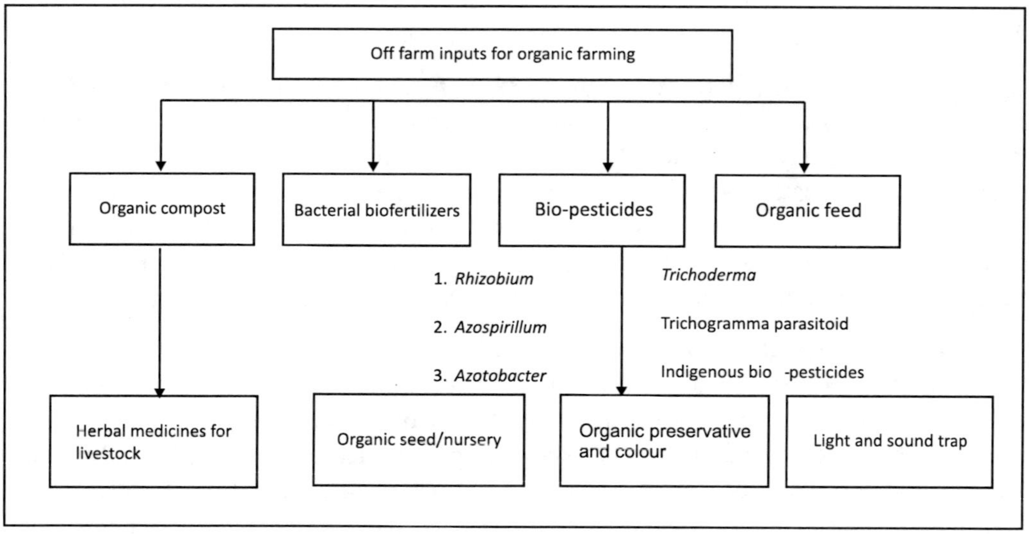

Fig. 2.1. Sketch diagram of off farm generated organic inputs

Nutrient Management in Organic Production of Fruits

Like all living beings, plants are made up of atoms derived from the planet Earth. Plants are autotrophic and have the unique ability to extract their nutrients from two sources i.e., the atmosphere and the soil. Bourguignon (2005) advocated that in fact thirty-four elements are necessary to plants and these are divided into two large groups, i.e., (i) those derived from the atmosphere and (ii) those derived from the soil as presented as under:

The atomic elements that constitute living organisms or participate in their metabolism.

Elements from the Atmosphere

These represent 92 to 98% of a plant's dry weight

- Four elements (all constitutive): C, O, H and N

Elements from the Soil

They represent 2-5% of a plant's dry weight

- Twelve vital elements of which two are non-constitutive–K and Cl ten are constitutive – P, B, Ca, Mg, S, Fe, Mn, Mo, Cu and Zn
- Eighteen other elements (of which the function of some is not known fully understood).

Critical review of above indicates that plants feed quantitatively from the atmosphere and qualitatively from the soil. A very few atmospheric elements make up the bulk of dry weight or vegetal matter and the many soil elements constitute only very little vegetal matter. The soil provides two types of elements – non-constitutive elements which the plants burrow and return at the end of the period of cultivation, and constitutive elements, which enter the vegetative organic molecular composition.

Carbon and oxygen, together constituting 88% of plant tissue, are derived from carbon dioxide. Photosynthesis allows the integration of carbon and oxygen from CO_2 within the structure of a sugar. Hydrogen, which contributes 6% of plant tissues, is derived from the break down of rainwater and hydrogen, is absorbed in the tissue of the plant. Nitrogen is derived from the atmosphere, through microbes that can transform gaseous nitrogen into ammonia. This ammonia directly enters in leguminous plants in the form of amino acids or is converted into nitrates by the soil microbes and then absorbed by other plants.

The constitutive elements of the plants derived from soil do not enter in the form of atoms, but in the form of oxides and chelates. These transformations are affected by the soil microbes, which render these elements assimilable by the plants from the mother rocks or humus. Thus all other soil elements can be absorbed by plants through microbial action. The oxidation of soil elements is often achieved by very specific micro flora. Fifteen oligo-elements can enter in plants in the organic form. This statement enables us to nullify the strong notion that plants have purely mineral diet.

As regards the availability of major organic nutritional inputs, i.e., N, P, K in India, the estimate of National Center of Organic Farming, (Bhattacharya and Kumar, 2005, Tandon and Narayan, 1990) is depicted in Table 2.1.

Table 2.1. Estimate of potential availability of nutrients from organic sources

S.no.	*Resources*	*Million tonnes*
1.	Crop residue	3.865
2.	Animal residue	3.854
3.	Green manure	0.223
4.	Bio-fertilizers	0.370

It is interesting to record that sizeable quantum of crop and animal residues are often burnt, or left as such in the villages. If the above residues are properly decomposed in the form of FYM/compost and utilized, they can meet the nutritional requirement. If applied through chemical sources, hardly it can meet 1/3 of the nutrient requirement. Besides in order to meet the nutrient requirement, they have to be used in large quantities, which may pollute the native soil, ground water and also the environment. In recent years, number of techniques of composting have been developed in different parts of the country. Some of the prevalent composting techniques like Indore, Bangalore, Coimbatore, NADEP, vermi, bio-dynamic, microbe mediated, phospho-compost, and accelerated compost are employed to enrich the nutrient availability.

Composting is an important technique for recycling (farming/industrial) waste and for improving the quantity and quality of organic manures. Significant improvement in the nutritive value of these composts is a common experience, which is evident from Table 2.2.

Table 2.2. Nutrient status of different composts

S.no	*Preparations*	*N (%)*	*P (%)*	*K(%)*	*Ca(%)*	*Zn (ppm)*	*Cu (ppm)*	*Fe (ppm)*	*Mn(ppm)*
1.	Vermi-compost	3.15	4.29	0.53	1.72	168	61	3545	252
2.	BD compost	1.68	0.17	1.23	1.20	96	45	357	3352
3.	NADEP compost	0.98	0.35	1.00	1.25	162	56	430	230
4.	MM compost	1.54	0.51	1.06	1.35	140	45	433	275
5.	FYM	0.70	0.19	0.37	0.24	75	34	222	3134

Liquid Manures/Pesticides

Liquid manures are prepared using different organic materials *viz.*; cow dung, urine, chopped leaves of legumes, fish/blood meal, sea weeds, etc. On an average, preparations of liquid manures take 2-3 weeks time. Proper stirring, twice in a day and incorporation of one BD set (BD-502-507) enhances the fermentation process and nutritive value of liquid manures as depicted in Table 2.3.

Table 2.3. Nutrient status of liquid manures prepared from cow dung, urine, plant parts and BD preparation

Liquid manures	Nutrient level				
	N (%)	*P* (%)	*K* (%)	*Zn* (ppm)	*Cu* (ppm)
BD Liquid manure with *Neem*	0.63	0.32	0.48	2.00	0.31
BD Liquid manure with *Subabul*	0.87	0.47	0.22	1.80	0.31
BD Liquid manure with *Lantana*	0.45	0.32	0.27	2.12	0.17

In general, requirement of potash is much high in horticultural crops. It is difficult to take care of potash being mined through these crops. Incidentally leaves of *Glyricidia*, *Calotropis* and water hyacinth, etc., are rich in potash. Leaves of these plants and sulphomag can be incorporated in preparation of composts; liquid manures or even these can be used as surface mulch. Liquid manures are diluted 2-3 times and frequently can be sprayed on plants to boost the growth, crop yield and quality. Decoction of nettle leaves is effective for the control of hard pests like mango hoppers, mealy bugs and mites.

These composts are rich in NPK content compared with ordinary FYM. Besides NPK, these are also good source of Ca, S, and micro-nutrients *viz.*; Zn, Cu, Mn, and Fe. Nutritive value of these composts can be further enriched by incorporation of Cow Pat Pit (CPP) biodynamic microbial rich formulations. Thimmaiah (2004) coined these as hybrid compost. Nutritive values of a few of these composts are depicted in Table 2.4.

Table 2.4. NPK content in different hybrid composts

Sl.No.	*Compost*	*N (%)*	*P (%)*	*K (%)*
1.	Biodynamic composting			
	(i) Standard BD composting (substrate cow dung)	2.43	1.43	1.32
	(ii) Biogas slurry as substrate at the place of cow dung	2.87	1.05	1.10
2.	NADEP composting			
	(i) Standard technique	1.05	0.71	0.85
	(ii) Standard technique + B D preparations	2.12	1.05	1.47
3.	Traditional heap method	0.7	0.38	0.43

It is interesting to record that only a fraction of the nutrients are removed through ultimate production and rest are available for further recycling. However, besides these wastes, a lot of leaf litters are available if properly recycled. These can be a good source of organic nutrition. In the case of mango and banana, nutrient removed by leaves and fruits can be taken as example for addition of nutrients for sustainable production (Chundawat, 1997) (Table 2.5).

Table 2.5. Quantity of various nutrients elements in various parts of mango and banana

Part	*Total quantity of nutrients elements*				
	N (%)	*P (%)*	*K (%)*	*Ca (%)*	*Mg (%)*
Fruit	1.23	0.60	2.14	0.91	0.52
Leaf	1.23	0.06	0.54	1.71	0.91
Banana (per 100g of fruit) (Lahav and Turner, 1983)					
Leaf	2.3	0.22	4.5	–	–
Fruit	3.78	3.78	7.76	–	–

Approximately 40–50% of nutrient requirement can be met after recycling of mango and banana leaves. Bananas is a gross feeder but besides bunch if other wastes are recycled, around 60% nutrient need can be met from these sources. Similarly, lot of litter and organic wastes can be procured from nearby forest or pond for nutrient availability.

Strategically selected crops and forest litters in a system can thus be a good source of biomass and nutrients. It is often argued that if the crop residues are needed as cattle feed or as fuel, then either the cattle excreta after composting or equivalent quantity of other non-economic biomass may be returned to land. In the one long-term experiment, Rupela *et al.,* (2004) reported production of 4.5 tonnes ha^{-1} (dry mass) of lopping from *Glyricidia* grown on the bund in the 5th year. Addition of lopping to a field meant application of 103 kg N and 6.7 kg P ha^{-1}. Besides *Glyricidia* there are large number of plants *viz.*; *Ipomea cornea, Pongamia pinnata, Cassia auriculata, Azadirachta indica, Thespesia populnea,* etc., which could be utilized as green manure. Similarly, crops like *Crotolaria juncea*, *Sesbania* spp, *Indigofera tinctoria*, *Calapogonium muconoides*, *Phaseolus trilobus*, *Stylosanthus hamata*, *Lablab pupureus*, etc., are excellent crops which could be grown to check the soil erosion and can be recycled as green manure/through composts as surface mulch.

PEST AND DISEASE MANAGEMENT

In pollution free atmosphere and fertile soil, incidentally pest and disease infection are minimum. Crop rotation, solarization, organic seed treatment, trap crops, predators, parasites, use of bio-pesticides and peppering are the number of options available which can be utilized for pest control. Phelan *et al.* (1997) and Phelan (1995) investigated mechanism that could explain why insect pests can be lower on organic farms. Soils from organic and conventional farms were brought in the laboratory and treated with various conventional and organic amendments. Maize plants were grown in pots of these soils in the greenhouse and exposed to female Europeans corn borers (ECB). These insect preferred to lay eggs on plant grown in soil with history of conventional management. With respect to plant disease, lower incidence of some disease was observed such as Fusarium head blight on organic farm (Oerke *et al.*, 2001)

The top ten plants which posses pesticidal properties are neem (*Azadirachta indica*), custard apple (*Anona squamosa*), papaya (*Carica papaya*), giant milkweed (*Caliotropis procera*), pongamia (*Pongamia pinnata*), oleander (*Nerium indicum*), chaste tree (*Vitex nugundo*), snake root (*Aristolichia bracteata*), Indian tinospora (*Tinospora cordifolia*) and castor (*Ricinus communis*). Decoction prepared from these alone, or in combination with cow urine is effective for control of large number of pests. Garlic–onion–chilli acts as repellent and buttermilk spray is also effective for the control of caterpillar and spider mites. Spray of BD-501 brings immunity against fungal infection. BD liquid pesticides prepared from casuarina/horse tail leaves also provide tolerance to diseases.

ORGANIC FARMING SYSTEMS PREVALENT IN INDIA

In order to resolve above apprehensions, number of organic production systems *viz.* biodynamic farming, *rishi krishi*, *panchgavya krishi*, nauteco farming and natural farming have emerged over the years in different parts of the country. Brief account of these has been given below:

Biodynamic Farming

Biodynamic farming, under the present scenario appears to be one of the sound alternatives. It is based on systematic and synergistic harnessing of energies from cosmos, earth, plant and cow. Some of the of the preparations used in minute quantities *viz.*; BD 500–508, Cow Pat Pit, etc., show remarkable effects on growth, metabolism, crop yield and quality. It is interesting that these practices do not require sophisticated facilities and most of them can be created at the farm itself by simple training. These are components of biological farming, capable of affording long-term sustainability to farming and particularly to the ecosystem. Emphasis is made to adopt a farming calendar for crop cultivation along with some specific preparations. Efforts are made to restore soil fertility in the form of humus, increase the living system of soil by skilful application of appropriate crop rotation (Steiner, 1974).

Rishi Krishi

This system is very popular and practiced by the farmers in Maharastra, Madhya Pradesh, Rajasthan and now in Uttar Pradesh for many years. Thousands of farmers are now following the system for crop production. In this system, the following preparations are used to maintain soil fertility, crop yield and quality (Deshpandey, 2003).

Angara

Bhomi *Sanskar* is done to make the soil fertile, in which 15-20 kg rhizosphere soil of banyan tree (*Ficus bengalensis*) "*Vat vriksh*" is broadcasted on one acre of land. It has lot of earthworms and other beneficial microbes, which improves the soil fertility and its biological activity (Despandey, 2003).

Amritpani

Amritpani is prepared by mixing 10 kg cow dung, 250g ghee and 500g honey. All the ingredients are mixed, stirred and fermented and diluted to 200 litres. This is used after proper stirring to treat the seeds (*Beej Sanskar*) enrich soil (*Jala Sanskar*) and seedlings (*Vanaspati Sanskar*) by spraying on field and plants. This is also followed by mulching with locally available organic wastes.

Panchgavya Krishi

Panchgavya consists of five products from cow, i.e., cow dung (5 kg), urine (10 litres), milk (3 litres), curd (2 litres) and ghee (1 kg). They are suitably mixed along with sugar-cane juice, tender coconut water, ripe banana and toddy and incubated for 30–40 days and stirred daily. Three per cent solution is used to treat seed/seedlings and encourage vegetative and reproductive growth of the plants. A group of farmers in Tamil Nadu are practicing this for cultivation of number of cereals and horticultural crops. In order to address these issues in better way, Central Institute for Subtropical Horticulture, Lucknow started study on integration of these systems and develop a package of *Jaivik Krishi* system, which can be adopted by the common Indian farmers in commercial way. It is interesting to record that in all the systems mentioned above, cow plays an important role. This needs to be promoted in the country.

Seeds and Planting Material

Non-availability of organically produced seeds and planting materials is the key component in promotion of organic farming in fruit crops. Hence, major research and development efforts are required to ensure availability of quality seeds and planting materials required in organic production. According to "Manu", good seed and good land yields abundant produce. Quality of seed should be true, viable, free from pests and diseases, weeds, foreign matter, fairly priced, bold and uniform in size. Local seeds and varieties are better adapted to organic farming. Genetically modified seeds are not accepted in organic farming (Pathak and Ram, 2006).

CONCLUSION

The interest in organic farming in developing countries is growing because it requires less financial inputs and places more reliance on natural and human resource available. India is land of small holdings. Organic farming on such land holdings, especially under rainfed zones, tribal areas and North West to North East Himalayas still will go long way in promoting organic production of fruit crops in integrating these systems and develop package of *Jaivik Krishi* (JK), which can be promoted in different parts of the country by the common Indian farmers, which has been depicted in Fig.2.2. It is interesting to record that in all four systems "COW" particularly those with hump (indigenous breed) is one of the key components, hence provision of at least one cow per hectare need to be promoted for *Jaivik Krishi* activities in the country.

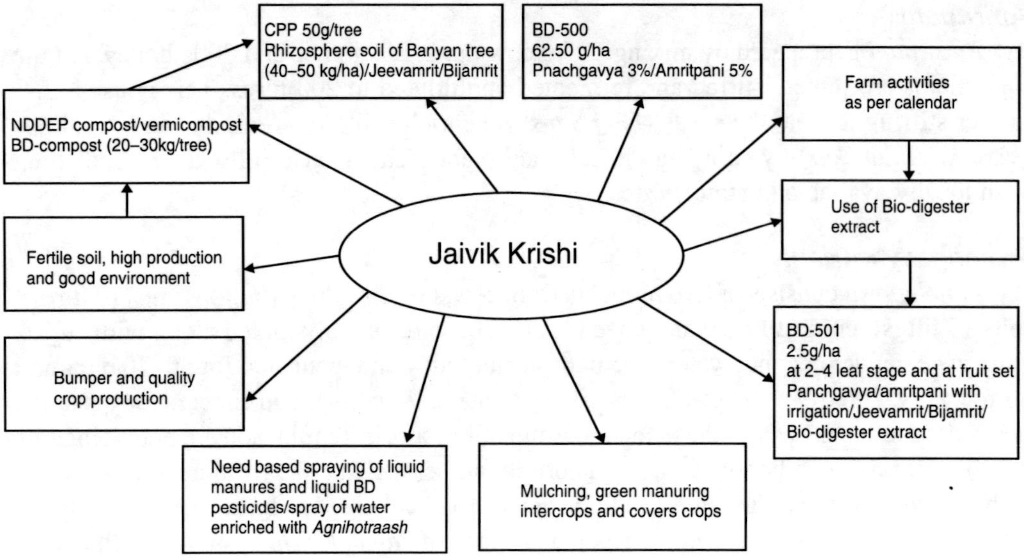

Fig. 2.2. Package of *Jaivik Krishi*

Strategies

- Number of organic production methods, *viz.*, biodynamic, *Rishi Krishi*, *Panchagavya*, are being practiced by a large number of farmers. There is need to inventorize these methods, study the possibility of their integration, validation and development of package of practice, specific to crop and agro-climatic situations in the name of *Jaivik* farming.
- Identification/selection of appropriate varieties adapted to organic production methods and agro-climatic situations.
- Research/extension activities on production of seed and planting materials need to be taken up vigorously in order to ensure quality seeds and planting materials.
- Controlled experiments to study the cosmic inputs, i.e., planting calendar and biodynamic preparations through science of astro-physics need to be initiated.
- Appropriate crop rotation and role of legumes as cover, intercrop as green manure, need to be investigated and included in the package.
- Development of techniques to enhance the nutritive value of composts through incorporation of various organic waste, rock phosphate, dolomite, lime, cakes, bio-fertilizers, ash, bone, blood, fish meal acceptable in organic production system.
- In order to minimize the impact of insect, pest, disease and weeds, various methods such as cultural, mechanical, peppering, use of predators, parasites, bio-pesticides, bio-agents, etc., need to be integrated and package developed.

- Besides the quantum of production, the nutritive value of produce (protein, amino acids, vitamins, micronutrients, antioxidants, etc.), taste, keeping quality and therapeutic value, etc., should be considered in organic production.
- There is need of continuous monitoring of soil health with respect to physico-chemical and biological properties and monitoring of ground water, environment and flora and fauna on conventional and organic farm.
- Integrated model involving various stakeholders such as farmer's organization (FIGs, FA's), NGOs, Corporate bodies, State/Central Government/ICAR/SAUs and International organizations in organic farming need to be developed and promoted.

References

Bhattacharya, P. and Chakraborty, C. 2005. Current status of organic farming in India and other countries. *Ind. J. Fert.*, 1(9): 111-123.

Chundawat, B.S. 1997. Nutrient management in fruit crops. Ed B.S. Chundawat, Agrotech, Publishing Academy, Udaipur.

Claude Bourguignon 2005. Regenerating the soil. Other India Press, G-8, St Briton Apartments Feira Alta, Goa. 62p.

Deshpandey, M.S. 2003. Organic farming wrt. Cosmic Energy - Non-Voilence *Rishi Krishi.* Mrs Kamal Prakashan Patil, Mrs Vijay Printers, Khede, Kolhapur.

Lahav, E. and Turner, D.W. 1983. *Banana nutrition.* Plant Bulletin No. 7. International Potash Institute, Berne, Switzerland.

Oerke, E.C., Meier, A.A., Steiner, U. and Denhe, H.W. 2001. Incidence and control of *Fusarium* head blight in intensive wheat production in Western Germany. Phytopathaology 91 (Suppl. 6), 567.

Pathak, R.K. and Ram, R.A. 2006. Organic Seeds: Planting Material Production and its Sustainability. Seed: A global Perspective. Eds, Kalloo, G. Jain, Alice, S.K., Jain, K. and Umesh Srivastawa, Associated Publishing Company, New Delhi, India, pp-160-63.

Phelan, P.L. 1995. Soil management history and the role of plant mineral balance as a determinant and host preference by European corn borer. Biological farming and Horticulture **15**: 25-34.

Phelan, P.L., Mason, J.F. and Stinner, B.R. 1997. Soil fertility management and host preference by European corn borer, *Ostrina nubilis* (Hubner), on *Zea mays* L: a comparison of organic and conventional chemical farming. Farming, ecosystems and Environment 56, 1-8.

Rupela, O.P., Gowda, C.L.L., Wani, S.P. and Ranga, Rao, G.V. 2004. Lesion from non-chemical input treatments based on scientific and traditional knowledge in a long-term experiment. International conference on Farming heritage of Asia, 6-8 Dec., 2004, Asian Agri. History Foundation, A.P.

Steiner, R. 1997. Theosophy. An introduction to the supersensible knowledge of the World and the Destination of man. Anthrosophic Press, Hudson, New York.

Tandon, H.L.S. and Narayan, P. 1990. Fertilizer in Indian Farming–Past, present and Future (1950-2000). FDCO-New Delhi.

Thimmaiah, A., Tewari, R.C., Tewari, K.N., Pathak, R.K. and Ram, R.A. 2004. Nutrient and pest management in organic farming: A case Study. National Conference on Organic Farming for Sustainable Production" held at NASC, Complex, New Delhi, March 23-25, 2004, p34.

3

Technological Advances in Organic Farming of Plantation Crops

George V. Thomas

INTRODUCTION

A range of plantation crops including tea, coffee, rubber, coconut, oil palm, cashew, arecanut, cocoa and spices are cultivated in the humid tropics and tropical belts of the country extending throughout the peninsular India comprising of Kerala, Karnataka, Tamil Nadu, Andhra Pradesh, Orissa, Bihar, West Bengal, Goa, parts of Maharashtra, and the north-eastern region. They are cultivated in wide range of soils ranging from sandy, sandy loam and laterite. These crops occupy an area of about six million hectares which is about 3.4% in the gross cropped area. Plantation crops serve a variety of human needs such as food, oil, industrial raw materials, beverages and confectionary items. Coconut, arecanut, cashew and cocoa are the major small holder's plantation crops cultivated in India. The major socio-economic features in which these crops are cultivated include predominance of small and marginal holdings, medium to resource poor farm environment, less marketable surplus and marketed surplus.

Plantation Crop Based Cropping/Farming Systems

Cropping/farming systems by raising compatible subsidiary crops and/or integrating with livestock enabled to increase the productivity and net returns from unit area of coconut and arecanut plantations. Farm resources like land, labour, sunlight, water and nutrients were effectively utilized in the system and higher productivity was achieved as a result of synergistic interaction among the crop and crop-livestock components. Crop diversity involving a number of annual, biennial or perennial crops as inter/mixed crops in perennial stands of coconut and arecanut promoted the productivity and sustainability of the system.

The beneficial effects were also reflected on non-monitory inputs such as enhanced soil fertility status, microbial activity and nutrient transformations which were conducive for sustainable crop productivity.

The coconut farmers in India, especially in the homestead gardens of Kerala state are accustomed to grow a variety of crops in the interspaces of coconut palms to meet this food, fodder, fuel, wood and timber requirements in addition to generating supplementary income. Many of this traditional cropping system mimic mature and create well-established agro-ecosystems with minimum dependency on external inputs to sustain production. The scientific research resulted in the development of a number of highly productive economically viable cropping system models. Coconut based High Density Multispecies Cropping Systems developed at CPCRI involves growing a diversity of crops at very high population per unit area to maximize production and economic returns. The inter crops include black pepper, clove, banana and pineapple (Bavappa *et al.,* 1986). Mixed farming system involves cultivation of forage grasses as feed for dairy cows which are maintained in the system and recycling of wastes back to the system as a source of nutrients for coconut palm. This system is a typical example of utilization of soil–plant–animal relations in a man made agro-ecosystem (Maheswarappa *et al.,* 2000).

Need for Sustainable Organic Production Systems in Plantation Crops

The low level of adoption of management practices particularly chemical fertilizers by the farming community is the major factor responsible for low productivity in farmers' gardens particularly in small holders' plantation crop like coconut. Resource constraints is also leading to low level of adoption of technologies. The soil conditions and climatic factors including high level of precipitation and high temperature result in the loss of major portion of chemical fertilizers applied to the soil by leaching and de-nitrification resulting in low nutrient use efficiency in tropical soils. Soil degradation is taking place due to the continuous application of chemical fertilizers without adequate organic inputs and the farming practices which leads to over exploitation of natural resources. Pollution of water bodies is taking place as a result of leaching of chemicals from the agricultural systems. Reduced crop and soil health because of poor organic matter content and micro-nutrient deficiencies are the other constraints in the production systems experienced now. All these factors underline the need for soil and plant health management to achieve sustainable production protecting the environment and safeguarding the natural resources. There is also a need to reduce the cost of inputs to enhance competitiveness in the international market under the changing global scenario.

Plantation Crops – Amenability to Organic Farming

Plantation crops are highly amenable to organic farming as they produce large quantities of waste biomass which, if recycled, can meet the nutrient demand of the crops to a great extent. The availability of waste biomass from a well managed coconut garden with 175 trees/ha has been estimated to be 14 tonnes/ha/yr in the form of leaves, spathe, bunch waste and husk. A considerable portion of husk is used for extraction of coir fibre. Coir dust, the by-product of coir processing factories, is usually dumped without any use often causing environmental pollution. It has been estimated that 7.5 million tonnes of coir dust is available in the country from the various coir defibering units. The total availability of waste biomass from 1.93 million hectare of coconut plantation in the country has been estimated as 14.36 million tonnes annually (Table 3.1). The natural decomposition of these wastes and the nutrient release are very slow due to high lignin content and the nature of lignocellulose complex of the coconut waste materials. Substantial saving in terms of fertilizer input is possible through effective recycling of the waste biomass.

Table 3.1. Availability of on-farm wastes/ by products in plantation sector in India

Crop/Crop waste	*Quantity available*
Coconut (excluding coir pith)*	14.36 million tonnes
Areca leaves	0.13 million tonnes
Areca rachis	0.08 million tonnes
Areca husk	0.22 million tonnes
Cocoa shed leaves	360.03 tonnes
Cocoa pruning	12056.33 tonnes
Cocoa pod husk	32900 tonnes
Coir pith	7.5 million tonnes
Coffee husk	0.18 million tonnes
Tea waste	0.22 million tonnes

* includes spadices, bunch wastes, sheath, inflorescences and husks

The organic wastes can be directly utilized as mulch or converted into composts by employing earthworms or microbial cultures. Addition of nitrogen through biological nitrogen fixation can be favoured by growing green manure legumes and use of biofertilizers based on asymbiotic and associative symbiotic nitrogen fixing bacteria.

On-farm Waste Recycling through Vermi-composting

When the organic wastes available from a plantation are recycled, it can supply the major portion of nitrogen and a part of other nutrients required by the palms. Studies conducted at CPCRI, Kasaragod have revealed that coconut plantation wastes could be effectively converted into rich vermi-compost using a local species of epigeic earthworm belonging to *Eudrilus* sp. The *Eudrilus* sp. was located from a heap of coconut palm wastes at CPCRI farm. Among the local earthworms tested for their ability to multiply and produce vermi-compost from coconut palm wastes, this strain of *Eudrilus* was the best and hence more studies were conducted. The local strain of *Eudrilus* is different from *Eudrilus eugeniae*, the African night crawler, which has its origin in tropical and subtropical Africa. The strain of *Eudrilus* located at Kasaragod differs from the African strain in its colour, vigour, behaviour, cocoon colour and reproductive rate on coconut wastes. The local strain is more darkly pigmented and vigorous and the cocoon case is thick and dark. Interestingly, it is also capable of burrowing in coarse soils such as sandy loam and in soft bedding materials. It is also capable of actively feeding on soil and digested soil in the form of pellets is deposited around burrow holes.

The low cost vermi-composting technology developed at CPCRI enables production of high quality organic manure from coconut leaves in a period of 60–75 days (Prabhu *et al.*, 1998; Thomas *et al.*, 2001). Vermi-composting can be done ideally in permanent cement or brick tanks constructed under shaded conditions to maintain appropriate quantity of substrate, optimum moisture and temperature which are necessary for efficient vermi-composting. This will also provide an opportunity to provide protection for the worms from predators like rodents, ants, birds and wild boars. The weathered wastes obtained during rainy seasons are preferred. These organic wastes are to be treated with cow dung at the rate of 10% by weight in the form of slurry and must be allowed to undergo a preliminary decomposition for about 2-3 weeks. The earthworms at the rate of 1000 worms per tonne of coconut leaves are to be introduced. The compost bed should be mulched properly using any locally available plant material or gunny bags and has to be protected from direct sun light. Watering is to be done to maintain enough moisture. In about 60–75 days compost will be ready, leaving behind only a portion of undecomposed material. Composting was

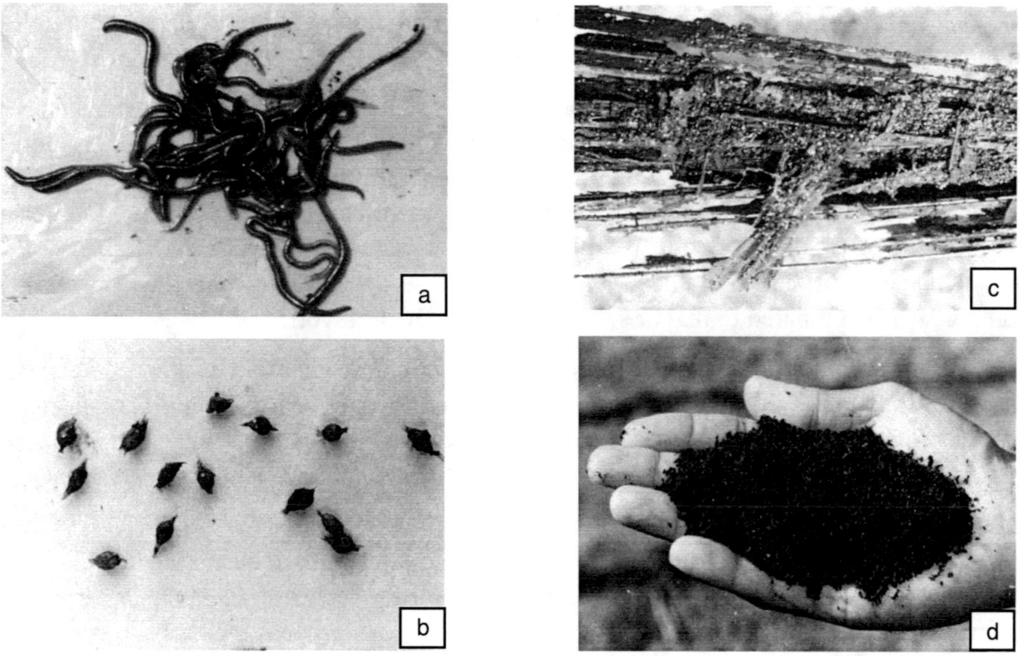

Fig. 3.1. Vermicomposting of coconut leaves: a. *Eudrilus* sp., b. Cocoons of *Eudrilus* sp., c. Coconut leaves undergoing composting, d. Vermicompost

faster in heaps and the earthworms were seen feeding and casting from all sides may be due to better aeration and exposure. On an average, 70% recovery of vermi-compost was obtained. The average nutrient composition of vermi-compost recovered was found to be 1.8% N, 0.2%P, 0.2% K, 17.84% organic carbon and C: N ratio of 9.95:1. The composted palm wastes also contained higher levels of micro nutrients such as Fe, Zn, Cu and Mn when compared to that of the untreated substrate. The same technology for vermi-composting was also tested in large pits taken in the inter spaces of four coconut palms in sandy loam and coastal sandy soils and was found to work well. When composting is done in the field itself, lot of labour required for transportation of the biomass and compost can be saved. This technology is also suitable for plantations with very limited irrigation facilities, as only a limited number of pits or trenches need to be watered.

The C : N ratio of the organic matter ingested by the earthworm decreases and bound nutrients are converted into easily available forms during the vermi-composting process. A small portion of this digested organic matter is absorbed by the earthworm and the left over, which is relatively stable humus-like organic matter is egested in the form of pellets or granules, frequently referred to as vermi-castings. The gut passage of organic matter results in enhanced availability of nutrients, increased counts of microbes and enrichment with a number of bio-active compounds. Two types of active nitrogen fixing bacteria not commonly isolated from soils have also been found regularly associated with vermi-casts. Thus, vermi-compost increases soil fertility through addition of plant nutrients, growth hormones, increased level of soil enzymes and important micro-organisms as they are rich in microbial diversity, population and activity.

The changes in the occurrence and population of 15 types of beneficial micro organisms were studied during different stages of vermi-composting. There was significant variation in the population levels of different groups of micro-organisms in the substrate, during the composting process in the gut of earthworm and vermi-compost. The levels of fungi, asymbiotic nitrogen fixers, nitrosomonos, nitrobacter, ammonifiers, fluorescent pseudomonas, phosphate solubilizers and silicate oxidisers were significantly high in vermi-compost when compared to that of substrate and gut of earthworm.

Coir pith, a byproduct obtained after extraction of coir fibres from husk, accumulates as waste around the coir fiber extraction units causing environmental hazard. Though coir pith has many beneficial attributes such as high moisture holding capacity and plant nutrient content, the use of raw coir pith in crop production is not recommended as manure due to the high C: N ratio of more than 100:1.

Research efforts on the utilization of this valuable resource indicated that it can be converted into compost either through the use of local species of *Eudrilus* or through microbial degradation, and the final product can be used to improve soil physical properties and moisture holding capacity to a great extent (Thomas *et al.*, 2001). The composted biomass was found to be of superior quality with respect to content of major and micro-nutrients and plant beneficial micro-organisms particularly the bacteria involved in nutrient transformation and growth promotion.

Co-composting coir pith with solid poultry manure with and without lime and rock phosphate amendment revealed that the composting process facilitated by poultry manure amendment brought about bioconversion of coir pith to a final product in 45 days and the final product possessed physico-chemical characteristics required for quality organic manure. C : N ratio, which is considered as a maturity index of composting process, got reduced during the composting process to 21.42. The composted coir pith can be used as manure in coconut plantations and can increase the capability of soils to store moisture and nutrients.

Fallen leaves, bunch wastes husk are the major source of organic wastes in arecanut garden. About 0.1317 million tonnes of dried areca leaves, 0.0832 million tonnes of dried arecanut bunch waste and 0.2224 million tonnes of dry arecanut husk are estimated to be available annually in India. The non-marketed produce in arecanut garden viz., areca-shed leaves, sheath, husk and arecanut bunch waste supplies approximately 94.6g, 10.0g and 109.63g of N, P_2O_5 and K_2O, respectively per palm per year. These available arecanut wastes have potential to supply 5,260, 1,337 and 6,230 tonnes of N, P and K, respectively to the agricultural system annually. Cocoa is grown as mixed crop in coconut and arecanut plantations and crop is cultivated in an area of 30,000 ha in the country. Cocoa wastes could contribute 540, 72 and 1,244 tonnes of N, P and K per year, respectively.

Crop Residue Availability and Recycling in Cropping/Farming System

The coconut based High Density Multi Spices Cropping System comprising clove, black pepper, banana and pineapple yielded 23.51 t/ha/year of recyclable biomass from the system. The extent of nutrient recycling from biomass was estimated to be 130 kg N/ha, 8.5 kg P/ha and 121 kg K/ha (Palaniswami *et al.*, 2007).

Vermi-composting trials undertaken in large cement tanks to evaluate the effect of mixing crop residues of component crops with coconut leaves on vermi-compost production by *Eudrilus* sp. revealed that recovery of compost was 60–65 % in the mixture of coconut leaves + banana pseudostem and leaves, coconut leaves + glyricidia leaves (3:1 proportion) compared to 70% when coconut leaves alone was tested. (Table 3.2). The C: N ratio of vermicompost produced varied from 12.27–13.20.

Table 3.2. Recycling of waste biomass of coconut based cropping system by vermicomposting

Treatments	*Recovery (%)*	*Organic carbon (%)*	*Nitrogen (%)*	*C:N ratio*	*Phosphorus (%)*	*Potassium (%)*
Coconut leaves + banana (3: 1) pseudostem	60	19.96	1.51	12.27	0.24	0.29
Coconut leaves + Glyricidia (3:1)	65	15.84	1.23	13.20	0.26	0.48
Coconut leaves alone	70	17.15	1.34	12.49	0.22	0.41

The coconut based mixed farming system comprising of dairy, poultry and silk worm rearing contributed sizable quantities of biomass viz. 14 tonnes of cow dung, 295 kg of poultry bedding materials and droppings, 12 kg of silk worm waste (all on dry weight basis) and 50,000 litres of cowshed washings. This waste biomass on recycling could meet 74 % N, entire quantity of P, and 82 % of K requirement of both coconut and grass (Thomas and Palaniswami, 2005).

The recyclable biomass production from sole arecanut was estimated at 8.7t ha^{-1} and it can be considerably increased to 14.3 tonnes ha^{-1} with adoption of High Density Multi Spices Cropping System (Bavappa e*t al*., 1986). Though the quantity of recyclable biomass is reduced after 15 years of adoption of Arecanut Based Cropping System, still there is scope for efficient and economical recycling (Bhat and Sujatha, 2007). Recyclable biomass produced from different component crops in arecanut based cropping system varied between 8,724–10,354 kg ha^{-1} $year^{1}$. The recyclable biomass can be converted into vermi-compost with 82–87% recovery. The estimated nutrient content in vermicompost was 1.71% N, 0.21% P and 0.43% K (Table 3.3).

Table 3.3. Nutrient composition of vermicompost made out of cocoa and arecanut leaves

Nutrients	*Cocoa*		*Arecanut*	
	Dried leaves	*Vermi-compost*	*Dried leaves*	*Vermi-compost*
OC (%)	47.10	24.4	OC(%)	47.10
N(%)	1.27	1.65	N(%)	1.27
P(%)	0.17	0.19	P(%)	0.17
K(%)	0.27	0.32	K(%)	0.27
C:N ratio	37.00	14.78	C:N ratio	37.00
Cu(ppm)	32.66	83.60	Cu(ppm)	32.66
Fe(ppm)	1157.41	2593.0	Fe(ppm)	1157.41
Zn(ppm)	228.39	367.7	Zn(ppm)	228.39
Mn(ppm)	363.1	679.84	Mn(ppm)	363.1
pH		7.5	pH	

Vermi-wash for Growth Promotion and Quality Produce

Another important product of vermi-composting technology is the vermiwash, which is rich in nutrients in readily available form, plant growth promoting hormones, beneficial micro-organisms and can be used for improving the productivity of crops. Vermi-wash has been found to be effective as foliar spray for growth promotion and bio suppression of pathogens in crop plants. All the physiologically active water soluble components of vermi-compost such as humic acids, plant growth regulators, amino acids, vitamins, micro-nutrients and microbial cells are extracted in water and is known as vermi-wash. The water soluble components from vermi-compost may be collected by passing water slowly through the worm beds or by simple suspension of vermicompost in water. This is used for foliar applications as such or after sufficiently diluting, based on the need. Vegetables and ornamental plants have been found to respond very well to this treatment.

Vermi-wash produced from actively vermi-composting substrates of coconut leaf + cow dung by *Eudrilus* sp had (Table 4.4) an alkaline pH, contained major and minor nutrients, growth hormones, humic acid and plant beneficial bacteria (Murali Gopal *et al.,* 2008). Application of appropriately diluted coconut leaf vermi-wash (CLV) increased germination and seedling vigour index of cowpea and paddy seeds in laboratory bioassays. Field trials with cowpea, maize and bhendi in the Institute farm showed its capacity to enhance the biomass and yield of the crops accompanied by higher soil microbial activities. Field trials were conducted in farmers' plots with bitter gourd, cowpea, amaranthus, and chillies. Observations indicated that application of vermi-wash had resulted in yield of crops on par or slightly lower than the plots that received regular fertilizer inputs (Murali Gopal *et al.,* 2008). However, the farmers listed many other important points viz. healthy plant and root hair growth, lesser pest and disease damage, larger and deep leaf colours, longer ability of plants to stay without wilting in field as well as longer time of remaining fresh in case of amaranthus after harvesting, more freshness of fruits etc.

Table 3.4. Characteristics of Vermi-wash from coconut leaf vermi-compost

Nutrients	*Conc. (ppm)*	*Biochemical constituents*	*Conc. (µg/ml)*	*Micro-organisms*	*Population (cfu/ml)*
N	2.8	Total sugars	61.6–111.2	Bacteria	2 x 10^3
P	10.2	Reducing sugars	41.9–88.4	Fungi	85
K	205	Free amino acids	20.8–32.3	Actinomycetes	9 x 10^3
Ca	37.9	Proteins	615–890	Phosphate solublilizers	25 x 10^2
Mg	6.5	IAA	0.52–1.15	N_2 fixers	15
Fe	Traces	Gibberellic acid	0.23–1.61	Fluorescent pseudomonads	8 x 10^2
Cu	Traces	Total pheno	10.2–14.8		
Zn	0.07	pH	7.6–8.9		
Mn	0.17				

Utilization of Efficient Legume-Rhizobium/Bradyrhizobium N_2 Fixing Systems

Nitrogen fixing legumes possessing symbiotic association with efficient *Rhizobium/ Bradyrhizobium* strains can be important inputs in organic cultivation of coconut (Fig 3.2). The technique for utilization of leguminous cover crops as green manures to supply biologically fixed nitrogen and easily decomposable biomass to coconut was standardized at CPCRI (Thomas and Shantaram, 1993). It involves cultivation of leguminous creepers such as *Pueraria phaseoloides, Mimosa invisa* and *Calopogonium mucunoides* in coconut basins during the monsoon period from June—October and incorporation of legume biomass in respective basins. During a growth period of 140–150 days, the promising legumes generated 15–28 kg of biomass and 102–197g of nitrogen in the basin of a coconut palm (Table 3.5).

Fig. 3.2. Basin management of coconut with *Calopogonium mucunoides*

Table 3.5. Biomass and N contribution by green manure legumes in coconut basins

Legume species	*Laterite soil*		*Sandy soil*	
	Biomass(kg/basin)	*Nitrogen(g/basin)*	*Biomass(kg/basin)*	*Nitrogen(g/basin)*
Calapogonium mucunoides	27.21	186.53	14.71	102.61
Mimosa invisa	24.67	197.55	17.00	153.19
Pueraria phaseoloides	28.45	196.19	19.43	121.29

Generation of large quantities of nitrogen rich biomass is also possible through the cultivation of the fast growing leguminous tree crop, *Glyricidia* in the coconut plantations. It can also be grown in littoral sandy soils where no other green manure can establish. The planting density and pruning frequency to obtain higher biomass yield from *Glyricidia sepium* when grown as green manure crops in coconut plantation in littoral sandy soil was standardized (Subramaniam *et al.*, 2000). The highest fresh biomass yield of 8t/ha/year was obtained in the treatment with three rows of *Glyricidia* and three prunings per year (Table 3.6). The contribution of major nutrients through the loppings of *Glyricidia* was about 77.6 kg N, 5.7 kg P and 26.8 kg K per ha annually. The cultivation and incorporation of green leaves of *Glyricidia* helps in improvement in physical, chemical and biological properties of soil which are beneficial for the plant growth.

Table 3.6. Biomass yield and nutrient contribution of *Glyricidia* alley cropped in coconut gardens (kg/ha)

Glyricidia treatment	*Fresh biomass*	*Dry biomass*	*N*	*P*	*K*
2 rows, 3 prunings	6194	1816	61.5	4.6	20.4
2 rows, 3 prunings	5479	1594	54.8	4.0	18.4
3 rows, 3 prunings	7970	2296	77.6	5.7	26.8
3 rows, 4 prunings	7211	2011	69.5	5.1	24.0

Living Soil as Basis for Organic Farming

Living soil concept is basic to organic farming. The role of soil flora and fauna is well-recognized in this system of production as the prime drivers for various processes resulting in enhanced physical, chemical and biological health of soil. The inadequacy of nutrient and energy sources in the tropical soils is the limiting factor for the low biological activity and dormancy of large number of beneficial micro-organisms in soil. The concept of organic farming dwells on the realization of the biological potential of soil microbes and their capability to support the production process if the soils are fed with sufficient organic matter. The focus in organic farming is on development and application of innovative agro-technologies based on soil biodiversity to enhance the natural nutrient cycles and utilization of specific soil-plant-microbial associations in managing the soil in organic production. More soil biodiversity leads to stability and sustainability of production systems. Biological systems of regulation promoted by the soil microbes should function at the highest level to achieve closed nutrient cycles with less nutrient losses.

Microbial Resources and their Utilization

Biofertilizer formulations containing living cells of beneficial micro-organisms multiplied in suitable carrier can form low cost ecofriendly inputs for organic crop production. A range of micro-organisms possessing beneficial traits such as biological nitrogen fixation, phosphorus mobilization and production of plant growth promoting substances have been found to be closely associated with coconut palms (Thomas *et al.*,1991).

The diazotrophic bacteria isolated from coconut roots include different species of *Azospirillum*, *Herbaspirillum* sp., *Azoarcus* sp., *Burkholderia* sp., *Arthrobacter* sp., *Pseudomonas* sp. and *Bacillus* sp. (Prabhu and Thomas, 1999 Thomas *et al.,* 2001), *Beijerinckia indica* was the predominant asymbiotic N_2 fixer in coconut soils (Table 3.7) (Thomas *et al.*, 1985). Phosphate solubilizing microbes of coconut soil include *Pseudomonas* sp., *Bacillus* sp. and *Micrococcus* sp. (Thomas and Shantaram,1986). Efficient strains of nitrogen fixing bacteria and phosphate solubilisers were used for preparation of biofertilizers employing locally available materials such as vermicompost and coir pith as carrier materials. Inoculation trials indicated that the biofertilizers were effective in enhancing the vigour of coconut seedlings raised in polybags with 1:1 coir pith - soil mixture as substrate.

Table 3.7. Diversity of beneficial microbes associated with coconut palm

Type of association	*Group of organisms*	*Important genera*	*Beneficial traits*
Symbiotic	Arbuscular mycorrhizae	*Glomus* spp.	• Uptake of immobile elements·
		Gigaspora spp.	• Uptake of water under moisture stress condition·
		Acaulospora spp.	• Biosuppression of root pathogen
		Sclerocystis spp.	
		Scutellospora spp.	
Associative symbiotic	Diazotrophs	*Azospirillum* spp.	• Fixation of atmospheric N_2·
		Herbaspirillum spp.	• Production of plant growth promoting substances (PGPS)
		Azoarcus spp.	
		Burkholderia spp.	
		Arthrobacter spp.	
		Xanthobacter spp.	
		Pseudomonas spp.	
Endophytic	Plant Growth Promoting	*Bacillus* spp.	• Production of plant growth promoting substances·
	Rhizobacteria (PGPR)	*Pseudomonas* spp.	• Biosuppression
Rhizospheric association	Diazotrophs	*Beijerinckia* spp.	• Fixation of atmospheric N_2
		Azospirillum spp	• Production of PGPS
	Phosphate mobilisers Fungi	*Aspergillus* spp.	• Production of polysaccharides helping in soil aggregation
		Penicillium spp.	
	Bacteria	*Bacillus* spp.	• Solubilisation of insoluble phosphates·
		Micrococcus spp.	
		Pseudomonas spp.	• Production of PGPS

Recent studies at CPCRI resulted in the isolation of several strains of fluorescent pseudomonads from rhizosphere of coconut and cocoa with plant growth promotion properties (Litty Thomas, 2009; Priya George *et al.*, 2009). Screening of the isolates revealed that 90% of the isolates from coconut exhibited phosphate solubilizing capability, siderophore production and ammonification. IAA production and ACC deaminase activity was noticed in 85% of the isolates. Five percent of the isolates produced HCN and 2% of the isolates showed growth in N-free media. Antagonistic activity against pathogens of coconut such as *Ganoderma sp.* (causing basal stem rot disease) and *Thielaviopsis paradoxa* (causing stem bleeding disease) was shown by 8–16% of the PGPR isolates. Enormous possibilities exist for utilization of these efficient PGPRs for development of biofomulations and their applications in crop production.

Symbiotic associations with coconut roots is formed by several species of AM fungi belonging to five genera and the association provides benefit to the host plants in terms of crop nutrition, disease suppression and increased growth and yield of crops.

Soil Health Benefits Under Organic Farming

Soil health benefits due to organic farming is evident from the results obtained in the coconut based mixed farming system involving cultivation of fodder grass in the interspaces of coconut palms and integration of animal husbandry involving dairy, poultry and pisiculture as components to enhance the income from coconut gardens. The system was maintained with three treatments including 100% organics, 100% inorganic and 50% organic + 50% inorganic. The population density of soil microbial groups and function specific micro-organisms of beneficial micro-organisms were very high in the organic treatment when compared to the other treatments. The increase in the population of fungi was considerably high in 100% organic treatment compared to the other treatments (Table 3.8). Proliferation of micro-organisms was remarkably low in the treatment that received 100% chemical fertilizer.

Table 3.8. Microbial distribution in organic, inorganic and integrated nutrient management in coconut - fodder grass system

Treatments	*Bacteria (10^5 cfu/g soil)*		*Fungi (10^3 cfu/g soil,)*		*Actinomycetes (10^5 cfu/g soil)*	
	Coconut	*Interspace*	*Coconut*	*Interspace*	*Coconut*	*Interspace*
Monocrop + RF	13.45	10.00	6.86	4.16	9.62	6.16
MF + 50% organic + 50% inorganics	18.22	21.64	6.6	7.23	7.83	8.35
MF + 100% organics	23.17	29.22	18.16	19.31	11.33	14.26
MF + 100% inorganics	11.76	14.66	7.12	8.35	7	12.94

RF – Recommended fertilizers dose; MF – Mixed farming

Increase in microbial biomass and activities of soil enzymes (phosphatase and dehydrogenase) indicated that the soil biological activity was more in the farming system under organic system of cultivation than under inorganic or integrated treatment (Table 3.9). Microbial biomass was significantly high in the system where organic recycling was practiced. The microbial biomass is considered among the most labile pools of organic matter and thus, serves as the reservoir of plant nutrients.

Phosphatase activity relates to hydrolyzing P compounds and liberating P from insoluble sources for absorption by plants. The dehydrogenase activity was remarkably high in organic treatment which indicated higher level of microbial oxidative activity.

Table 3.9. Microbial biomass and enzyme activities in coconut based mixed farming systems

Microbial parameters	*Microbial biomass (µg C g⁻¹ soil)*		*Phosphatase (µg p-nitrophenol g⁻¹ soil h⁻¹)*		*Dehydrogenase (µg formazon g⁻¹ soil h⁻¹)*	
Treatment/crop	*Coconut*	*Interspace*	*Coconut*	*Interspace*	*Coconut*	*interspace*
Monocrop	190.79	153.56	45.07	22.49	7.53	5.34
50% organic	213.51	264.26	49.87	31.79	10.57	15.77
100% organic	292.04	312.18	62.15	51.98	16.66	22.99
100% inorganic	142.07	193.03	32.97	19.92	5.73	12.46
CD (P = 0.05)	11.19	4.85	2.69			

The quality of an agricultural soil is a measure of its capacity to sustain crop productivity while preserving environmental quality. Organic farming had positive contribution to soil functions to sustain crop growth and productivity. The augmented microbial build up and activities, biochemical changes and nutrient transformations mediated by soil micro-flora and microbial interactions enabled to achieve sustainability of the system.

CROP YIELD REALISED UNDER ORGANIC FARMING

Organic cultivation of coconut was found feasible with integrated treatments utilizing organic and bioinputs including recycling of waste biomass by vermicomposting, *in situ* cultivation and incorporation of leguminous cover crops and biofertilizers of *Azospirillum* and *Bacillus*. In a field trial conducted in an adult coconut plantation, the yield obtained with West Coast Tall cultivar of coconut under organic cultivation was 83 nuts/palm/year while in Dwarf X Tall hybrid (Chowghat Orange Dwarf X West Coast Tall), the yield was 93 nuts/palm/year.

Coconut based farming system with inter cropping of fodder grass in the inter-spaces of coconut with dairy and poultry as components (Fig. 3.3) was maintained with organic inputs including the recycling of cowshed wastes and vermi-composting of the biomass generated in the system. The yield of coconut in the organic cultivation was 108 nuts/palm/year and fodder yield was 106 t/ha/year. The system was highly remunerative with net return of Rs. 1,37,164/ha/year achieving economic and environmental advantages under organic farming.

Field experiments conducted under the All India Coordinated Research Project on Palms (AICRP, 2000) in different coconut growing tracts of the country had organic treatments such as composted coir pith to supply 100% N and neem cake + bone meal + ash. The yield under organic cultivation ranged from 80 nuts/palm/year to 145 nuts/palm/year in different locations (Fig. 3.4). This data revealed that economically viable coconut cultivation is feasible when palms are managed with organic inputs alone and organic cultivation of coconut is a viable proposition for profitable cultivation of coconut in a sustainable environment friendly manner.

Fodder grass intercropping

Dairy unit

Poultry unit

Fig. 3.3. Coconut based farming system at CPCRI, Kasaragod

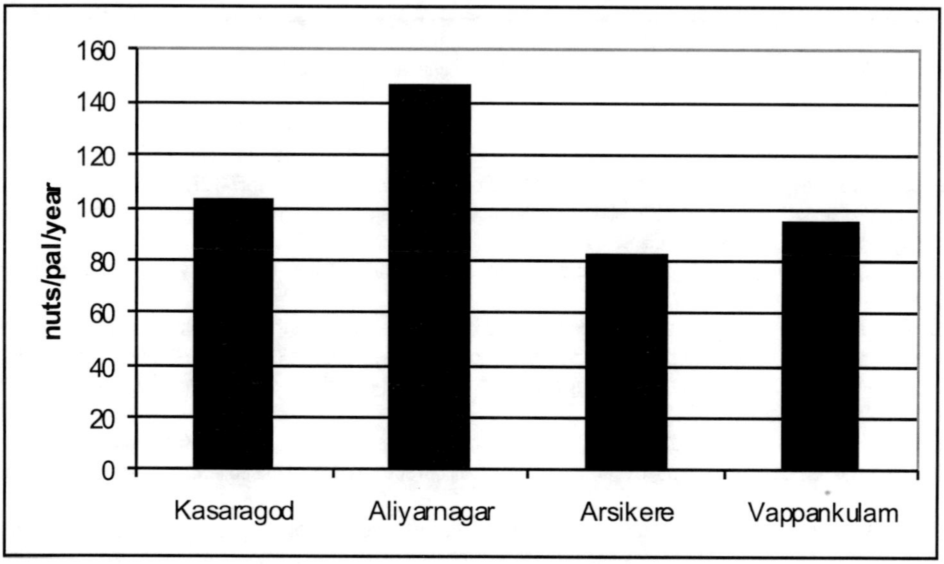

Fig. 3.4. Effect of organics on coconut yield

Under organic farming with application of vermicompost to supply 100 % of recommended N in arecanut the yield obtained was 1.70 kg dry nut (chali)/palm/year which indicated that arecanut crop can be cultivated with organic inputs to realize higher economic yield.

CONCLUSION

There have been significant advancements in the development of agro-technologies based on the principles of organic farming in plantation crops. The growth habit and planting methods of crops like coconut and arecanut make it highly amenable for organic farming through the low cost technologies based on the local resources. These agro-technologies integrate ecological principles into intensification process and ensure that plant nutrients are in constant and close cycling within the soil and plant compartments. They rely on wide variety of biological strategies adopted by the soil systems and are governed by the soil microbial dynamics.

Technologies are available for production of sufficient quantities of organic matter, efficient recycling and generation of high quality compost and growth promoting liquid formulations and utilization of microbes to supply the plant nutrients without any ecological damage as is the case with chemical agriculture. Enhancement of biodiversity is the primary principle used to evoke self-regulation and sustainability of the agro-ecosystems. When biodiversity is restored, a number of complex interactions between soil, soil organisms, plants and animals are established, giving stability to the system to form viable and sustainable organic production systems.

References

AICRP. 2000. All India Coordinated Research Project on Palms, annual Report, 2000. Central Plantation Crops Research Institute, Kasaragod, Kerala (India).

Bavappa, K.V.A., Kailasam, C., Khader, K.B.A., Biddappa, C.C., Khan, H.H., Kasturi Bai, K.V., Ramadasan, A., Sundararaju, P., Bopaiah, B.M., George V. Thomas., Misra, L.P., Balasimha, D., Bhat, N.T. and Bhat, K.S. 1986. Coconut and arecanut based high density multispecies cropping systems. *J. Plantation Crops* **14:**74-87.

Bhat, Ravi and Sujatha, S. 2007. Soil fertility status as influenced by arecanut based cropping system and nutrient management. *J. Plantation Crops* **35:**158-65

Biddappa, C.C., Upadhyay, A.*K.,* Hegde, M.R. and Palaniswami, C. 1996. Organic matter recycling in plantation crops. *Journal of Plantation Crops* **24:** 71-85.

Bopaiah, B.M. and Shetty, H.S. 1991a. Soil microflora and biological activities in the rhizospheres and root regions of coconut-based multistoreyed cropping and coconut monocropping systems. *Soil Biology & Biochemistry* **23**(1): 89-94.

Bopaiah, B.M. and Shetty, H.S. 1991b. Microbiology arid fertility in coconut-based mixed farming and coconut monocropping system. *Tropical Agriculture* **68**(2): 135-38.

CPCRI 2003. In *Annual Report* 2002-03, pp. 178, Central Plantation Crops Research Institute, Kasaragod, Kerala.

Ghai, S.K. and Thomas George, V. 1989. Occurrence of *Azospirillum* spp. in coconut based farming system. *Plant and Soil* **114:** 235-41.

Khan, H.H., Sankaranarayanan, M.P. and Narayana, K.B. 1978. Characteristics of coconut soils of India I. Morphology, some physico-chemical characteristics and taxonomy, Proceedings *PLACROSYM-J;* E.V. Nelliat (ed.), pp.54-79, Indian Society for Plantation Crops, CPCRI, Kasaragod, Kerala.

Liyanage, L.V.K. and Jayasundera, H.P.S. 1988. *Coconut Bulletin* **5**(1):1-4.

Maheswarappa, H.P., Anithakumari, P., Kamalakshyamma, P.G. and Shanvas, M. 2003. Cowpea *(Vigna unguiculata)* as low cost input green mananure crop for basin management under root *(wilt)* garden. *Indian Coconut Journal* **34**(5): 11-13.

Maheswarappa, H.P., Sairam. C.V., Dhanapal, R., Vidhan Singh, T. and Hegde, M.R. 2000. Coconut based mixed farming systems and its economic advantages. *Indian Coconut Journal,* **31** (3): 17-19.

Merilyn, V.J. and Thomas George, V. 1992. Distribution of nitrogen fixing *Beijerinckia* in the Rizosphere of coconut. *Journal of Plantation Crops* **20:**149-55.

Nair, S.K. and Subba Rao, N.S. 1977. Microbiology of the root region of coconut and Cocoa under mixed cropping. *Plant and Soil* **146:**511-20.

Palaniswami. C., George, V., Thomas, R., Dhanapal, P., Subramannian, H.P., Maheswarappa and A.K. Upadhyay 2007. Integrated Nutrient Management in Coconut based Cropping System. Technical bulletin No. 49. CPCRI, Kasaragod, Kerala, India, 24p

Pillai, N.G. and Davis, T.A. 1963. Exhaust of micronutrients by the coconut palms: A preliminary study. *Indian Coconut Journal* **16:** 81-87.

Prabhu, S.R., Thomas George, V. and *Baldani,* J.I. 1998. Occurrence of the associative symbiotic nitrogen fixing *Azospirillum amazonense* in the root environments of coconut palm *(Cocos nucifera* L.). 39th Annual Conference *Association for Microbiologists of India,* held at College of Fisheries, Mangalore, 5-6 Dec. 1998.

Prabhu, S.R., Thomas George, V., Nierzwicki-Baner, S.A. and Prasad, T.G. 2000. GA-like substances producing endophytic Gram positive bacteria are associated with coconut palm *(Cocos nucifera* L.). National Seminar *Recent Advances in Plant Biology,* held at CPCRI, Kasaragod, 3-5 Feb. 2000.

Subramanium, P., Biddappa, C.C., Maheswarappa, H.P., Dhanapal, R. and Palaniswami, P. 2000. Growing *Glyricidia* as a green manure crop in coconut garden. *Journal of Planation Crops* **28**:212-17.

Thomas George, V. 1988. Vesicular-arbuscular mycorrhizal colonization in coconut in relation to root (wilt) disease and inter/mixed cropping. *Indian Journal of Agricultural Sciences* **57:**145-47.

Thomas George, V. and Prabhu, S.R. 2003. Association of diazotrophic and plant growth promoting rhizobacteria with coconut palm *(Cocos nucifera* L.). Paper, VIth International *PGPR Workshop,* M. S. Reddy, M. Anandaraj, S. J. Eapen, Y.R. Sarma and J.W. Kloepper (eds.), pp. 20-26, held at Indian Institute of Spices Research, Calicut, Kerala.

Thomas George, V. and Shantaram, M.V. 1984. *In situ* cultivation and incorporation of green manure legumes in coconut basins: An approach to Improve soil fertility and microbial activity. *Plant and Soil* **80:**373-80.

Thomas George, V. and Shantaram, M. V. 1993. Biomass production and nodulation of green manure legumes in coconut basins in laterite soil and their influence on soil fertility. *CORD* **IX:**38-47.

Thomas George, V., Prabhu, S.R., Reeny, M.Z, and Bopaiah, B.M. 1998. Evaluation of lignocellulosic biomass from coconut palm as substrate for cultivation of *Pleurotus sajor* caju (Fr.) Singer. *World Journal of Microbiology* and *Biotechnology* **14:** 879-82.

Thomas George, V., Prabhu, S.R., Subramanian, P. and Iyer Rohini 2001. *Organic Farming Technologies in Coconut,* ATIC Series Publication No.4, CPCRI, Kasaragod. 14p.

Thomas George, V.and Palaniswami, C. 2005. Effect of different components of farming system and on-farm organic recycling on coconut crop and soil fertility. *In: Proceedings of National Seminar on More remunerative coconut based farming system* 12th December 2005, Dr. Balasaheb Sawant Konkan Krishi Vidyapeeth, Dapoli, Maharashtra, pp 49-61.

4

Technological Intervention for Organic Production of Cashew

N. Yadukumar, M.G. Nayak and M.G. Bhat

INTRODUCTION

Cashew is grown over an area of 8.54 lakh ha with a production of 6.2 lakh tonnes in India. During the year 2006-07, the total quantity of 1.14 lakh tonnes of processed kernels was exported with earnings of Rs. 2289 crore (NRCC Vision 2025, 2007). Cashew is generally grown on least fertile soil. Added to this, cashew is a neglected crop resulting in non-exploitation of full yield potential. Nutrient management in cashew is very much crucial as cashew being totally neglected crop, does not receive manure at all. If fertilizers are applied significant increase in yield is realized. Considering average annual removal of major plant nutrients at the present level of production by six or more years old cashew trees and total productive area under cultivation in India, the total nutrient removal from soil is up to the tune of 185 thousand tonnes of N, 40.24 thousand tonnes of P_2O_5 and 120 thousand tonnes of K_2O annually. Actual estimated addition of the nutrients through fertilizer application is meager for the reason that only 20% of the total area (1, 43,250 ha) under cultivation receives fertilizers. If we take this into account only 11.17 thousand tonnes of N, 2.99 thousand tons of P_2O_5 and 2.7 thousand tonnes of K_2O are applied through fertilizers. This much quantity of fertilizer application is too low compared to any other horticultural crop.

Though cashew trees shed 20–40 kg leaves and apple per ha plantation annually, 50% of this is used for composting and application to other important crops like arecanut, coconut and rice in west coast and east coast region where cashew is generally grown in addition to the above crops. There is a local practice of collecting cashew leaf litter at the time of harvest of the crop which coincides with summer season and applying the same to cattle shed where it undergoes half decomposition due to presence of cow dung and cow urine and trampling of the biomass. Such half decomposed biomass, instead of applying to cashew gardens, goes to other important crops like arecanut, coconut and rice. Occasionally cashew leaf litter is also collected for fuel purpose. These practices lead to general depletion of nutrients in cashew garden every year there by reduction in productivity. If the above mentioned huge quantity of biomass is properly utilized for application to cashew gardens, considerable quantity of nutrients can be retained for replenishment of nutrients. Appreciable quantity of nutrients present in the crop residue is often not recycled into the system in a systematic manner. The potential nutrients in recycling of biomass generated in cashew has been quantified as 33.8 kg N, 17.6 kg P_2O_5 and 20.6 kg K_2O/ha/year and considerable quantity of other essential elements including micronutrients (Anonymous, 2003). These organic forms of nutrients improve soil health through their effect on soil physico-chemical properties and microbial actions. All the same, use of chemical fertilizers is the easiest way of boosting crop yield. However, due to its increasing cost and removal of subsidies to the farmers and environmental problems, the farmers are inclined to prefer organic manure to inorganic manures.

The farmers of the developed and developing countries are beginning to realize the harmful effects caused by excessive use of chemical fertilizers and pesticides on the food crops produced. Those who are already conscious of the problem are now focusing attention on cost-effective and nature-friendly technologies, partly owing to the reason that present agriculture based on chemical fertilizers is becoming less and less competitive and partly also in response to the changing consumer preferences for farm produces free from chemical contaminants. Such farmers are evincing interest in leaning towards organic agriculture as an alternative farming system. Though integrated nutrient management involving application of both organic and inorganic manuring is suggested to be the best, it is not catching up to the expected level. Technological interventions for organic production of cashew are discussed in this chapter.

Table 4.1. Nutrient status of soils in cashew growing tracts (Nutrient kg/ha)

Locations	*N*	*P_2O_5*	*K_2O*	*Nutrient status*
Coastal Karnataka	120	3–43	100	Low to high in N, Low to high in P_2O_5 and low in K_2O
Coastal Kerala	60–120	6–13	186–345	Low in N, Low to medium in P_2O_5 and medium to high in K_2O
Maharashtra	156–479	07–2.12	79–208	Low to medium in N, Low in P_2O_5 and medium to high in K_2O
Andhra Pradesh (Bapatla)	54–71	4–6	150–190	Low
Orissa	38–80	15–28	100–150	Low in N, medium to high in P_2O_5 and medium in K_2O

Source: Collected from reports of AICRP on Cashew

Varieties Recommended for Organic Farming

Only mid and late season varieties are best suited for organic farming. The reason being that in early season varieties the crop flushes early and flowers early, which attracts maximum infestation of tea mosquito. Because it is during this period (November—January) that the climate is cool and experiences dewfall during early morning followed by cloudy day resulting in maximum spoilage of the flowers due to the highest infestation by the tea mosquito bug (TMB). Mid and late season (February—April) varieties escape this. Due to increased temperature during flowering and fruiting in mid and late varieties the TMB population comes down thereby the crop damage due to it is minimum or nil. As such, control of TMB does not arise under such situations. However, under extreme changed weather conditions favorable to sudden outbreak of TMB, controlling this pest even with chemical means is also difficult. Some of the most promising mid and late flowering varieties released are given below:

***Mid-season flowering varieties*:** Bhaskara, Dhana (H1608), Dharashree, Amrutha (H1597), Priyanka (H1591) BPP-8, V-4, V-7.

***Late season flowering varieties*:** Ullal-1, Chinthamani-1 and Madakkathra-2

Production of Vegetatively Propagated Cashew Grafts Organically

The pot mixture used for raising seedlings meant for grafting should contain organic manure (cow dung), rock phosphate and biofertilizers. The potting mixture should have 1/3rd soil, 1/3rd cow dung and the remaining 1/3rd sand and 5 g each of *Azospirillum*, *Pseudomonas* and *Aspergillus* biofertilizers and 5 g of rock phosphate. Bags filled with potting mixture are kept for raising seedlings organically.

Planting Pit and Spacing

Planting grafts in bigger pits of size one cubic meter after filling 2/3rd depth with top soil, farmyard manure of 5 kg or 2 kg poultry manure or 1kg castor cake/pit is recommended. Planting of cashew grafts should be at 2/3rd depth and after planting, it should be properly mulched. Normal soil and water conservation measures like terracing with crescent bund (Rejani and Yadukumar, 2007) should be taken within three years after planting. Spacing maintained can be 8 m x 8 m, 10m $\times$ 5m under normal density planting system and 5m $\times$ 5m, 6m $\times$ 4m under high density planting systems.

Organically Recyclable Biomass Waste (Crop Residue) Available in Cashew Garden as Potential Nutrient Source

In general every year considerable quantity of leaf litter and cashew apple are available at the base of the crop if the same is not taken away from the location. The cashew leaf litter, apple waste and weed biomass collected at the base of the plant is to the tune of 40 kg/tree/year. When it is decomposed, 65% recovery can be expected. This amounts to 26 kg of biomass. This half decomposed biomass contains around 0.65% N, 0.41% P,

0.45% K, 0.22% Ca, 0.19% Mg, and 369 ppm Fe, 14.6 ppm Cu, 16.5 ppm Zn and 283 ppm Mn (Yadukumar and Nandan, 2005). The total macro-nutrients added to the soil when above quantity of biomass is incorporated will be approximately 169 g N, 107 g of P_2O_5 and 117g K_2O/tree. Soil mineralisation also takes place in the system, thereby, releasing the nutrients in available form. In addition to the utilization of native soil nutrients cashew requires additional quantities of nutrients for economical yields.

Results of the Organic Manuring Trials

In an experiment conducted at NRCC, Puttur (2000-03), it was observed that application of 20 kg poultry manure alone with soil conservation technique (with crescent bund) increased the yield by 80 % compared to control plot (Table 4.2).To understand the performance of the crop for different combinations of organic and inorganic manures a trial was laid out in 2002 and continued upto 2003 on six and seven years old cashew trees (NRCC Selection 2 variety). The results indicated that application of bio-fertilizers-*Azospirillum* with the compost of organically recyclable biomass (ORB) available in cashew garden to meet 100% N requirement gave the highest yield of 1.053 t/ha on the seventh year whereas in the previous year it gave minimum yield. This indicated that in the long run this treatment is good. Next higher yield obtained was from the plot receiving 50–75% N through inorganic fertilizers and remaining 25–50% N through compost of organically recyclable biomass (1.02 and 1.03 t/ha, Table 4.3). Compost of ORB available in cashew garden was prepared by adding 20% of weight of total biomass with cow dung slurry. The compost so prepared contained 1.6% N, 0.54% P and 0.4% K and considerable amount of other micro-nutrients. The procedure to convert cashew leaf litter and waste apple available in cashew garden into suitable compost material in a large scale is standardized at NRCC, Puttur.

Table 4.2. Influence of soil and water conservation techniques on nut yield of 13 years old cashew orchard (kg/ha)

*Treatments**	*Sub plot treatments*			*Mean*	*% of control*
	1	*2*	*3*		
Individual tree base terracing with catch pit	818.5	946.7	1068.3	944.5	145.2
Individual tree base terracing with crescent bund	1051.1	1212.3	1261.7	1175.0	180.6
Staggered trenches between two rows	831.1	1128.9	1032.0	997.3	153.3
Control plot without any soil and water conservation	560.4	614.4	776.4	650.4	100
Mean	815.3	975.6	1034.6		
CD (P = 0.05) for Soil conservation measures	289.7				
CD (P = 0.05) values for manures	565.47				
CD (P = 0.05) values for interaction	NS				

*All the treatments received recommended organic and inorganic manures

NB: 1 = Control (No manure application), 2 = Recommended doses of fertilizers (500g N, 125 g P_2O_5, 125 g K_2O/tree + 10 kg poultry manure/ tree/ year), 3 = 20 kg poultry manure

Table 4.3. Effect of manurial doses on yield-five and six years after planting

Treatments (g/tree/year)	*Nut yield (kg/ha)*		***Total for 2 years***
	5YAP	*6YAP*	
No fertilizers	228	669	877
100% N RFD	243	951	1194
Azospirillum + 100% N-RCB	206	1053	1259
Azatobactor + 100% N-RCB	212	831	1043
25%N RFD + 75% N-RCB	112	942	1054
50%N RFD + 50% N-RCB	300	1024	1324
75%N RFD + 25% N-RCB	247	1031	1278
25%N RFD + 75% N-PM	258	795	1053
50%N RFD + 50% N-PM	315	718	1033
75%N RFD + 25% N-PM	222	989	1211
CD (P=0.05)for manurial doses		546	

RFD = Recommended Fertilizer Doses. RCB = Recyclable Cashew Biomass available in garden. PM = Poultry Manure. YAP = Years After Planting

PREPARATION OF VERMI-COMPOST

Attempt was also made to prepare vermicompost from the recyclable biomass available in cashew garden especially cashew leaf liter and apples. Cashew leaf litter and apples are allowed to decompose for 20 days after mixing with 15% cow dung slurry. Then the material is fed to earthworms that were developed previously in plastic bucket by feeding cow dung material. Vermi-compost chambers can be prepared as per the following details. Simple structure of chamber with 1.5 m height, 5 m long with 1.5 m width is erected. At the bottom the chamber, there should be a cement channel to fill water to prevent entry of ants. The chamber should be shaded during summer and protected from rain during rainy season by covering with shaded nets and silpauline sheet or any local thatching materials like coconut or areca nut leaves spread over the roof constructed specially for the purpose. For drainage of vermiwash a hole should be provided which is connected to a tank with PVC pipes so that the vermiwash can be utilized as liquid manure. Percentage recovery of compost will be 65. The compost should be sprinkled with sufficient quantity of water as and when moisture level goes below 40 %. For this, the water should be sprinkled once in four days so that the biomass is always in wet condition. Nearly 750 kg compost can be obtained from one tonne cashew recyclable biomass. The vermi-compost contained 1.21% N, 0.898% P, 0.59% K, 2.75% Ca, 0.82% Mg, 29.66 ppm Zn, 24.26 ppm Mn, 12.23 ppm Cu and 162 ppm Fe. Population of useful bacteria, fungi, actinomycetes, azospirillum were 39×10^6, 39×10^5, 28×10^5, and 0.52×10^4 cfu (colony forming units), respectively (Yadukumar, 2007).

Nutrient Management in Organic Farming

In the case of high density planting system it is proved beyond doubt that the recommended doses of nutrients are not required six to seven years after planting since the crop canopy increases every year and covers the ground completely. It is at this stage that the crop does not perform in terms of yield. Every year if we add lot of external nutrients as per the recommended doses, it causes unwanted increase in vegetative growth and overlapping of branches giving heavy shade and consequent decreased flowering and fruiting resulting in decreased yield. In high density planting system, the leaf deposit/unit area is more than two times compared to normal density planting system. This huge deposit of leaf litter and apples contribute natural build up of nutrients in soil, which serves as rich source of nutrients. Every year external nutrient application has to be reduced. Studies on nutrient management indicated that 1/3rd of recommended doses will be enough for maintaining cashew plantation when grown in high density planting system.

In view of growing demand for organically grown cashew, strategies need to be developed for organic farming. Some of the commonly available organic farming manures like poultry manure, FYM, vermicompost, oil cakes, coir pith-organic manures mixture hold promise in cashew in organic farming and the same is under investigation at National Research Center for Cashew. There is high potentiality to utilize cow dung slurry in organic farming. Approximate quantities of cow dung available in different states are given in Table 4.4.

Table 4.4. Annual outturn of wet dung from cattle and buffalo (in million tones) and nutrient potential of bovine excreta

States	*Million tons of cow dung*
Andhra Pradesh	77.389
Kerala	4.538
Maharashtra	107.986
Karnataka	54.806
Tamil Nadu	70.027

Source: Gaur *et al., 1992.* Organics in soil health and crop production, Edited by P.K. Thampan.

Green leaf is in abundance in Kerala, Coastal Karnataka and Maharashtra, where mostly cashew is grown. The green leaf can serve as good organic manure to cashew. Green manure crops like glyricidia and pongamia can be grown very well under rain fed conditions on the border or inside cashew plantations to produce green manure. Approximate nutrient contents in each kind of organic manure are given in Table 4.5.

While considering the quantity of nutrients available in soil, the approximate quantity of manure (organic manure) required to meet the recommended dose of N of 500g/tree is presented in Table 4.5. The recommended dose of N has been arrived based on the trials conducted in various Research Stations. In organic manure, availability of nutrients normally stretches for longer period and in addition, the loss of nutrients through leaching is minimum. Hence, we need not apply the quantity of organic manure to the extent of actual nutrients requirement, which we otherwise calculate, when we consider applying nutrients in the form of chemical fertilizers.

Table 4.5. Nutrient status of organic manure (g/kg of manure)

Manure	*N*	P_2O_5	K_2O
Poultry manure	28.7	10.3	13.0
Farm Yard Manure (FYM)	8.0	4.1	7.4
Vermicompost	16.0	22.0	6.7
Green Manure -Glyricidia	39.8	7.1	11.8
Cashew waste compost or compost of recyclable cashew biomass	10.5	5.3	4
Biogas slurry	18.0	9.2	8.0
Water Hyacinth compost	20.0	10.0	23.0
Castor cake	58.0	18.0	16.0
Groundnut cake	78.0	15.0	14.0
Pongamia cake	34.0	10.0	12.0
Neem cake	52.0	10.0	14.0

Source: Organics in Soil Health in Crop Production 1993 by PK Thampan, Pee Kay Tree – Crop Development Foundation, Cochin

Due to nutrient loss through seepage, runoff and denitrification, hardly 50% of N and 75% of the applied K_2O will be made available to the plants if we apply these nutrients through chemical fertilizers ,whereas, such nutrient loss is negligible in the case of organic manures. Detailed study on this aspect need to be taken up for suggesting options to farmers like the exact quantity of different organic manures to be applied/tree depending upon the availability of these manures in the respective local markets.

The quantity of organic manures to meet the full requirement of P_2O_5 and K_2O in all cases except in the case of green manure and groundnut cake is mentioned in Table 4.6. When groundnut cake and green manures are used, bone meal and wood ash may be suggested to supplement P_2O_5 and K_2O requirements if the farmer prefers producing cashew organically, which fetches almost 30 % increase in price.

Green Manuring

Leguminous crops such as *Peuraria javanica*, *Calapogonium muconoides* and *Centrosema pubescens* and *Glyricidia maculata* add organic matter, enrich the soil nutrients, prevent soil erosion and conserve soil moisture. Seeds of these cover crops can be sown in the beginning of rainy season with a seed rate of 12 kg/ha in the interspaces of cashew orchard. Beds of 30cm × 30cm are prepared in sloppy degraded soils by loosening the soil and mixing compost or cow dung.

Table 4.6. Quantity of organic manures required by adult cashew tree (kg/tree)

Manure	*Quantity required* (kg/tree)	*Nutrients supplied (g/tree)* N	P_2O_5	K_2O
Poultry manure	17.4	500	180	226
Farm Yard Manure (FYM)	62.5	500	256	462
Vermi-compost	31.3	500	689	210
Green Manure Glyricidia	12.5	500	89	183
Biogas slurry	27.8	500	256	222
Water Hyacinth compost	25.0	500	250	575
Castor cake	8.6	500	155	138
Groundnut cake	6.4	500	96	90
Pongamia cake	14.7	500	147	191
Cashew waste compost	33.0	500	250	200
Neem cake	9.61	500	96	134

Note: Calculated based on nutrient concentration available in each kind of manure as worked out at NRCC

Presoaking of the seeds in water for 6 hours ensures better germination. Seeds sown in the beds should be covered by a thin layer of soil. Fencing is necessary to avoid cattle grazing. In addition to utilization of recyclable cashew biomass waste, glyricidia can be grown in the inter space between two rows of cashew and yearly with three cuttings 60kg green biomass of glyricidia can be produced which can be added to base of the cashew plant (Yadukumar *et al.*, 2007). From one ha area of cashew plantation nearly seven tonnes of dried green biomass can be produced and applied to cashew that is equal to application of nearly 186 kg N, 40 kg P_2O_5 and 67 kg K_2O/ha/year. This will meet entire nutrient requirement of cashew and it was found that the same yield could be achieved compared to the plot receiving recommended Package of Practices in the case of nutrients.

Inter and Mixed Cropping

Leguminous crops like horse gram, cowpea, and other crops like groundnut; tapioca, vegetable and fodder crops, etc., are grown as inte-rcrops in cashew plantations. Out of the annuals, biennials, fruit crops and tree species when grown as inter/mixed crops pineapple has been found to be the best intercrop in cashew garden during the first seven years (Yadukumar, 2007a).

Three trenches (1 m width 0.5 m depth and of convenient length) are to be opened across the slope between two rows of cashew. In each trench, two rows pineapple suckers should be planted at 60 cm between rows and 40 cm between two suckers within a row. In each trench for 1 m length half basket of compost should be added and mixed with soil before planting the suckers. One-hectare cashew orchard can accommodate 15,000 pineapple suckers. Pineapple starts yielding from 2nd year and from 4th year it should be replanted in a new trench dug out by the side of existing trench and the same can be retained till 7th year of cashew plantation. After seven years, because of heavy shade of branches of cashew over the pineapple and due to difficulty in picking raw cashew nuts fallen over pineapple plants it may not be feasible to grow this inter-crop economically. Nearly 40 tonnes of pineapple fruits can be expected for seven years with a net return of about Rs. 2 lakh. Expenditure would be Rs. 1 lakh and hence the net profit from pineapple alone will be Rs.1 lakh/ha of cashew garden. Apart from gaining a net profit of Rs.1 lakh from pineapple 30 % increase in cashew nut production was observed mainly due to better soil and water conservation in trenches where pineapple is planted across the slope between two rows of cashew.

Other crops like turmeric, ginger and elephant foot yam can be grown as intercrops organically for the first five years of cashew plantations. The profit realized from these crops may range from Rs. 20,000 to 50, 000/ha. Ample scope is there to grow vanilla and pepper as mixed crops with cashew if irrigation facility is provided. *Garcinia indica* (kokum) can be grown as mixed crop by maintaining same population as that of main crop-cashew (200 each/ha). All the above crops are to be grown organically as mixed crops.

Weed Management

Generally weeding in cashew can be done twice a year. One weeding in the month of August during manure application and another weeding can be done just before the start of flushing and flowering (October or November). The weed biomass can be effectively recycled as that of cashew biomass waste explained earlier.

Management of Tea Mosquito Bug (TMB)

Tea mosquito bug can cause yield reduction to the extent of 30–40% by damaging tender shoots, inflorescences and nuts. Both adults and nymphs of TMB suck the sap from tender shoots, panicles and immature nuts and apples which results in formation of black lesions. These lesions on shoots and panicles coalesce causing shoot blight or blossom blight. The plants can escape TMB attack if the new flush is delayed. Planting mid-season or late season flowering varieties would be the right strategy to escape TMB. Even early flowering varieties also flower 10–15 days late if the plants are grown organically compared to chemical fertilizer applied ones (Table 4.7). The delay in flowering naturally minimizes the incidence of TMB by escaping multiplication of its population. Clean cultivation is another method to check multiplication of TMB considerably. All weeds should be removed and heaped as mulch. Encouraging multiplication of weaver ants (*Occoyphylla smargdina*) is another way to check the TMB. The red ants also feed on the nymphs of TMB. In the initial stages, placing small pieces of jaggery at the cracks and crevices of the bark can attract the red ants.

Table 4.7. Time of flowering in different varieties during 2007-08 (50% completion of flowering)

Varieties	*Organic farming*	*Conventional farming*
Bhaskara	Dec. 2nd week	Nov. 4th week
Ulal-3	Dec. 2nd week	Nov. 4th week
VRI-3	Nov. 3rd week	Nov. 1st week
Madakkathara-2	Jan.1st week	Dec. 3rd wek
K22-1	Jan. 3rd week	Jan. 1st week
NRCC Selection 2	Nov. 3rd week	Nov. 1st week

At National Research Centre for Cashew, plot of 2.5 ha is maintained for the last four years and here though initially TMB damage was noticed in small pockets, no economic loss of yield was observed. Studies on the effect of various organic pesticides like Phalada, Phytozeal and Phalini in late variety (NDR2-1) indicated that combination of Phytozeal and Phalada resulted in maximum yield with minimum tea mosquito damage (Table 4.8). Phalada IIIC is a herbal based liquid insecticide certified as 100 % organic insecticide by SKAL International (Certifying agency based at Netherland). The contents of this herbal preparations are combination of water and extracts of *Saptora japonica* roots (2%), *Annona reticulata* seeds (15 %), *Adathoda vasika* (leaf-12 %)), *Sapindus trifolatus* (30 %), *Pongamia pinnata* (20%). Phytozeal is a herbal based liquid insecticide combined with vermiwash which serves as good nutrient source also. Phytozeal contained vermiwash (90 %), plant extracts of *Lantana camera* flower (0.5%), *Cynadon dactylon* grass (0.5 %), seaweed (5 %), *Calotropis procea* (2 %) and broths of *Pseudomonas fluorescence* (0.1 %) and *Trichoderma viridae* (0.1 %). Phalini is diluted form of cashew nut shell oil mixed with water. In another observational, trial it was found that tea moasquito damage score is more in both Bhaskara and Ullal-3 varieties in conventional farming compared to organic farming (Table 4.9).

Table 4.8. Effect of organic pesticides on the performance of four years old cashew tree (Madk-2)

Treatments	*Yield (kg/tree)*
Phytozeal 5.0 ml/l + phalada IIIC 1.0 ml/l	3.07
Phytozeal 7.5 ml/l + phalada IIIC 1.5 ml/l	3.52
Phytozeal 10.0 ml/l + phalada IIIC 2.0 ml/l	4.17
Phalini (1:30) + Phytozeal 5.0 ml/l	3.03
Phytozeal 5.0 ml/l	3.15
Control	1.81

Table 4.9. Tea mosquito damage score in different farming systems

Varieties	*Organic farming*	*Conventional farming*
Bhaskara	0.21	0.51
Ullal-3	0.30	2.10

Damage score (1-4), 1 = least damaged, 4 = severely damaged (panicle completely dried)

Management of Cashew Stem and Root Borer

Older cashew trees are more prone to infestation by CSRB. Adult female beetles lay eggs in the bark crevices of the stem (near to the collar region) or on the exposed roots. After hatching, the young grubs tunnel immediately into the bark of the trunk or root regions. Due to extensive feeding by irregular tunneling in the bark, the translocation of the nutrients is hampered leading to drying of leaves and twigs. In the infested portions, the gum and frass (chewed fibres and excreta pellets) start oozing out from the tree, which is the initial symptom of pest attack. Curative measures should be taken up immediately which are given below.

The infested portion having frass below the bark on either the main stem or root portion (after suitably digging the soil to a depth of 60cm) should be carefully chiseled to inflict minimum damage to the bark and tunnels should be tracked towards the fresher frass to locate the CSRB grubs, which should then be removed and killed. In case white powdery fibres are seen, it indicates that the grubs have entered to the heartwood for pupation. In such cases a pliable wire (for example, automobile gear wire) should be inserted deep into the pupation hole and pushed in till a slushy sound is heard/milky white fluid flows out indicating damage to the grub or pupa. Later chiseled portion of the bark should be swabbed with 10% Bordeaux paste. When a paste of cow dung and ash is applied to the injured portion of the bark it helps in curing of wounds.

Trees having more than 50% of the bark circumferences damaged by insect or with yellowing of the canopy hardly ever recover. However, the existence of different pest stages in such trees will act as inoculums for the next season. Hence, phyto-sanitation should be adopted by uprooting such trees, which are beyond recovery and disposed of immediately, along with removal of pest stages in other infested trees to prevent build up of pest inoculums. As the pest incidence occurs during the period of nut collection (Feb–May), the infested trees can be marked and curative measures must be adopted immediately, so that the grubs do not inflict more damage.

Yield

In the experiments conducted at National Research Centre for Cashew, Puttur on the organic farming the yield realized was to the tune of 1-1.3 tonne per hectare (Tables 4.10 and 4.11) under high density planting system in the second harvest (four years old plantation).

Table 4.10. Yield performance during third and fourth year under organic farming (Mdk2)

System of planting	*3rd year*	*4th year*
HDP with irrigation	2.9 kg/tree 1200 kg/ha	3.38 kg/tree 1352 kg/ha
HDP- without irrigation	2 kg/tree 800 kg/ha	2.5 kg/tree 1000 kg/ha
NDP-with Pineapple	3.2 kg/tree 640 kg/ha	4.7 kg/tree 940 kg/ha
HDP with turmeric	1.37 kg/tree 548 kg/ha	2.83 kg/tree 1132 kg/ha

HDP = High Density Planting NDP = Normal Density Planting

Table 4.11. Variety-wise performance of cashew grown organically during third and fourth year (up to second harvest)

Variety (HDP)	*3rd year(kg/ha)*	*4th year(kg/ha)*
Ullal 3	710	1208
Ullal 2	638	1064
Sel 2	546	544
Bhaskara	506	1212
VRI 3	593	1300
V 4	284	1200
Dhana	206	1088
Kanaka	313	1092
Madakkathara 2	256	1288
K 22-1	273	1432
Priyanka	330	1244
Ullal 4	250	1028

References

Anonymous, 2003. Develpoping integrated production packages for enhancing productivity of cashew, 2003, NATP Final Report, NRCC, Puttur, DK, Karnataka: 55 p.

NRCC, Vision – 2025. 2007. NRC Cashew Perspective Plan, National Research centre, Puthur, Karnataka, 46 p.

Rejani, R and Yadukumar, N. 2007. Response of cashew plantations grown in slope areas of West Coast region of India for soil and water conservation measures. Proceedings of the National Seminar on Research, Development and Marketing of Cashew, 20-21st November, ICAR, Research Complex, Goa, pp: 63-64.

Yadukumar, N, Rejani, R. and Nadan, S.L. 2007. Green manuring for high density cashew orchards. Proceedings of the National seminar on Research, Development and Marketing of Cashew, 20-21st November, ICAR, Research Complex, Goa, pp: 59-61.

Yadukumar, N. 2007. Organic farming in cashew. Proceedings of the National Seminar on Research, Development and Marketing of Cashew, 20-21st November, ICAR, Research Complex, Goa, pp 36-46.

Yadukumar, N. 2003. Organic farming practices in cashew plantations. Recent advances in organic farming technologies in plantation crops, short course:18-27, Nov 2003. edited by George V Thomas and Palanisami, CPCRI, Kasaragod, Kerala.

Yadukumar, N and Nandan, S L. 2005. Recycling of organic waste of cashew plantations by aerobic composting- *J. Plantation Crops.* **33**(2): 99-102.

Yadukumar. N. 2007. Development of suitable cashew based cropping systems – inter and mixed cropping with cashew. Proceedings of the National Seminar on Research, Development and Marketing of Cashew, 20-21st November, ICAR, Research Complex, Goa, 62 p.

5

Prospects of Organic Vegetable Cultivation in India

M. Prabhakar and S.S. Hebbar

INTRODUCTION

Vegetable farming is generally an intensive but profitable venture since per hectare yields are comparatively more and the income is large. In order to minimize risks involved in vegetable production, heavy fertilization as well as pesticide application is commonly practiced. To make it a sustainable and profitable system, reducing usage of these harmful chemicals and slowly changing over to non-chemical farming is the concern of the day. In this direction for change over or shift from the present heavily chemical oriented farming practices to environmentally friendly but profitable production systems require developing appropriate package of practices that depend on natural inputs and it is better if they are sourced within the farm. Although vegetable crops vary substantially as compared to other field crops, they all require skilled management to produce a high quality product without use of chemicals. These crops rely on intensive management of soil, water, nutrient resources and pest infestation organic way. Organic farming in vegetable crops needs to be standardized more in crops which are consuming maximum chemicals and pesticides in their cultivation such as cauliflower, cabbage, okra, tomato and also in vegetable crops which have high export potential like onion including rose onion, gherkins, okra, beans, peas, etc.

Although, as yet infancy, Organic Vegetable Farming is becoming important in the agriculture sector in India. In its most developed form, organic vegetable farming is both a philosophy and a system of agriculture. It includes all agricultural systems that promote environmentally, socially and economically sound production of vegetables. These systems take local soil fertility as a key to successful production. By respecting the natural capacity of plants, animals and landscape, it aims to optimize quality in all aspects of agriculture and the environment. Organic vegetable agriculture reduces external inputs by refraining from the use of chemo-synthetic fertilizers, pesticides and pharmaceuticals. Instead it allows the powerful laws of nature to increase both agricultural yields and disease resistance. We have exploited our natural resources beyond all limits to realize high productivity and production. During the recent past, the use of chemicals for the management of pest and diseases and improve the soil fertility has been increased. Some of these chemicals do not degrade easily and enter into human body leading to health hazards. Therefore, in several countries around the world including India, consumers demand for organically produced vegetables. In this direction it is endeavour of all people concerned with organic or safe vegetable production to develop feasible and economically viable package of practices based on the information available and technology developed which will be useful for farmers interested in organic vegetable farming. The principal objectives of organic vegetable farming are to develop a sustainable farming system for guaranteed adequate vegetable production in the foreseeable future, to develop self-sufficient agriculture system which would rely as much as possible upon resources from within its own resources and to develop an alternative strategy over chemical farming which would be a guideline for working of biological processes in natural eco-systems.

Organic Farming Practices in Vegetable Crops Production

Climate: Each vegetable crop has specific climatic requirements and the season selected for the cultivation should be the one during which period the crop growth, yields and produce quality are the best. Some disease and insect problems are especially prevalent at specific, predictable times of the year and vegetable crops can be planted at dates that protect them from these high-risk periods. The following sowing schedules can be generally followed:

Kharif: Brinjal, Chilli, Onion, Okra, Drumstick

***Rabi*:** Carrot, Cauliflower, Cabbage, Capsicums, French Beans, Onion, Peas, Tomato, Gherkins

***Summer*:** Bottle gourd, Cucumber, Pumpkin, Ridge gourd, Watermelon

It is better to avoid tomato cultivation during rainy period as *Alternaria*, *Phytophthora* foliar and fruit rot diseases are severe. During winter okra (*bhendi*) growth is retarded and during high temperature period virus diseases are severe in okra, chilli, tomato and beans and in capsicum, fruit set is a problem.

Soil: The ideal soils for growing vegetables are well drained, fairly deep, and relatively high in organic matter with a C:N ratio between 15–20:1. Well-drained loam soils rich in organic matter (1.5–2%) with gentle slope of 2–3% are better for organic vegetable production. Healthy and productive soil helps crops to develop good root systems and reduces crop stress caused by drought or excess rainfall. If drainage is a problem, plant on raised beds to promote drainage and faster warming of soil. The better job done in matching crop and soil, the greater chance of producing a successful crop. On the whole most of the vegetable crops grow well in pH range of 5.5 – 6.6 and below 5.5 liming is required at the rate of 5 t/ha. Salinity problems can be solved by frequent irrigations before planting to leach out salts or effect can be reduced by adopting drip irrigation, while the sodicity can be checked by the application of gypsum.

Soil fertility management: Any definition of sustainable or organic agriculture includes a commitment to proper soil management and organic vegetable production systems are guided by an overriding philosophy of "feed the soil to feed the plant." This basic precept is implemented through a series of approved practices designed to increase soil organic matter, biological activity, and nutrient availability. Over time, adding organic materials such as green manure, crop residues, and composts to cultivated soils builds levels of soil organic matter. As soil organic matter increases, the ability of the soil to supply nutrients to crops also increases. The ultimate goal is a healthy, fertile, biologically active soil with improved structure and enhanced nutrient.

Many soil amendments and fertilizers commonly approved for organic production systems have appreciable amounts of nutrients, but only a portion of these nutrients are available to the current crop. Organic soil fertility programmes are designed to maintain adequate levels of nutrients in the soil nutrient pool and to augment the pool as needed. Management practices strive to optimize diverse biological processes in the soil to create a complex environment that ensures adequate nutrition to the crop.

Role of Organic Matter and Humus

The increase of soil organic matter to optimum levels is a key aspect of any organic production system. Native organic matter levels are relatively low in Indian soils, generally less than one percent, and organic matter levels fall when the soil is brought under cultivation. Studies have shown that it is unreasonable for a grower to expect to increase soil organic matter by more than about one percent, but a relatively small increase can dramatically improve the soil fertility environment in a given field.

Soil organic matter improves cation-exchange capacity and serves as a reservoir of nutrients for the growing crop. Incorporation of organic matter, also improves soil aeration, drainage, and water-holding capacity. Green manure crops are an economical means for elevating soil organic matter and providing nitrogen for the succeeding crop. They also reduce soil erosion and may offer benefits related to pest and disease suppression. The decomposition of organic matter in soils can provide much of the N, P and S needed for crop nutrition. A portion of the N from many organic amendments is converted readily into available mineral forms. P from organic amendments reacts quickly, is bound to soil minerals, and moves very little from where it is placed. K, Ca and Mg are relatively soluble from plant residues or soil organic matter fractions and also contribute to the soil pool of these nutrients. Organic matter is also a valuable balanced source of many minor elements. Organic matter releases nutrients as it decomposes and provides slow, constant availability.

Management of Soil Biological Characteristics

A complex array of soil-dwelling plants and animals decomposes organic matter, mineralizes organic forms of nutrients, and fixes nitrogen. These beneficial biological activities enhance the soil's ability to release nutrients needed for plant growth and to break down plant residues. Tillage practices that aerate the soil enhance biological activity by providing oxygen and mixing organic matter throughout the tilled area. On the other hand, broad-spectrum fumigants such as methyl bromide kill beneficial soil organisms along with soil pests.

Bacteria and Other Micro-organisms

Both ammonium and nitrate are readily available sources of nitrogen to plants and soil micro-organisms. A variety of soil microorganisms convert organic nitrogen to ammonium, but only specific (nitrifying) bacteria convert ammonium to nitrate (nitrification). Nitrifying bacteria are most effective in well aerated soils at soil temperatures of 80–90° F, when the C:N ratio is low to medium and when soil moisture is adequate for plant growth. High activity of these bacteria increases N availability, but the nitrates released by nitrifying bacteria are highly soluble and subject to leaching. Thus, if not absorbed by crop or covers, nitrate may be lost from the field. In addition, other soil bacteria (denitrifiers) reduce nitrates to elemental nitrogen or nitrous oxide which are lost into the atmosphere as they are volatilized. Denitrification is most likely to occur in poorly drained soils where oxygen levels are low (anaerobic conditions).

Not only do naturally occurring micro-organisms play an essential role in the nitrogen cycle, they have also been reported to decrease some populations of pathogenic bacteria and fungi. For example, the fungus *Gliocladium virens,* which tends to increase with increased levels of organic matter, controls damping-off pathogens. *G. virens* is also available commercially. *Agrobacterium,* a naturally occurring soil micro-organism, reportedly restricts growth of *Fusarium,* a fungal pathogen causing a number of diseases in vegetable crops. Beneficial micro-organisms in the root zone, particularly species of *Pseudomonas,* have been associated with a decrease in take-all decline of wheat. Over time, two species of soil-dwelling amoebae in Canada reduced populations of two root pathogens.

Earthworms

Earthworms break up organic matter, combine it and humus with soil particles, and enhance microbial activity. Soils with high earthworm populations or soils containing earthworm castings generally have a greater ability to hold water and plant nutrients. Cation-exchange capacity, exchangeable calcium, magnesium, and potassium, and available phosphorus are generally higher in soils with earthworms. Earthworm activity has also been associated with reduced severity of several diseases. In Australia, *Rhizoctonia solani* caused less damage to wheat seedlings when earthworms were present. Mechanical soil disturbance also reduces disease severity, and it is possible that the physical mixing of soil by the worms reduces pathogen severity. The presence of earthworms also enhanced the movement and colonization of wheat roots by the biological disease control micro-organism *Pseudomonas corrugata.* When earthworms are introduced to the soil, they spread rapidly from the inoculation point up to 12 m per year. In most soils, however, if organic matter increases, earthworm populations will increase rapidly. Continuous cropping without adding organic matter usually reduces earthworm populations to a very low level, as do copper fungicides, soil fumigation, and frequent tillage.

Source of Nutrition in Organic Farming

Cover crops/green manure crops**:** Cover crops can be beneficial for intensive organic vegetable production in a number of ways. If leguminous cover crops are grown, soil nitrogen can be increased through nitrogen fixation. Grasses are particularly helpful in promoting soil structure and soil aggregate stability because of their fibrous root systems. Weed suppression for subsequent crops may be another benefit. Furthermore, cover crops can provide a favourable environment to attract and sustain beneficial arthropods. Growing a green manure crop that includes N fixing legume is the most economical way to provide nitrogen to succeeding crop. The green manure contributes 30–60 kg N/ha /crop depending upon the crop. Sannhemp and Dhaincha could be suitable catch crop in many agro-climatic situations. To realize some economic returns farmers who have irrigation facility can grow profitable vegetable crop such as french beans or cow pea in place of green manure crops. Green manure crops should be grown in the rainy summer and turned under well in advance of crop planting.

Compost**:** Compost is a relatively cost-effective organic source of nutrients. These composts can be prepared by crop residues, farm wastes. The C:N ratio of a compost is one indication of the maturity and N availability which should be less than 20:1.

Manure**:** Farm yard manure, bio gas slurry, poultry manure and other animal manures. These are the by-products in a integrated farming system, where output of one enterprise will become the input of the other activity and recycling is the key here.

Vermi-compost: Vermicompost production is economically remunerative due to its low cost of production. The quality and nutrient content depends upon the raw material used for production.

Other manures used in organic cultivation: A number of approved organic fertilizers are available for organic cultivation. Many of these nutrients are by-products of fish, meat and soybean processing industries. Other simple fertilizer materials that for most part offer only one macro-nutrient include: sodium nitrate (mined), blood meal, rock or colloidal phosphate, potassium sulphate (mined) and green sand (for K). Certain by-products of the meat processing industry, such as blood and bone meal have recently come under scrutiny because of food safety concerns and the potential for disease transmission. Mined Chilean sodium nitrate, a source of N, has been an imported component of organic fertilizer programmes because of its low cost. However, some certifiers restrict the use of sodium nitrate because of the long term adverse effect of sodium on soil. Minor element sources of organic fertilizer commonly contain one or more minor elements.

Special purpose fertilizers: Specific approved nutrient sources of K, Ca and Mg may be useful to an organic grower when a deficiency is indicated. Materials such is gypsum, lime and potassium magnesium sulphate have been in use.

Biofertilizers: The contribution of *rhizobium* bacteria as symbiotic nitrogen fixer is well-established. Also *Azotobacter* and *Azospirillum,* are also used as nitrogen fixers. The use of bio-fertilizers are also gaining popularity for solubulising phosphorus and VA for absorption of nutrients from soil.

Manuring and Fertility Management

1. Enrichment of FYM with *Trichoderma* and Bio-fertilizers: Well decomposed FYM is thoroughly mixed with *T. harzianum* , *Azatobacter* or *Azospirillum* and Phosphate Solubilizing bacteria (PSB), @ all1 kg/t of FYM, moistened with sprinkling water and covered with plastic sheet or dried coconut fronds and kept to incubate for 15 days. This enriched FYM should be mixed with remaining FYM before applying to the field.
2. About 35 to 40 tonnes of fully decomposed FYM and 1.5 tonnes of vermi-compost and 250 kg Neem cake having 8% oil content is added to the soil. Ridges for transplanting at recommended row spacing according to the crop to be grown is prepared after basal manure application.
3. For enhanced supply of nitrogen, green manuring crop can be grown and incorporated at least 3 weeks before transplanting as N source.
4. Concentrated organic cakes (4–7% N) like castor, soybean, cotton, etc. can also be used depending on the availability. At 30 days after planting, vermicompost (1.5 t/ha) and neemcake (250 kg/ha) application is done followed by earthing up operation.
5. As a source of P, Rock phosphate in combination with Phosphate Solubilizing Bacteria or Bone meal can be used , both contain about 20–22% $P_2 O_5$.
6. As a source of K, wood ash (2.5 to 3% K_2O) or Sheep manure (3- 4% K_2O) can be used.
7. Spray of panchagavya can be followed 4-5 times at 10 days interval to supplement nutrients and as a plant growth promoting tonic.
8. For legume vegetables, seeds need to be treated with *Rhizobium* cultures before sowing.

Selection of Varieties

For better performance select a variety which is resistant or showing field tolerant to important diseases and insects. Also choose varieties according to the season in which they need to be grown based on their performance. For example, in tomato for organic cultivation during *kharif* the major problem is *Alternaria* Leaf Spot, so better choose a cultivar known to have field tolerance to this disease and also if available a variety having field tolerance to ToSPO virus., while for summer season, TLCV resistant hybrids need to be selected. In cabbage and cauliflower photo and thermo insensitive hybrids are to be used. In tomato, bacterial wilt resistant varieties need to be grown. In French beans bacterial blight and rust resistant varieties need to be selected.

Tomato: Arka Abhijith, Arka Shreshta, (BW resistant), Arka Ananya (BW and TLCV resistant)

***Okra*:** Arka Anamika, Arka Abhay, Parbhani Kranti (YVMV resistant)

***Brinjal*:** Arka Keshav, Arka Neelkant, Arkaa Anand (BWR)

Cowpea: Arka Garima, Arka Suman (rust resistant)

***Cluster bean*:** Pusa Navbahar and Gouri

***Dolichos bean*:** Arka Jay and Arka Vijay

***French beans*:** Arka Komal, Arka Suvidha and Arka Anoop (rust resistant)

***Chilli*:** Arka Suphal, Arka Meghana, Arka Harith (powdery mildew and CVMV tolerant)

***Onion*:** Arka kalyan, Arka Niketan (purple blotch tolerant)

Organic Seedling Production

Raised-bed nursery: Beds of 10–15 cm height providing necessary drainage are prepared. During summer season following solarization of nursery beds will prevent the nursery diseases. FYM, vermicompost and trichoderma powder in the ratio of 10:1:0.1 is mixed and applied @ 0.3 kg/m^2. Neem cake @ 100 g/m^2 is also added to the nursery bed. Seeds are sown in lines with a spacing of 7.5–10.0 cm × 2.5 cm and covered by vermi-compost and straw. It is good to protect nursery beds with 40 or 50 mesh nylon net cover to prevent insect vectors. The seedlings will be ready in 25–45 days depending on the vegetable crop.

Raising seedlings in plastic pro or flat trays: Each flat or pro tray having 98 cells are used for raising seedlings of tomato, capsicum/ chilli, brinjal, cauliflower and cabbage. Cocopeat fully decomposed by the use of organic manure and enriched with phosphorus is the ideal media for filling the trays. Otherwise vermi-compost can be used as growing media. About 1-1.25 kg cocopeat is required to fill one tray. A small depression in the centre of the coco peat is made using finger and one seed is put per cell and covered with by cocopeat just enough to cover the seeds. Such filled and sown trays are stalked one over other till sprouting is observed. After sprouting, the trays are spread in net house and irrigated daily depending on the weather conditions. The seedlings of 20–25 days old in cabbage, cauliflower, tomato and 30–40 days in chilli, capsicum and brinjal are ideal for transplanting and better crop establishment.

***Transplanting/ direct sowing*:** BT (*Bacillus thuringiensis*) formulation (1g or ml/litre) is sprayed to the cabbage or cauliflower seedlings one day before transplanting. At the time of transplanting root portion of seedlings is dipped in 1 % Burgundy solution or seedling trays are drenched in it. Similarly bio-fertilizers such as *Azatobactor* or *Azatospirillum* and Phosphate solubilizing bacterial slurry can be used for root dipping of all vegetable seedlings. When seeds are used for direct sowing the above bio-fertilizers can be used for seed coating for non-leguminous vegetables while for leguminous vegetables rhizobium and PSB are recommended.

***Irrigation*:** Irrigation schedules recommended for commercial vegetable cultivation can be followed with only caution that foliage wetting is to be minimized in order to reduce foliar disease incidence. With drip irrigation plastic mulching helps to maintain better soil moisture regimes and also reduces pest and disease incidence in addition to saving of irrigation water through reduced evaporation losses as well by preventing weed growth.

***Weed Management*:** For the production of transplants from seedbeds, select a land area that does not have a history of troublesome weeds. Vegetable crops that germinate quickly and grow rapidly in the first few weeks after planting do well because they are more competitive with weeds than crops that initially grow slowly. Examples of rapidly-growing plants are transplants and seedlings of large-seeded vegetables such as sweet corn and beans.

Hand removal (i.e., pulling or hoeing) and cultivation are the most widely practiced method of eliminating weeds. You can use a properly set cultivator to remove small weeds quickly and effectively. Take care, however, to avoid damage to the crop roots. Use caution also to avoid cultivation in soil that is excessively wet (to prevent damage to the developing crop and soil compaction). Cultivation provides additional benefits to the crop, including soil aeration and improved water infiltration.

For weed management plastic mulch is a widely used ground cover in organic vegetable production especially for widely spaced crops such as cucumbers and melons and for greenhouse capsicum and tomato. The mulch provides a physical barrier between soil and plant surfaces and reduces the amount of disease inoculum splashed onto foliage, stems, and fruits during rainy periods. The amount of cucumber rot and buckeye rot of tomato may be reduced in this manner. Organic mulches, such as dry grass, straw, bark, banana sheath, composted sawdust, and similar materials, can also be used as physical barriers.

Organic Pest and Disease Management Practices

Organic disease management provides satisfactory protection from many vegetable diseases that commonly occur and these practices are based on non-chemical sanitation, cultural, physical, and biological means, as well as application of organically approved chemicals.

Used in an integrated programme, these practices reduce populations of fungi, bacteria, nematodes, viruses, and other pathogenic micro-organisms that cause vegetable diseases. A combination of practices is necessary, since no single practice is effective for all diseases that threaten production of a given crop. Some organic disease management practices should be carried out before the crop is planted such as soil solarization and others later in the season such as collection of debris and burning them. To eliminate fungus spores, mycelium, bacterial cells, and other forms of inoculums, you can disinfect stakes and cages by washing to remove soil, followed by dipping or spraying with a copper sulfate solution before reuse. Some of the plant protection measures that can be followed in organic farming are as follows.

It is necessary to use of resistant vegetable varieties wherever available. Clean seeds and transplants are a must in vegetable production. Seeds are to be avaoided from fields or areas where diseases are prevalent. Examples of seed-borne diseases are anthracnose of lima beans, mosaic virus of southern peas, bacterial blight of snap beans, black rot of cabbage, and leaf spots of turnip and mustard greens. Seed treatment with *Psuedomonas fluorescens* or *Trichoderma harzianum* or *Trichoderma viride* @ 10g/kg should be undertaken. Plant healthy and vigorous seedlings early in the season and keep the plants in a healthy growing condition by reducing stress by supplying adequate water and nutrients in a weed free environment. In crop rotation legumes are ploughed early so vegetation (green manure crop) has time to decompose before main crop planting. Use of mulch enables to keep vegetables from soil contact and rots. Keep out weeds which harbor insects and diseases. It is necessary to irrigate early in the morning to reduce time plants are wet. Remove and dispose of diseased plants to reduce the source of inoculums. It is essential to turn under crop refuse as soon as harvesting is completed. Growing of fodder maize or sorghum as a barrier crop all round the vegetable planted field helps to reduce incidence of virus carrying vector infestation on the main crop. Nursery beds are to be covered with 40 or 50 mesh nylon net to prevent insect vectors transmitting virus diseases. Nematodes can be managed by use of cover crops such as southern pea and hairy indigo which will reduce nematode build up. The use of mulch and keeping adequate soil moisture will reduce nematode populations. Solar heating the soil using large plastic covers reduces nematodes, soil-borne insects, some weeds, and some pathogens. Summer flooding, where soil type permits, will help control nematodes and other pests. Application of bio-agents like *Trichoderma* through FYM or seed treatment also helps in management of nematodes.

Bio-agents for Foliar and Fruit Diseases and Soil Borne Organisms

Trichoderma harzianum (10 g/l), *Trichoderma viride*, *Pseudomonas fluorescens* (10 g/l) *Bacillus subtillis* (10 g/l).

Bioagents and Botanicals for Management of Insect Pests

Beauveria basiana (10 g/l), *Verticillium lecani* (10 g/l), *Metarrhizum anisiplea* (10 g/l), Neem soap (10 g/l), Neem oil (8 ml/l), Neen seed Powder Extract (4%), Pongamia oil

(8 ml/l), Pongamia soap (10 g/l), NPV, *Bacillus thuringiensis*, on certain insects such as cabbage worms. Pyrethrins, rotenone, ryania, insecticidal oils and soaps.

Protected Cultivation of Vegetables

Protected cultivation is an important tool to overcome many biotic and abiotic stresses in organic cultivation of vegetables. In moderate to heavy rainfall area, rain shelter is proving effective in overcoming the problems of foliar diseases of vegetable crops like tomato. In crops like brinjal and capsicum where the control of certain insects like shoot and fruit borer management is difficult under organic farming, growing these vegetables under net house hold a bright promise.

Harvest and Post-harvest Management

Schedule harvest timings during cool hours of the day, either morning or evening hours. Immediately after harvest shift the produce to sheltered place and pack them in clean ply wood boxes or plastic crates. Pick the produce at right stage to retain firmness so the transport shocks are reduced. For example in tomato the fruits should be harvested at breaker stage or when the fruits are turning from green to light pink colour. The results of organic vegetable trials conducted at IIHR and as on farm trials in farmers' fields have indicated that the yield levels are 80/85% of conventional farming in cabbage, cauliflower, tomato, brinjal, chillies, onion and drumstick while they are on par even more in French Beans.

Marketing

The marketing of organically produced vegetables presents growers with a number of challenges. Total supply, consumer demand, pricing, perishability of the product, and market structure are all factors that contribute to a grower's ability to sell his/her product. Therefore, production and market risks both affect the profitability and economic viability of organically grown vegetables. In all cases, the risks associated with organic vegetable operations should not be minimized. Organic vegetables, like many fresh market commodities, are a highly perishable product. Growers must harvest, pack, and sell their products in an expedient manner to receive satisfactory returns. Therefore, from a marketing perspective vegetables carry a greater risk than such storable commodities as nuts and grains. Some organic vegetable growers may reduce this risk by planting crops such as winter squash, onions, garlic, and potatoes which may be stored for longer periods of time. Commodities that are produced organically can often be sold for a premium price over conventionally grown products. However, the industry is extremely competitive and returns to growers are dictated by the total supply, consumer demand, and the available organic outlets. Market saturation often occurs. Growers may then be forced to accept lower returns and/or market their product without the organic designation at conventional prices.

REVIEW OF RESEARCH WORK DONE ON ORGANIC VEGETABLE PRODUCTION IN INDIA

Crop nutrition

The effects of organic manures (FYM at 70% and oilseed cake + rock phosphate at 30%, *Azotobacter* at 20 kg/ha and oilseed cake + wood ashes+rock phosphate, and NPK fertilizer at 60: 70: 70 kg/ha) and biological pesticides (Neemax or an *Azadirachta indica*-based pesticide at 1 kg/ha, garlic extract at 10 l/ha and endosulfan at 1.2 l/ha) on radish was studied in Himachal Pradesh by Singh and Singh (2001). The highest yield (12.2 t/ha) was obtained with FYM and oilseed cake + rock phosphate combined with endosulfan compared to FYM and oilseed cake + rock phosphate coupled with garlic extract and other organic and inorganic treatments. Another field experiment was conducted to find out the effect of different sources and rates of organic manures (FYM, vermicompost and poultry manure) on yield of organically grown garden pea (Pandey *et al.*, 2006). The highest pod yields of 8.65–9.15 tonnes/ha was recorded with the application of FYM manure 15–20 tonnes/ha which were statistically significant compared to other treatments. Application of organic manures, irrespective of sources and rates, recorded significant improvement in physico-chemical properties of soil. No major insect-pests and diseases were recorded in the study. However, application of organic manures, irrespective of sources and rates, recorded significant reduction in the incidence of Fusarium wilt compared to control.

The effects of organic farming on the morphology of cluster bean and okara were investigated by Girija *et al*, (2006). Plant height, number of leaves, leaf length, leaf breadth, leaf area, number of branches, number of flowers, number of fruits, fruit length, and weight of fruits were greater under organic farming than under inorganic farming. Organic field experiment conducted at Coimbatore with an objective to evaluate the impact of organic and inorganic sources of nutrients, panchagavya spray and botanical spray on growth, yield and economics of okara indicated that conventional cultivation with monocrotophos spray recorded an yield of 10.39 t/ha and net returns of Rs. 47,250/ha. However, the yield was on par with okara raised with basal application of poultry manure @ 21 t/ha with panchagavya foliar spray (3%) four times at 30, 45, 60 and 75 days after sowing (10.27 tonnes ha^{-1} and net returns of Rs. 46,440/ha) (Boomiraj and Lourduraj, 2006). Organic farming in cowpea crop revealed that application of farmyard manure+biofertilizers gave 9.7% higher green pod yield than the treatment having recommended dose of chemical fertilizers but at par with treatment receiving poultry manure+biofertilizers and combination of farmyard manure, poultry manure, vermi-compost along with biofertilizers (Raghav and Shashi Kamal, 2007). The organically produced cowpea fruits possessed longer shelf life at room storage. Maximum uptake of nutrients was obtained from the treatment having the combination of farmyard manure, poultry manure and vermicompost along with biofertilizers as well as all the organic manures improved the soil health.

In a study to assess the performance of annual drumstick (cv. PKM-1) under organic culture for three consecutive years at Bangalore showed that the crop performance with respect to tree growth, yield and yield components were significantly influenced by organic treatments tested (Prabhakar and Hebbar, 2007). Tree trunk diameter was higher with farm yard manure (FYM) applied at 15 kg/tree or with vermicompost applied at 5 kg/tree along with bio-fertilizers (*Azospirillum* and phosphate solubilizing bacteria) at 5 kg/hectare as compared to other organic treatments such as green leaf manuring combinations or reduced FYM application rates. Fresh pod yield was higher (9.7 t/ha/year) with FYM applied at 15 kg/tree followed by green leaf manuring supplemented with rock phosphate and wood ash (8.5 t/ha/year). These yields were on par with conventional treatment receiving recommended dose of chemical fertilizers and manures (8.6 t/ha/year). Rashmi-Srivastava *et al.* (2007) made an effort to see the effect of arbuscular mycorrhizal fungi (AMF) and *pseudomonas* as the microbial inoculants in vegetable based cropping systems under organic farming practices. A significant increase in yield was observed in the inoculated plots over the control. The findings indicated that microbial gene pool especially the key helpers for the maintenance of soil health residing in the vicinity of roots, was positively affected by using *pseudomonads* and AMF. The effects of organic and inorganic fertilizer application on the performance of yellow onion (cv. Phule Suvarna) were studied by Sankar *et al.* (2008) at Coimbatore. The highest number of leaves, average weight of bulbs and marketable bulb yield (17.5 and 20.8 t/ha in two seasons) were obtained with Panchakavya + NPK compared to several organic treatments. In organic rose onion trials conducted at IIHR farm, Bangalore the crop growth and yield and bulb quality parameters such as dry matter and TSS content were marginally superior in organic treatment receiving organic manures equivalent to 100 % of recommended Nitrogen equivalents (Prabhakar and Hebbar, 2007a). Only chemical fertilizer application resulted in drastic reduction in bulb yield to the tune of more than 30 per cent. In cauliflower the results of the study conducted by Hebbar *et al.* (2008) showed that 50 percent of recommended N equivalent through enriched FYM (in addition to 25 t/ha recommended FYM) gave a curd yield of 18.61t/ha, which was on par with the yield obtained under conventional practice (19.76 t/ha) where recommended chemical fertilizer and recommended FYM was applied. However, significantly lower yield was obtained when only recommended NPK was used for cultivation (14.12t/ha).

Soil Health and Produce Quality

Brinjal cv. Surya plants were grown with farmyard manure, poultry manure and chemical fertilizers equivalent to the NPK contents in organic sources (Prasanna and Rajan, 2001). The result indicated that after 5 and 7 days of storage, the highest number of unmarketable fruits was from plants treated with inorganic fertilizers, while the lowest were from plants treated with organic fertilizers. In another experiment, Kannan *et al.* (2006) studied the effect of application of farmyard manure, vermicompost and coir pith, applied at 50, 75 and 100%, with or without 2 kg *Azospirillum*/ha on the yield, nutrient uptake, and crop quality of tomato cv. PKM1. Application of 75% vermi-compost in combination with *Azospirillum* resulted in the highest yield, titratable acidity (0.72%), ascorbic acid content (23 mg/100 g), total solids (5.4%), pH (3.9), crude protein content (1.70%) and lycopene content (3.7 mg/100 g), and the lowest nonreducing sugar content (0.37 g/100 g). Singh (2006) studied the effects of different biofertilizers and biopesticides used in combination with organic nutrient on different vegetable crops, viz. cabbage, radish and brinjal. The result revealed that though the yield in organic produce was at par with produce obtained under conventional method, the produce was superior in protein and vitamin C contents with better shelf life. The nitrate content, which has an ill effect on health, was also quite low in organic treatments. A field study was conducted in tomato crop to determine the impact of organic sources on uptake, quality and nutrient availability (Sable *et. al.*, 2006). The result revealed that nutrient uptake by tomato plants, quality of tomato fruits and availability of nutrients in soil after the harvest of tomato crop was more in treatments with organic source of nutrients through neemcake and vermi-compost in variable combinations and alone. Organic tomato recorded higher, vitamin C with increase shelf life in the treatments where 50% N through neemcake and 50% N through vermicompost as well as 25% N through neemcake and 75% N through vermicompost were given together. The latter treatment was also more effective in increasing the nutritional value of soil after the harvest of tomato crop. Tomato fruits grown on organic substrates contained significantly more Ca and ascorbic acid and less Fe than did the fruits grown on inorganic media (Shweta *et. al.*, 2007). Phosphorus and K content did not differ between the fruits grown on organic and inorganic fertilizers.

Crop Protection

In an organic trial in chilli the results indicated that the yield of chilli on organic plot applied with FYM, bio-fertilizers and bio-pesticides was on par (502.8 kg/ha) with yield on conventional practices (612.7 kg/ha) applied with chemical fertilizers and pesticides (Ali *et al.*, 2002). The incidence of pests viz., thrips, mites and aphids was 16.9%, 8.8% and 4% respectively in organic plot compared to 26.3%, 12.8% and 3.2% respectively in conventional plot. The occurrence of natural enemies was high (9.5%) in organic plot. The incidence of little leaf and fruit rot diseases was low 11% and 4% respectively in organic plot compared to 16.2% and 12.5% respectively in conventional plot. Ghorai (2005) made an attempt to harvest borer-free brinjals using physical barriers like transparent polythene, brass paper and newspaper packets to protect the individual fruits from fruit and shoot borer damage. Through this eco-friendly method healthy, fresh, soft, speckless and quality fruits were harvested. Paper packets are cheap and biodegradable. Forty to fifty man-days were required to bag the fruits at 2-3 g stage in one acre land. This gives the scope of producing organic brinjals, valued highly in the domestic and international markets. Singh, (2006) reported that bio-pesticide *Bacillus thuringiensis* and Neem-based pesticide were found equally effective in controlling Diamondback moth of cabbage, leaf caterpillar of radish and shoot and fruit borer of aubergine as compared to chemical pesticide endosulfan.

Sithanantham *et al.* (2007) have reported that entomopathogens (like *Verticillium lecani*) have shown promise for augmentative biological control (of *Scirtothrips dorsalis*), and there is scope for identifying more adapted and virulent strains of the entomopathogens, besides developing improved formulations for extending their shelf and field life. Experiments on control of soil and tuber borne diseases of potato using environmentally friendly methods such as soil solarization, organic amendments and biological control were carried out by Arora *et al.*, (2007), so as to find safe alternatives to the use of hazardous pesticides. Results of these experiments suggested that soil solarization, organic amendments and biological control using *Trichoderma viride* can provide safe alternatives to the use of pesticides in organic farming. Zadda Kavitharaghavan *et al.* (2007) reported that application of organic sources of nutrients and amendments (FYM, poultry manure, neem cake, mahua cake, pungam cake and biofertilizers, i.e., *Azospirillum*, *phosphobacteria* and silica solubilizing bacteria) significantly enhanced the production of defensive chemicals viz., silica and phenols and thus exhibited induced resistance in terms of antibiosis by means of reduction in feeding rate, oviposition, longevity and population build up and prolonged the nymphal duration of brinjal pests. Brinjal plants, which received organics, registered less quantity of reducing sugars, proteins and leaf chlorophyll when compared to NPK as straight fertilizer applied plots, thus making the plants less prone to insect attack.

Gaps in Current Knowledge

Scientific evidence on differences in yields under conventional and organic agriculture is very meager.Specific information on organic vegetable production practices in India is scarce, and vegetable growers need sound information to guide their management decisions. While rigorous research is done elsewhere on organic vegetable production under protected cultivation, it is almost nil in developing countries including India.

Future Thrust: Research and Development Opportunities

Non-availability of suitable package of practices on organic farming of various Horticultural Crops is a constraint. Appropriate methods to enhance the nutrient content of the bulky organic manure need to be developed. There is a need to assess the nutrient availability pattern of different organic sources during the cropping periods under different agroclimatic situations. Development of efficient bio-pesticides specific to insect pests and diseases. Issues like poor marketing access and remunerative prices when grown on large scale is to be addressed. There is a need to determine quality standards for various organic inputs. Research on organic horticulture should focus on farming system approach. Simplification of production protocols by the certifying agencies for the domestic markets is necessary for promotion of organic farming. Organic seed production and distribution system need to be encouraged. There is a strong belief in the potential of organic production as an income enhancing option for farmers, particularly when seen in light of the cost differentials obtained on the world market. This optimism and interest, however, needs to be matched with concrete knowledge and recommendations on how to go about producing and marketing organic products, and this knowledge base is not very well-developed in either public or private sector agricultural support system. This calls for authentic data generation by conducting appropriate research experiments and validation of the same by carrying out adequate on-farm trials in different agro climatic regions.

CONCLUSIONS

Day-by-day demand for organically grown, fresh and processed vegetables is on the increase world over. Although information on conventional cultivation practices is available from many sources, comprehensive information on organic cultivation practices is difficult to find. Organic production differs from conventional production primarily in soil fertility, weed, insect, and disease management. While there have been varying notions of organic farming over the years, the growth of the organic farming system and the introduction of standards and certification have led to a clearer definition in recent years. That definition describes organics as a viable agriculture, based on sound farming practices that does not include synthetic chemicals. Keeping this in view, it is an endeavour of all those interested in protecting the soil sustainability by adopting eco-friendly within farm operations as much as possible and generate appropriate organic vegetable production technology to supply safe vegetables for human consumption.

References

Ali, M.A.A., Chandrasekar, S.S., Vardarasan, S., Gopakumar, B., Paramaguru, P., Ponnusamy, V. and Muthusamy, M. 2002. Impact of organic cultivation of chilli (*Capsicum annuum* Linn) on pests and diseases. Proceedings of the 15th Plantation Crops Symposium Placrosym XV, Mysore, India, 10-13 December. 451-56.

Arora, R. K. and Jyotsana Sharma, 2007. Eco-friendly alternatives to pesticides in management of soil and tuber borne diseases of potato in organic farming. Sustainable environmental management: *Dr Jayashree Deshpande Festschrift* Volume. pp: 1-6.

Boomiraj, K. and Lourduraj, A.C., 2006. Organic production of bhendi (*Abelmoschus esculentus*). *Journal of Ecobiology* **19**(4): 389-96.

Ghorai, A.K, 2005, Avoidance of brinjal fruit borer (*Leucinodes orbonalis Guenn.*) through eco friendly physical barriers. *Journal of Entomological Research* **29**(2): 127-29.

Girija, I.S., Britto, and A.J. de, 2006. Comparative study on the effect of organic and inorganic farming methods on cluster bean and bhendi. *Environment and Ecology* **24S**(Special 4): 1115-18.

Hebbar, S.S., Prabhakar, M., and Ganesha Murthy.A.N. 2008. Organic farming in cauliflower. Presented in National conference on organic horticulture held at CPCRI, Kasaragod during 15-18 October, 2008. (Abstracts Book, P32).

Kannan, P. Saravanan, A. and Balaji, T. 2006. Organic farming on tomato yield and quality. *Crop Research Hissa* **32**(2): 196-200.

Pandey, A.K., Gopinath, K.A., Chattacharya, P., Hooda, K. S., Sushil, S. N., Kundu, S., Selvakumar, G., and Gupta, H. S. 2006. Effect of source and rate of organic manures on yield attributes, pod yield and economics of organic garden pea in north west Himalaya. *Indian Journal of Agricultural Sciences* **76**(4): 230-34.

Prabhakar, M. and Hebbar, S.S. 2007. Studies on organic production technology of annual drumstick in a semi-arid agro-ecosystem. *Acta Horticulturae* **752**: 345-48.

Prabhakar, M. and Hebbar, S.S. 2007 a. Studies on organic production technology of rose onion (*Allium cepa*) in semi arid agro-ecosystem of India. Paper Presented in 5th International ISHS Symposium on Edible Alliaceae, held from 29-31 October 2007 at Dronten, The Netherlands.

Prasanna, K.P. and Rajan, S. 2001. Effect of organic farming on storage life of brinjal fruits.*South Indian Horticulture* **49**(Special): 255-256.

Raghav, M. and Shashi Kamal, 2007, Organic farming technology for higher and eco-friendly cowpea production in Tarai region of Uttaranchal. *Acta Horticulturae* **752**: 469-471.

Rashmi Srivastava, Roseti, D. and Sharma, A.K., 2007. The evaluation of microbial diversity in a vegetable based cropping system under organic farming practices. *Applied Soil Ecology* **36**(2/3): 116-123.

Sable, C.R. Ghuge, T.D. Jadhav, S.B. and Gore, A. 2007. Impact of organic sources on uptake, quality and availability of nutrients after harvest of tomato. *Journal of Soils and Crops* **17**(2): 284-287.

Sankar, V. Veeraragavathatham, D. Kannan, M. Subbiah, K. and Prakasam, V. 2008. Studies on organic farming practices in yellow onion for the production of export quality bulbs. *Journal of Maharashtra Agricultural Universities* 33(2): 255-257.

Singh, S.R. and Singh, A.K. 2001. Effect of organic farming technology on yield and quality of radish (Raphanus sativus L.) under mid hills of H.P. *Agricultural Science Digest* **21**(2): 115-117.

Singh, S.R. 2006. Response of bio-fertilizers and pesticides on yield and quality of cabbage, radish and brinjal in vegetable based rotation system. *Applied Biological Research* **8**(1/2): 33-36.

Shweta, Sonal and Deepika Sharma, 2007. Enhancing nutrients in tomato (*Lycopersicon esculentum Mill.*) through vermi-composting. *Phytomedicines* 623-627.

Sithanantham, S. Varatharajan, R. Ballal, C.R. and Visalakshy, P.N.G., 2007. Research status and scope for biological control of sucking pests in India: case study of thrips. *Journal of Biological Control* **21**(Special): 1-19.

Zadda Kavitharaghavan, Ragjendran, R. and Vijayaraghavan, C. 2007. Induced systemic resistance to major insect pests of brinjal through organic farming. *Crop Research Hissar* **34**(1/3): 125-129.

6

Potentials in Organic Farming of Spices

V. A. Parthasarathy, V. Srinivasan and R. Dinesh

Spices in Indian Economy

India produces about 4 million tonnes of spices and exports around 180 spice products to more than 150 countries. The world export of spices during 2005 was 3592. 48 ('000 t) and import was 3454.40 ('000t) (FAO database 2007). In 2007-08, Indian share of the world trade in spices was about 35% in volume and 37% in value and is planning to increase its share in the world spices trade to 65% both in volume and value by 2017. The country's exports touched a record $1.1 billion in 2007-08 and likely to raise to US $1.3 billion in 2008-09 due to good demand for chilli, pepper and cumin seed. Organic spice exports have been growing rapidly in recent years, recorded 25 % increase in quantity, and 11 % increase in value. The share of organic spice is negligible compared to conventional spices trade. The main organic spices traded by volume are pepper (39%), ginger (19%), turmeric (11%), cardamom (8%), chilli (7%), cloves (7%), nutmeg (7%), and others (7%). The largest producers and exporters of organic spices in the world are Sri Lanka (28%), India (12%), Indonesia (10%), Madagascar (8%), Tanzania (10%) and Guatemala (8%). Sri Lanka is the largest producer of organic pepper followed by India. India exported around 700 tonnes of different varieties of organic spices in 2002-03 and Spices Board, Ministry of Commerce, Govt. of India expects to export organic spices worth Rs. 240–260 crore by 2012. The current market size is small but demand is expected to grow at a rate of 15–20% in the coming decade. Given the ever-growing shift towards chemical-free organic food items in western world, the potential for export of organic spices is high, and India is placed at an advantageous position, being a major exporter of spices to the global market.

Organic Spices

According to the ITC and UNCTAD/GAT, more than 130 countries produce certified organic foods. Internationally, there is a shift towards traditional / ethnic system of medicines. Since spices form part of many of these medicines, the demand for organically produced spices is increasing considerably. Organic pepper is produced and exported to international markets by India Indonesia and Sri Lanka. Organic pepper has established its market share better than other pepper products like dehydrated green pepper, pepper in brine, freeze-dried pepper or frozen pepper. The country at present exports around 50 tonnes of different varieties of organic spices. Exports will get a significant boost in the coming years as more farmers switch to organic methods. The important markets are the European Union, USA and Japan. The current market size is small but demand is growing at an annual rate of 15–20% and may grow to 57,000 tonnes in the next 10 years.

Relevance of Organic Farming in Spices

Farmers in our country have always been practicing traditional ways of using indigenous technologies and inputs mostly in line with modern organic farming principles. India has great potential to capture the expanding global organic spice market as per capita consumption of fertilizers and pesticides in our country is far below that of developed countries and it is very easy for Indian farmers to embrace organic spice farming in its true sense. With global retail sales of organic products fast picking up, India could emerge as a major player by exploiting its own strengths like traditionally followed agriculture systems, diverse agro-climatic regions with self-sustaining agriculture systems.

Most of the North Eastern states of India are having virgin land without any commercial cultivation that are very much suitable for organic cultivation of spices and major spices like black pepper, ginger, turmeric and chillies have larger area in this tract, enabling it to become the "organic spices hub of India". Most of these spices are cultivated under shade trees, which are available naturally as evergreen forest and left out Jhum land. The North Eastern states viz., Meghalaya and Arunachal Pradesh ranks 2nd and 3rd in production of ginger and Assam holds 4th position in area of turmeric in India. Besides, large cardamom is cultivated in Arunachal Pradesh in an area of 2,192 ha with a production of 512 tonnes and productivity of 500 kg/ha. In the world map of large cardamom production, India shares 54%, in which 87% is contributed by Sikkim and the remaining 13% by Darjeeling (West Bengal). In large cardamom, Sikkim claims the major share in the country with an area and production of 26,734 ha and 5401 tonnes, respectively with a productivity of 242 kg/ha. India is the leader in large cardamom in the world market and there is a vast scope to produce organic large cardamom as well as to export it.

Organic standards, in which the definition is set out for practical application stipulate not only the prohibition of use of certain inputs but dictate a range of practices to be followed that will ensure that a farm maintains its sustainable productivity. Large-scale use of high analysis fertilizers and pesticides pose a number of environmental hazards and imbalances in soil nutrient levels. An unfavourable ratio of fertilizer consumption and losses of nutrients resulting from excessive erosion and heavy leaching under heavy rainfall in spice growing areas threaten the environment also. Excessive and indiscriminate use of fungicides/pesticides could result in pesticide residues in the produce, harm non-target organisms, develop resistance and environmental pollution. Negligence of the biological aspect of soil has led to depleted population of soil organisms such as earthworms and their allies responsible for healthy plant growth. Moreover, since spices are high-value and export-oriented in nature, it is also imperative to keep the levels of pesticide residues below tolerance limits in view of the stringent standards set by the importing countries.

Mixed Cropping Systems

Multiple cropping or farming systems helps efficient use of natural resources, buffer against price fluctuations, generates more employment opportunities and production of commodities for varied end uses and thus ensures crop diversity. Unlike other field crops, spices are highly amenable for inter-cropping owing to their ecological niche. As the pressure on land under cropping is increasing, area expansion for spice cultivation is questioned. The best alternative would be cultivation of spices as a component crop in the existing coffee, tea, coconut and arecanut plantations. Mixed cropping of black pepper with rubber, cocoa, orange, etc., are common in other countries like Brazil.

Vines like black pepper and vanilla and rhizomatous crops like ginger, turmeric, cardamom and galangal are ideally suited for mixed/ inter cropping under tree plantations with partial light. In West Bengal, ginger was found to be highly remunerative as inter-crop in coconut. The profit was Rs. 40,250/ha whereas, it was Rs. 23,450/ha from turmeric. An increase in net returns to the tune of Rs.13, 300/ha can be obtained in mixed cropping black pepper in coconut gardens (Sarma *et al.*, 2003). In a cropping system the addition of organics due to litter fall would also result in organic matter build up and enhanced nutrient availability in soils (Khan *et al.*, 2002). The minimum tillage practice would ensure no or minimum damage to root systems and the associated microbial load would be ideal. It is reported that there is great scope of bringing tea and coffee estates under black pepper production by covering all the shade trees with pepper. Area under coffee and tea in South India can accommodate about 3.8 and 2 million standards for pepper, respectively and can result in additional pepper yield up to 20,500 t/year.

Plantation crops including black pepper and cardamom, being perennial in nature contribute large quantities of waste by-products, which by composting and recycling will meet the nitrogen requirement of crop and partly other nutrients. The multiple cropping system results in the continuous uptake, addition of biomass and higher level of nutrient recycling that would lead to a positive influence on the physico-chemical and biological properties of soil.

The microbial biomass, the organic content, total N, P and K would be higher in the root region soils of multistoried cropping system when compared to the monocrop. The beneficial effects are reflected through enhanced soil fertility status, enhanced microbial activity and better utilization of natural resources for the benefit of plant growth and sustainable crop productivity.

The organic addition brought through leaf litter recycling and microbial activity would lead to integrated nutrient management for enhanced system productivity. In south India where coconut is a major plantation crop, abundance of waste materials, about 7.5 million tonnes of coir dust annually from coir based industries provides good scope for recycling the organic materials into the system. Composts made out of leaf and sheath wastes and husks of these plantation crops are valuable sources of plant nutrients (Table 6.1).

Although organic farming is better than conventional chemical farming systems, its crop yield is observed to be much lower in the developing countries as compared to that of the latter. Therefore, for enhancing the yield potential technologies like inoculating soils and plants with beneficial micro-organisms for creating a favourable micro-environment for plant growth are being promoted. The presence of *Azospirillum* reported in the spices based cropping is of great relevance and needs to be exploited. So also the P solubilisers present in several perennial cropping systems (Bopaiah and Thomas, 1993). Combination of bio-fertilizers *viz., Azospirillum*, Phosphobacteria and VAM in the nursery mixture has significantly enhanced the root and shoot growth of nursery cuttings. Field application of *Azospirillum* @ 20g/bed along with coir compost and FYM has significantly increased the yield of black pepper, ginger and turmeric.

Several PGPRs (plant growth promoting rhizobacteria) isolated from the rhizosphere of black pepper and other spice crops were found growth productive, disease suppressive and incidentally are efficient P solubilizers. The role of *Pseudomonas fluorescens* mediated nutrient flux in the soil microcosm in plant growth promotion was studied with the higher uptake of nutrients by the bacterized plants (Diby *et al.*, 2003). PGPRs and their effectiveness has been established on foot rot and slow decline of black pepper and rhizome rot of ginger and clump rot of cardamom.

Table 6.1. Nutrient content of different organic farm wastes

Organic waste	*N*	*P*	*K*	*Zn(mg kg^{-1})*
		(g /100g)		
Composted coffee pulp	3.20	0.16	2.3	12
Arecanut leaf	2.13	0.12	0.88	17
Cardamom dried leaves	1.85	0.11	1.80	18
Shade tree leaves	1.68	0.07	1.10	19
Leaf compost	1.10	0.84	0.78	76
Coir compost	1.20	0.60	1.20	16
Vermi-compost	2.00	0.80	1.20	44

The efficient isolates from black pepper with high rhizosphere competency and adaptability can be used in a cropping system involving black pepper, ginger and cardamom.

The use of EMs (Effective Microorganisms) containing a consortium of beneficial microbes that acts as microbial inoculants and anti-oxidants in soil thus improves soil health and promotes healthy environment for plant growth. The benefits of using EMs in agriculture are (Unnikrishnan Nair, 2006).

- Promotes germination, flowering, fruiting and ripening.
- Improves physical, chemical and biological environments of the soil and suppresses soil borne pathogens.
- Enhances photosynthetic capacity of crops.
- Increases the efficacy of organic matter as fertilizers.

Organic Farming Strategies

Definite principles, basic standards of production, documentation, inspection and certification guidelines are approved by the National Standards Committee constituted by the members of IFOAM in India. Experiments conducted at Indian Institute of Spices Research, Kozhikkode revealed that application of composts *viz.,* leaf and vermi-composts and FYM improved significantly the soil physico-chemical properties, nutrient availability, yield and quality in black pepper and cardamom. Leaf compost and vermi-compost treatments supported significantly higher population of free-living N-fixing bacteria, phosphate solublizing bacteria populations.

Converting coir dust to enriched coir pith compost and applying it to the spice basins helped in increasing the production and quality of spices and also reduced the amount of chemical fertilizer need in to half. Trials conducted through AICRP on Spices at various locations revealed that sustainable yield up to 6.5 kg vine^{-1} could be achieved in black pepper by application of FYM (10 kg) + rock phosphate (40g) + wood ash (2 kg) per vine along with *Azospirillum* on par to the conventional practice in nutrient supplementation.

Similarly in ginger and turmeric, combined application of different organic sources like FYM + pongamia oil cake + neem oil cake + stera meal + rock phosphate + wood ash has yielded on par to the conventional practice in addition to high quality. The quality parameters of ginger (high oleoresin and low fibre) and turmeric (curcumin and starch) are found to be high under organic farming (IISR Annual report, 2004-05).

Eco-friendly management schedules have been developed at IISR, Kozhikkode using biological control of pathogens and pests of black pepper that have great significance especially in the concept of organic farming. Soil solarization and fortifying the nursery mixture with *Trichoderma harzianum* and VAM to raise the nursery rooted cuttings of black pepper and cardamom gives protection from foot rot pathogen, *P. capsici* and nematodes, *R. similis* and *M. incognita*. Organic manures such as a neem cake, coffee husk, FYM and tea waste can be effectively used to mass multiply biocontrol agents for better field establishment and serve as nutrient supplement. Effective control of the foot rot disease of black pepper under field condition can be achieved by adopting cultural practices like phytosanitation, minimum tillage and better drainage and application of biocontrol agent *T. harzianum* @ 50 g/ vine mixed with 1 kg of neem cake or 5 kg of FYM in the basin.

Other biocontrol agents namely *Paecilomyces lilacinus* and *Verticillium chlamydosporium* are also found to be effective in suppressing nematode infestation in black pepper. Various plant products like leaf extracts of *Chromolaena odorata* and *Strychnos nux-vomica* and seed extracts of *Azadirachta indica* and *Anona squamosa* are promising against pollu beetle. The biocontrol agents of major insect pests of black pepper, ginger and turmeric were documented to develop biocontrol strategies for the management of these insect pests.

Certification

The most significant factor distinguishing organic farming from eco-farming or perma culture is the existence of production standards and certification procedures. Quality of the produce is of the prime importance. Quality parameters follow stringent standards like micro-biologically clean, absolutely residue free with authentic analytical data from approved laboratory.

Any grower who has inclination for organic farming can apply for organic certification. The tariff on inspection and certification varies between agencies and between the types of farmers. The farmer has to keep all the details with respect to cultivation like input-output details, farm map and organic system plan and reports etc. documented. Similarly, processing/ packing and exporting units are also needs to be certified before it deals with organically certified exports. Group certification programmes are also in vogue for organized group of producers, processors and exporters with similar production systems located in geographical proximity. The criteria used for group certification is (Kithu, 2005),

Marginal farmers	- below one hectare
Small farmers	- between 1and 1.99 ha
Semi medium farmers	- between 2 and 3.99 ha
Medium farmers	- between 4 and 9.99 ha
Large farmers	- above 10 ha

Certification and labeling are usually done by an independent body to provide a guarantee that the production standards are met. The certification bodies existing globally under the guidance of IFOAM, Germany charge exorbitantly for certification. In this connection Govt. of India has taken steps to have indigenous certification system to help small and marginal growers and to issue valid organic certificates through certifying agencies accredited by APEDA/Coffee Board/Tea Board/Spices Board. Guidelines for organic production of export oriented spices have also been developed and published by Spices Board, Cochin. A number of farmer's organizations representing organic growers have already entered into the export of spices with the help of Spices Board. Among them are the PDS, WSS (Kerala), SAMANVITA (Orissa) DWUS (Manipur) to name a few.

Spices Board also launched a brand ***Flavourit*** for commercially marketing a range of high quality premium spices. National Programme for Organic Production (NPOP) having standards for products and labeling as '**India Organic**' was started in May 2000 and Spices Board has accredited the following certifying agencies.

- M/s. Indian Organic Certification Agency (INDOCERT), Thottumugham, Alwaye, Kochi, Kerala – 683 105. (info@indocert.org).
- LACON GMBH, Germany C/o Renewable Energy Center, Chunanganveli, Alwaye – 683 105, Kerala (laconindia@indiatimes.com).

FUTURE STRATEGIES

The main strategies for organic farming should be on promoting adoption of biofertilizers, biocontrol agents, cover crops and ecofriendly inputs, adhering to strict phyto-sanitary measures and upgradation of post-harvest technology, direct and indirect effects of organic farming to be quantified on long term basis, strengthening research base on the, economic viability of the system, creating awareness among the farmers about benefits, market promotion and adequate publicity for increasing the demand for organic spices even in the local market.

References

Bopaiah, B.M. and Thomas, G.V. 1993. Occurrence and activities of phosphate solubilizing micro-organisms associated with coconut based cropping systems. *J. Plantation Crops* (Suppl.) **21**:104-07.

Diby, P., Srinivasan, V., Anandaraj, M. and Sarma, Y.R. 2003. *Pseudomonas fluorescens* mediated nutrient flux in the black pepper rhizosphere microcosm and enhanced plant growth. Abstract and short papers: 6th International PGPR Work-shop, 5-10 October, 2003, IISR, Calicut, p. 18-24.

Garibay, S.V. and Jyoti, K. 2003. *Market Opportunities and Challenges for Indian Organic Products*, Study funded by Swiss State Secretariat of Economic Affairs, February 2003.

Khan, H.H., Upadhyay, A.K. and Palaniswami, P. 2002. Integrated Nutrient Management in Plantation Crops, Plantation crops research and developments in the new millennium. In: proceedings of PLACROSYM XIV, National Research Centre for OilPalm, Pedavegi, Andhra Pradesh. pp. 9-22.

Kithu, C.J. 2005. Certification – the crux in organic farming. *Spice India*, **18**(8):7-10.

Lampkin, N. 1990. Organic Farming. Farming press Books, Ipswich. U.K. p. 701.

Sarma,Y.R., Krishnakumar, V. and Anandaraj, M. 2003. Scope and role of PGPR in cropping systems in plantation crops and spices in relation to their agronomy and pathology. Oral Presentations: 6th International PGPR Workshop, 5-10 October 2003, Calicut, India. pp. 3-19.

Unnikrishnan Nair, U.G.S. 2006. Benevolent microbes. *Science Reporter* **43**(1):20.

7

Organic Production of Seed Spices

S.K. Malhotra

The seed spices are used not only for flavouring and seasoning of food, but also for various cosmetic, perfumery and pharmaceutical purposes. The growing awareness of health globally has led to demand of organically produced seed spices as they are free from chemical residues and health hazards. India has been exporting seed spices and its value-added items to 70 countries in the world and is earning a foreign exchange of more than Rs. 362 crore annually. Though, contribution towards export is only 10% of the total spices exported, but there is tremendous scope to enhance it to twice if value added products of seed spices are also exported from India. The major value added products prepared from seed spices are essential oil, oleoresin, spices powder, curry powder and organic seed. In the recent past the focal point of export has shifted towards organic products of spices. The organic seed spices production in India is in infancy and major credit of organic spices export goes to main group of spices produced, particularly in Kerala and adjoining parts. The production of seed spices is concentrated mainly in the arid to semi-arid regions covering Rajasthan, Gujarat and Madhya Pradesh. A small group of farmers in these states have started organic cultivation, but official information regarding its production, area and export is not available. The organic production of seed spices crops have not gained much momentum due to lack of scientific information on suitable production technology, appropriate plant protection measures and specific varieties for organic production.

World over, people are becoming more conscious of health problems due to consumption of agricultural produce contaminated with pesticide residue and the adverse impact of modern agriculture on the environment. Organic farming is the viable alternative to these burning problems. Major advantages of organic farming are that organic spices are highly remunerative because of export demand and moreover the organic produce is of high quality, safe, nutritive and eco-friendly. The related benefits accrued are that long-term fertility of soils is protected as it is based on the principle of sustainable farming and the insoluble nutrient sources can be utilized through action of micro-organisms.

The production of seed spices in arid region is organic by default and it is rather easier to shift towards certified organic farming of seed spices. Some of obvious constraints identified towards such type of production system for seed spices crops are cultivation of seed spices, traditionally in marginal lands with low fertility and inherent nature of slow germination also leads to poor stand of the crop. Moreover, there is lack of specific varieties for organic cultivation and responsiveness to organic nutrient sources. Some of seed spices viz., coriander, ajowan, dill and fenugreek are grown in rainfed farming system for the technology available is insufficient. These crops experience high biotic pressure of wilt, blight, powdery mildew and root rot diseases and heavy infestation by aphids and store grain pests without complete organic control measure.

Procurement Management for Organic Farming of Seed Spices

A minimum of two years is required as conversion period for organic cultivation and undergoing the process of certification by the accredited agency. The general guidelines for selection of farm land, buffer zone, certification, etc., are required to be followed. The procurement management which includes operationalization of activities for production of organic seed spices crops and creating conducive environment for required kind of quality and quantity of produce by way of adopting pre- and post-harvest practices and deciding marketing system suitably. It shall ultimately be guideline to the grower/processor/exporter in planning the production/supply of organic seed spices for export is given here under:

i) Pre-harvest Activities

- Selection of site/area in terms of soil and crop conditions, irrigation facility and time suitability for procurement
- Selection of varieties in terms of requirement of overseas markets, resistance to biotic and abiotic stresses, in terms of ultimate use of the produce (herbs or seeds), local infrastructure availability
- Adoption of fields/farmers for better agronomic practices: Use of organic manures/ composts/bio-fertilizers/other permittable nutrient sources, etc., optimum to low irrigation intensity, adoption of fields as per crop rotation and suitable crop mixing.
- Designing supply scheme
- Assessment of harvesting/maturity as per market requirements.

ii) Harvest and Post-harvest Activities

- Stage of harvest and inspection of field.

- Harvesting and threshing according to requirement of particular country.
- Post-harvest operations like cleaning, drying, grading, packing, storing, transportation.
- Quality testing and labelling.

General Guidelines for Organic Spices Production

The general guidelines have been prepared by combining the information available from various sources (major part from Spices Board) which are in line with the principles and practices of organic cultivation for spices crops.

1. A minimum area of 0.40 ha is the eligibility requirement for organic spices growers and applicant should possess a valid organic certificate issued by an international agency.
2. The farmers are free to convert a portion of the farm but it is advisable that entire farm is converted to organic, truthfully and systematically.
3. It is essential that all the crops in the field, where organic spices are produced are also maintained using organic methods of production.
4. Crop rotation involving a legume in the cycle should be followed and crops belonging to the same family should be avoided.
5. Mixed farming system integrating crop husbandry and livestock is the most ideal.
6. The organic farm is to be maintained as a closed ecosystem limiting the net loss of nutrients through soil wash and run off from the system.
7. In order to avoid contamination of organically cultivated plots from neighbouring plots an isolation distance of at least 25 m width is to be left from all round the conventional farm.
8. A minimum of two years is required as conversion period for organic cultivation. In case of cultivation on virgin lands, the conversion period can be relaxed.
9. Varieties cultivated should be adapted to the soil and climatic conditions and possibly be naturally resistant to pests and diseases of the region. The organic basic seed material should be used. In case of non-availability initially conventional planting material shall be used.
10. Well-rotten farm yard manure or compost @ 10–12 t/ha for fennel and 4-5 tonnes may be used for coriander, cumin and fenugreek as basal application.
11. Aphids are major problem in case of seed spices crops, regular surveillance is necessary for effective management of this pest. Application of fish oil rosin soap, neem seed kernal extract (3%), neem oil (50ml/10 water mixed with 50ml soap solution) and tobacco decoction (0.05%) may be done for control of insects.

12. Restricted use of Bordeaux mixture 1% for control of root rot in fenugreek, wilts in cumin and coriander, blights in fennel and cumin can be practiced. Dusting of sulphur as a restricted use can be done for control of powdery mildew.

Major Inputs for Organic Farming

The major inputs which can be used for organic seed spices production are:

***Organic manures*:** FYM, sheep manure, crop residues, poultry manure, oil cakes and other farm wastes, composts-coir pith compost.

***Green manures*:** Sunhemp, Dhaincha, legumes are used for green manuring.

***Vermi-compost*:** It is the compost prepared by earth worms. Biologically degradable and decomposable organic wastes are used as earthworm feed.

***Biofertilizers*:** Natural fertilizers containing carrier based micro-organism viz *Rhizobium*, *Azotobactor, Azospirillum*, Blue-green Algae, Azolla, Mycorrhizae and phosphobacteria.

Herbal Extracts for Crop Production

***Biological agents for controlling insects*:** Protozoa are available as biological agents for controlling the population of insects. The natural enemies spiders, insects, mites, nematodes, birds, fungi, bacteria, virus may be the biological agents.

***Bio-Plants growth promoters*:** The decayed plant extracts are used as liquid manure for promoting plant growth (eg. Eupatorium weed and Glyricidia).

***Biodynamics to Agriculture*:** The Indigenous and Biodynamic Preparations such as Cow Horn manure, horn silica, compost preparations BD 502-508 can be used in organic nutrition management.

Specific Technology Available

The seed spices crops *viz.* coriander, cumin, fennel and fenugreek organically have its ready market mainly in Germany, The Netherlands, USA, and Japan. In addition to common cultural practices already standardized for seed spices crops, the general and specific guidelines available must be used for organic production. Such practices should involve Good Agricultural Practices (GAP), Good Management Practices (GMP) and Good Hygiene Practices (GHP) in value chain system of production, post-harvest handling and processing. The specific production technology available so far, for production of organic seed spices is given here crop-wise.

CORIANDER

Coriander (*Coriandrum sativum* L.) belonging to family *Apiaceae* is a herbaceous plant widely grown in most of the states of India as a spices crop. The stems, leaves and fruits have a pleasant aromatic flavour and thus used in flavouring different kind of foods and beverages worldwide. Coriander is a native of southern Europe and Asia minor. Coriander is commercially cultivated in India, Morocco, Russia, Hungary, Poland, Romania, Bulgaria, Czechoslovakia, Guatemala, Mexico and USA. In India, it is cultivated largely in Rajasthan, Madhya Pradesh, Gujarat, Tamil Nadu and Andhra Pradesh. The major importing countries of Indian coriander are USA, Saudi Arabia, Germany, Canada, Bahrain, Indonesia, South Africa, Singapore, etc.

Cultural Requirements

The cultural requirements including practices as per procurement management for organic coriander production are described here.

Climate

Coriander a tropical crop, requires a cool and comparatively dry and frost free climate. It is cultivated successfully during winter season in frost free weather in plains and as summer crop in temperate areas. Germination of coriander is severely reduced at temperature above 30 °C and below 10 °C. Heavy rains are harmful for the crop and continuous cloudy weather invites diseases and aphids. For green leaf, it can be grown throughout the year except extreme summer months provided moisture is made available. Coriander is cultivated either as a pure crop or mixed crop under irrigated as well as rainfed conditions.

Soil

Coriander is grown in a wide range of soils from medium to heavy loam in nature. Deep and fertile loamy soils are best suited for quality production of greens and fruits under the irrigated production system. But owing to high moisture retention, the heavy soils are ideally suited for rainfed cultivation of coriander and is thus grown successfully in areas under black cotton soils in southern and northern parts of India, as rain fed crop with conserved soil moisture. However, coriander as an irrigated crop can be cultivated in all type of soils containing sufficient organic matter.

Cropping System

Coriander can be grown as mixed or intercrop provided all the other crops are grown following organic farming methods. It is desireable to include leguminous crops or other suitable crops exhibiting lesser competition. Likewise rotation crops should also be grown following organic farming methods. Following crop rotations under the Punjab conditions have been suggested: Maize–Potato–Coriander, Summer moong–Coriander and, Rice–Coriander. Some of the other cropping systems suggested for seed spices under semi-arid conditions can be: Green gram/black gram–Coriander, Cluster bean/cowpea–Coriander and Cluster bean-Coriander-Summer Maize.

Cultivated Varieties

Varieties selected should have adaptation to the soil and climatic conditions and preferably have resistance/tolerance to pests and diseases. Efforts should be made to use organic seed of recommended variety for organic farming. When certified organic seed of coriander is not available, initially conventional seeds can be used. The description of some of the important cultivated varieties recommended for different states are given in Table 7.1.

Table 7.1. Recommended varieties of coriander for different states

Variety	*Salient characters*
	Rajasthan
RCr- 41	• Suitable for irrigated areas. • Highly resistant to stem gall and wilt but moderately tolerant to powdery mildew. • Matures in 130-140 days and average yield is 9.2 q/ha.
RCr-20	• Suitable for cultivation as rainfed crops or limited moisture conditions. • Moderately tolerant to powdery mildew, wilt as well as stem gall. • Matures in 120 days and average yield is 9 q/ha.
RCr-435	• Adapted to irrigated conditions under high fertility soil. • Moderately resistant to root knot nematode and powdery mildew. • An average yield of 10 q/ha can be obtained in 135 days.
RCr-436	• Suitable for cultivation in heavy soils of Rajasthan as rainfed crop. • Moderately resistant to root knot nematode and less susceptible to aphid. • Matures in 90-100 days and average yield is 11 q/ha
RCr-684	• A variety from medium maturity group with bold grains • It is resistant to stem gall and is suitable for irrigated cropping system. • Average yield - 990 kg/ha in crop duration of 130 days.
RCr-446	• Medium in maturity bears bold grains. • Suitable for irrigated cropping system • Moderate resistance to stem gall and wilt. • An average yield of 12 q/ha can be obtained in crop duration of 130 days.
NRCSS-ACr-1	• It is a long duration variety for irrigated conditions. • Dual purpose variety suitable for leaves and seeds production, • The grains are medium to small, having essential oil content up to 0.5% • Average yield 1250 kg seeds per hectare. • Resistant to stem gall and have tolerance to powdery mildew. • Suitable for growing as greens during off-season also.
	Gujarat
GCr-1	• Suitable for cultivation as early crop under irrigated conditions. • It has moderate tolerance to wilt and powdery mildew.

(*Contd.*)

Variety	*Salient characters*
	• Matures in 112 days with an average seed yield of 11 q/ha.
GCr-2	• Suitable for early sowing under rainfed areas.
	• It is moderately tolerant to wilt and powdery mildew.
	• It matures in 120 days with an average yield of 14 q/ha.
	Tamil Nadu
CO 1	• Suitable for leaf and seed production.
	• Has less water requirement and suited for rainfed areas of Tamil Nadu.
	• It matures in 100-120 days and average yield is 4 q/ha.
CO2	• It is a dual purpose variety and is suitable for saline, alkaline and drought prone areas of Tamil Nadu.
	• The variety matures in 90-100 days with an average grain yield of 5.2 q/ha and green yield of 10 q/ha in 40 days after sowing.
CO3	• Plants are dwarf with medium-bold, oblong, brownish-yellow grains.
	• Suitable for both *kharif* and *rabi* season cultivation in Tamil Nadu
	• Tolerant to wilt, powdery mildew and grain mould.
	• Matures in 86-104 days and average yield is 6.5 q/ha.
CS 287	• Early maturing variety, is suitable for cultivation under both rainfed and irrigated conditions of Tamil Nadu. · Field tolerance to wilt and grain mould.
	• Plants are early-maturing with medium-sized, oblong, straw grains
	• Matures in 78-97 days and average yield is 6.0 q/ha.
	Bihar
Rajendra Swathi	• An early variety and semi erect type is suitable for intercropping.
	• Resistant to stem gall and moderately resistant to wilt, aphids and weevil.
	• It matures in 100 days and average yield is 9.09 q/ha.
Rajendra Sonia	• It is a medium tall, high yielding variety.
	• Resistant to stem gall, moderately resistant to wilt, aphid, weevil and tolerant to fruit fly.
	• Average yield is 12 q/ha in crop duration of 110 days.
	Andhra Pradesh
Sadhna	• A semi-erect variety is early maturing and dual purpose type.
	• Suitable for cultivation under rainfed conditions of A.P.
	• Field tolerance to white fly and mites.
	• Matures in 100 days and average yield is 10.3 q/ha
Swathi (CS 6)	• Semi-erect early maturing variety and is suitable for rainfed condition and late sown crop.
	• It has field tolerance to fruit fly.

(*Contd.*)

Variety	Salient characters
	• The seed yield is 855 kg/ha in 95-105 days crop duration.
Sindhu	• Suitable for intercropping and rainfed farming system.
	• Tolerant to wilt, powdery mildew, and aphids.
	• Average seed yield of 10 q/ha in 102 days cropping period.
	Haryana
Hisar Sugandh	• Medium maturity group.
	• Suitable for dual purpose cultivation both as leaves and seeds.
	• This variety possesses resistance to stem gall. The average yield of 14 q/ha
Hisar Anand	• It is a dual purpose high yielding variety has wider adaptability to different soil conditions
	• Average yield 14 q/ha
	Uttar Pradesh and Uttarakhand
Pant Haritama	• A dual purpose variety, late maturing and high yield potential (15 q/ha)
	• It is tolerant to stem gall disease.
Azad Dhania- 1	• Tolerant to moisture stress· Tolerant to powdery mildew and aphids
	• Maturity in 120–125 days
	Madhya Pradesh and Chhatisgarh
Gwalior No. 5365	• A medium seed sized variety from medium maturity group
	• This variety produces 8–10 q/ha seed in 100-113 days cropping period.

Preparation of Land

Soil is brought to fine tilth by 2-3 ploughing with harrow or *desi* plough. Stubbles of the previous crop should be collected and removed from the field. Clods should be broken and field should be levelled with the help of plank. Beds of convenient size with provision of irrigation channels should be prepared before sowing of seed. Well-rotten farm yard manure or compost is to be incorporated into the soil preferably 3 weeks before sowing.

Sowing Time

Coriander is grown throughout the year for leaves except sever summer months. The main crop of coriander for grains/leaves as spices is grown during October-November and as *rabi* season crop in north India. In southern states Andhra Pradesh, Tamil Nadu and part of Karnataka coriander is cultivated both in *kharif* and *rabi* season. The *kharif* crop is sown during June-July while the *rabi* crop in October-November. The rainfed crop of coriander is also grown in Northern Rajasthan, M.P, Tamil Nadu and Andhra Pradesh during mid-September to October.

Sowing Method

Coriander is propagated through seeds (botanically fruits called schizocarp). Being bilocular each schizocarp has two mericarps and thus before sowing fruits are rubbed to split into two halves. The seed rate required is 10–12 kg per hectare for irrigated and 20 kg for rainfed cropping system. The crop is usually sown by broadcasting of seeds but line sowing is better. Soil depth should not exceed 2.5 cm. A line to line spacing of 20-30 cm and plant to plant spacing of 15 cm should be maintained. Soaking seeds in water for 8-12 hours before sowing hastens germination.

Seed Treatement and Manurial Requirement

The use of bioinoculant *Azospirillum or Azotobacter* as seed treatment and soil treatment in combination with 10 t sheep manure/ha could give seed yield of 9 q/ha and thereby reducing the incidence of stemgall to 8% under semi-arid conditions. The fusarium wilt incidence in coriander can be better managed by seed pelleting with *Trichoderma viride* (with CFU10^6 @ 4g/kg of seed) plus neem cake application (150 kg/ha) reduces wilt incidence.

The manurial recommendations have been made from different states. In general the well-rotten farm yard manure of 10–15 t/ha for heavy soils and 15–20 t/ha for light soils with low organic matter, is added while preparing the field. The application of vermicompost at 5 t/ha can also be made.

Irrigation

Coriander is cultivated both as rainfed and irrigated crop but in the latter production system, about 4–6 irrigations are required after germination, depending upon the soil and climate. Initial moisture is essential for good germination and crop stand. The *kharif* crop being rainfed, requires no irrigation but the *rabi* crops should be irrigated at 12–15 days interval. In general the crop requires 4-5 irrigations after germination at 0.8 IW/CPE. Safe irrigation water should be used for irrigation, use of sewage water is strictly prohibited.

Inter-cultural Operations

Coriander is slow growing plant initially and therefore thinning, first hoeing and hand weeding should be done 30 days after sowing. Two hand weedings followed by hoeing should be practiced 30 and 60 days after sowing, and if required, third weeding can be done. During first weeding, the plants should also be thinned to the desired distance.

Crop Protection

Coriander is prone to attack by some of the serious diseases and insect-pests. The plant protection measures under organic farming system should involve emphasis on crop management practices such as time of sowing, balanced nutrition, crop rotation, green manuring, etc., for reducing the incidence of diseases and pests, selection of resistant varieties and adoption of bio control measures.

Pest Monitoring

Pest monitoring is an important part for organic cultivation of coriander and should be done once in seven days to work out economic threshold level (ETL). It will help to keep close watch on the appearance of insect pests, diseases and bio-control fauna. For sucking pests, population should be counted on three leaves, i.e., top, middle and lower per plant. For cut worms and defoliators, per cent damage assessment can be made by counting total number of plants and affected plants. Management strategy suggested are setting up of yellow pan/sticky traps for monitoring of aphids/whitefly @ 10 yellow pans/sticky traps per hectare. Locally available empty yellow palmoline tins coated with grease/vaseline/castor oil on outer surface may also be used. The larvae of cutworm and leaf defoliators should be collected and destroyed.

Aphids (*Hyadaphis coriandri Das.*)

Both nymphs and adults suck sap from the tender leaves thus reducing marketability. The severe infestation affects the yield and quality of leaves badly. Moreover, secretions from aphids invite attack of sooty mould. For the control, clip off the heavily infested plant parts and spray the crop with pressurized water. Application of fish oil, resin soap or neem seed kernel-extract (3%), neem oil (2%) or tobacco decoction (0.05%) is effective measure. Spray of neem oil (1%, 5 litre/ha) or NKE (3%, 15 litre cow urine + 1.5 kg neem kernel extract/ha) are effective in controlling aphids (*Hyadaphis coriandri* Das) when sprayed three times at 15 days interval.

Mites (*Petrobia latens*)

The mites frequently attack the coriander crop and whole plant becomes whitish yellow and appears sickly. Plants get stunted at severe infestation. For the control, spray neem oil (1%).

Coriander wilt (*Fusarium oxysporum* f.sp. *corianderii*): The disease is more prevalent in light soils and affected plants show yellowing, drooping of leaves and terminal branches followed by the withering and drying of the plant. In the severe incidence of disease, plant dies and can be pulled out easily of the ground. Pathogen being coriander-specific can be managed through crop rotation. The occurrence of disease can be reduced by the application of oilcakes in the field with high soil pH 8.2. Solarization of soil during summer and use of disease free seed minimizes the disease incidence. The antagonistic fungi *Trichoderma viridae* 4 g/kg seed treatment is effective in reducing the wilt incidence to 5.3 %. Seed treatment with *Pseudomonas fluorescens* @ 10 g/kg seed + soil application of neem cake @ 150 kg/ha is effective biocontrol method. Extreme dry climate condition should be avoided as it favours the disease. Adoption of field sanitation measures and cultivation of wilt resistant variety suiting to the production area are recommended.

Powdery mildew (*Erysiphe polygoni* D.C.): The symptoms appear as white powdery mass on the leaves, twigs and fruits of the plant. For control, the application of sulphur dust, wettable sulphur suspension (0.6%) and sulphur spray and the restricted use of sulphur is permitted and described in guidelines from Spices Board. Foliar spray of onion extract

(5%) is effective in reducing the severity. The incidence is more during February-March, thus sowing of coriander should be completed in first week of October, as late sown crop has more severe damage. Uprooting of all affected plants followed by burning or burying, early planting manages the disease well and growing resistant/tolerant varietiesare recommended.

Stem gall (*Protomyces macrosporus* Unger): The symptoms appear as galls on the stem, leaf stalk and peduncles. The seeds are deformed and the yield is reduced. For the control, it is necessory to uproot and burn all the infected plants completely out of the field. Besides, solarization of soil and use of resistant variety RCr 41 can also be followed. Wet growing conditions should be avoided. Use of healthy seed is the best remedy. Follow long term crop rotation.

Blight (*Alternaria poonensis*)

The disease appears in the form of dark brown spots on the stem and the leaves. The disease mostly appears under cloudy weather. The measures suggested for control are avoiding excessive irrigation and high humidity, using disease free seed, adopting field sanitation measures and using resistant variety.

Frost protection: Frost causes damage to coriander crop at flowering and fruit setting stage. During severe winter months when there is danger of frost, the crop should be irrigated and smoke environment should be created before sunset in the field.

Harvesting and Yield

Coriander crop matures in about 100–150 days of sowing depending upon varieties, climate and geographical location. The stage of harvest depends upon the market requirement. For green coriander seeds, the crop is harvested when seeds are green and have attained full size. The green seed crop is dried under shade for retaining the green colour. Keeping in view, the market requirement, the crop can be harvested at yellow or brown colour of seeds. An average seed yield of 4.5 to 6.0 q/ha can be obtained from rainfed crop, whereas average seed yield of 8–12 q/ha.

Post-Harvest Management

Coriander is mostly grown for green leaves as vegetable, dehydrated leaves and seed. The post-harvest operations are followed as per the type of commodity grown. The fresh greens consisting edible portion are the tender leaves and stem, which are cut to length of about 5–7 cm above ground. The yellow, diseased, and damaged leaves are trimmed off and weed plants and straw is culled during cleaning and dressing. Healthy and disease free leaves are tied into small bunches for the convenience in handling, transportation and marketing. The leaves of coriander are sun-dried or dehydrated in a suitable dehydrator, for further use in off-season. The dried coarse powder or rubbed leaf powder is mostly used for flavouring by sprinkling on the various food items.

Storage

The fresh coriander leaves are highly perishable and deteriorates rapidly at temperature above 5 °C. Therefore, the fresh leaves bunched should be marketed soon after harvesting. The leaves can be sorted only for about 24–36 hours after harvesting under ambient condition, however in cold store at 0 °C temperature and 90% relative humidity, the storage period can be extended for one week. The dried seed is filled in plastic bags or in gunny bags lined with plastic sheet. Each bag is sealed and stored under clean, dry and ventilated place.

Processing

The mature dried seed of coriander are distilled to obtain the essential oil. Hydro or steam distillation method is generally used for extraction of essential oil. On an average, the dried seeds of coriander yield 0.1–1.7% volatile oil depending upon variety and location. Good quality oleoresin extracted from coriander seed has great demand in importing countries.

Coriander is sold as whole coriander, ground coriander, essential oil, oleoresin, coriander herb oil, coriander dhana dal. The volatile oil is used chiefly as a flavouring agent for liquors, cocoa and chocolate industry and it is also valuable ingredient in perfumes.

Quality Specifications for Export

Whole seed

The quality specification for coriander can be classified into three categories *viz.* commercial requirements, cleanliness and health specifications.

Commercial Specifications

These requirements vary from country to country and crop to crop. It depends on many factors like customer needs, acceptability and country specifications. The commercial specifications of seed spices crops are colour, appearance, taste, pungency, texture, shape, volatile oil and packaging. The general commercial and general requirements for coriander seed are:

Table 7.2. General commercial requirements

Coriander	Yellow-brown or green colour, small seeded, large seeded, free flowing, high volatile oil (min. 0.25%)

Cleanliness Specifications

The permit table cleanliness specifications as per ASTA, ESA and ISO are given in Table 7.3.

Table 7.3. Cleanliness specifications

ASTA cleanliness specifications for coriander

Crop	Whole insect dead (by count)	Excrets mammalian (mg/lb)	Excreta other (mg/lb)	Mould (% wt)	Insect defiled/infected (%/wt)	Extraneous foreign matter (%/wt)
Coriander	4	3	10	1	1	0.5

Other quality standards as per ESA and ISO

Crop	Ash level % w/w (min.)	Acid insoluble ash % w/w (max.)	Moisture content % (max)	Volatile oil % (Min.)
Coriander	7	1.5	12	0.3

Health Specifications

The micro-contaminants such as pesticide residues, microbial counts, aflatoxins, heavy metals, etc., are the health specifications. These contaminants make the food unsafe. The importing countries have limited the maximum level of these contaminants and limits are; Maximum Residue Level = 0.05 to 0.2%, Microbial counts = 1×10^2/g to 1×10^6/g (*Salmonella, E. coli*, yeast, moulds), Aflatoxin = 5 ppb to 10 ppb (Max.) and Heavy metals = Arsenic (5 mg/kg), copper (20 mg/kg) lead (10 mg/kg), Zinc (50 mg/kg).

Essential Oil

The specifications for coriander oil are: Colourless or pale yellow colour, specific gravity (20 °C, 0.870 to 0.885), optical rotation (20 °C), + 8 to +13, refractive index, 1.463 to 1.471, solubility (70% ethanol), 2-3 vols, aldehydes (Linalool) 45 to 70%.

Coriander Oleoresin

The coriander oleoresin should be prepared with recommended organic solvents followed by the subsequent removal of the solvent as per specifications of importing countries.

Coriander Powder

The coriander powdered hygienically through permittable devices, should at least 95% of the ground product pass through a U.S. Standard No. 30 sieve.

CUMIN

The cumin (*Cuminum cyminum* Linn.) belonging to family *Apiaceace* is an important seed spice believed to be indigenous to northern Egypt, East Mediterranean region, Iran, Turkey and India. It is popularly known as safed jeera and is one of the most common spices of daily use in the Indian sub-continent. It is mainly grown in India, Egypt, Turkey,

Iran, Pakistan, Syria, and Italy. In India it is extensively cultivated in Rajasthan, Gujarat and in some parts of Madhya Pradesh and Uttar Pradesh as *rabi* crop.

Cultural Requirements

The cultural requirements including practices as per procurement management for organic cumin production are described here.

Climate

Cumin is a tropical crop and is successfully cultivated in moderately cool and dry climate. It does not prefer humidity in atmosphere during flowering and seed setting stage and is prone to attack severely by diseases under such climatic conditions. It is also susceptible to frost injury during flowering and early fruit setting stage. Therefore, the cultivation of the cumin is restricted during *rabi* season in the area in which atmospheric humidity is low and winters are not severe. At the time of maturity cumin needs dry climate. The frequent rains during flowering or early fruit setting period are highly undesirable. High humidity associated with cloudy weather, is harmful and increases incidence of powdery mildew and blight. Frost is the most dangerous and limited factor for cumin cultivation.

Soil

The cumin can be cultivated in wide range of soils, however sandy loam to medium heavy soils having plenty of organic material with better fertility status are most suitable The soil should have better drainage facility because stagnated water and excessive moisture both are very harmful to cumin crop. In sandy soils, incidence of wilt disease has been reported more. The soil should be free from stones, clods, stumps and weed plants.

Cropping System

Cumin is not recommended for growing as mixed or inter-crop. However, the rotation crops required to be grown, should be based on organic farming methods. Some of the suggested cropping systems for cumin growing areas could be: Dhaincha-Cumin, Dhaincha-Cumin-Green gram, Maize-Cumin-Summer Moong and Pearl Millet-Cumin.

Cultivated Varieties

Varieties selected should be adapted to the soil and climatic conditions and should preferably have resistance/tolerance to pests and diseases. Efforts should be made to use organic seed of recommended variety for organic farming from certified organic farms or own seed rose organically. When certified organic seed is not available, initially conventional seeds can be used. The description of some of the important cultivated varieties developed for different states with information on salient characters are given in Table 7.4.

Table 7.4. Recommended varieties of cumin for different states

Variety	*Salient characters*
	Rajasthan
RZ 19	· The plants are erect in growth behaviour bear pink coloured flowers and bold pubescent grains.
	· Takes 120 –140 days to mature and average yield is 5-6 q/ha.
RZ 209	· The variety has shown tolerance to wilt.
	· The crop matures in 140-150 days and average seed yield is 6.5 q/ha.
RZ 223	· The variety possesses resistance to wilt.
	· The essential oil content is 3.23 % and gives seed yield of 6.0 q/ha.
	Gujarat
GC 1	· The plants are erect with pink flowers and bold, linear, oblong ash brown colour grains.
	· The variety is tolerant to wilt disease.
	· It matures in 105-110 days and average yield is 7.0 q/ha.
GC 2	· The plants are bushy with good branching habit, attractive grains.
	· Matures in 100 days and average seed yield is 7.0 q/ha.
GC 3	· The variety is resistant to wilt.
	· It matures in 100 days and average yield is 7.0 q/ha
	· The essential oil content is 3.5%.
GC 4	· It gives an average yield of 8.75 q/ha
	· It is resistant to Fusarium wilt

Cultural Practices

Preparation of Land

Soil is brought to fine tilth by 2-3 ploughing with harrow or *desi* plough. Stubbles of the previous crop should be collected and removed from the field. Clods should be broken and field should be levelled with the help of plank. Beds of convenient size with provision of irrigation channels should be prepared before sowing of seed. Well-rotten farm yard manure or compost is to be incorporated into the soil preferably 3 weeks before sowing.

Sowing Time

The time of sowing may vary a little according to place. However the optimum time of sowing is mid October—mid-November. Cumin can be sown in Pali district of Rajasthan throughout October but the second week of October is the optimum time of sowing to get higher seed yield.

Sowing Method

The quality seed of 12–15 kg is required for sowing of cumin crop in one hectare area.

Sowing of cumin is done by two methods, i.e., line sowing and broad-casting. Traditionally farmers sow cumin by broadcasting method but sowing done in lines facilitates inter-cultural operations. Cumin seeds are small in size, therefore sand or screened friable soil is mixed with seed just before sowing so that uniformity is maintained when seeds are dropped in shallow opened furrow. The line to line distance should be kept at 25 cm. The seed should not be sown deeper than 1.5 cm and the same should be covered with a thin layer of soil. Since germination is a problem in cumin, therefore soaking of seed in water for 8 hours before sowing improves germination. The soaked seed should be dried in shade before sowing.

Manurial Requirements

In Rajasthan, 6-8 tonnes of FYM should be applied at the time of seedbed preparation. The requirement of fertilizer depends upon the fertility status of the soil. Therefore, fertilizer should be applied based on soil testing report. In order to maintain good soil structure and biological property, apply 10 t / ha FYM or 5 t/ha compost three weeks before sowing of the crop. Inoculation with VAM fungi, *(Gigaspora calospora, Glomus fasciculatum, Glomus mosseae* and *Acaulospora laevis*) not only reduced the incidence of wilt in cumin but also enhanced nutrient uptake. The use of bioinoculant *Azospirillum* or *Azotobactor* as seed treatment and soil treatment in combination with 5t sheep manure/ha could give higher seed yield under semi-arid conditions.

Irrigation

The water requirement for raising cumin crop is comparatively less than many spices crops. Irrigation should be given judiciously using uncontaminated water. After sowing a light irrigation should be applied and initial moisture helps in getting good germination. The germination of the seed will be visible after 10–12 days of sowing. The crop is irrigated at an interval of about 30 days depending upon the weather condition and soil type. It is, however, important to keep in mind that irrigation should be avoided when the crop is at maturity stage, as this may affect seed quality adversely.

Inter-cultural Operation

The crop should be kept free from weeds for proper growth and development of plants. Generally 2-3 hand weeding are required to keep the crop weed free and reducing the crop-weed competition for growth resources like nutrients, moisture, light and others. The first hoeing and weeding is done at the time when the plants are 5 cm tall. The subsequent inter-cultural operations should be done on appearance of weeds. The weed like *Zeeri* (*Plantago pumila* L) is a serious in cumin and is difficult to distinguish at early growth phase but can be identified easily at flowering stage. It is suggested to keep cumin field free from weeds till 60 days after sowing for lowering weed intensity and realizing higher seed yields.

Crop Protection

Cumin is prone to attack by some of the serious diseases and insect-pests. The plant protection measures under organic farming system should involve emphasis on crop management practices time of sowing, balanced nutrition, crop rotation, green manuring, etc., for reducing the incidence of diseases and pests, selection of resistant varieties and adoption of biocontrol measures.

Pest Monitoring

Pest monitoring is an important part for organic cultivation of cumin and should be done once in seven days to work out economic threshold level (ETL). It will help to keep close watch on the appearance of insect pests, diseases and bio-control fauna.

For sucking pests: Population should be counted on three leaves, i.e., top, middle and lower per plant.

For cut worms and defoliators: Per cent damage assessment can be made by counting total number of plants and affected plants. Management strategy recommended include setting up yellow pan/sticky traps for monitoring of aphids/whitefly @ 10 yellow pans/sticky traps per hectare. Locally available empty yellow palmoline tins coated with grease/vaseline/castor oil on outer surface may also be used. The larvae of cutworm and leaf defoliators should be collected and destroyed.

Insect pest

Aphids: Aphids attack cumin crop at flowering stage causing serious damage. For the control, clip off the heavily infested plant parts and spray the crop with pressurized water. Application of fish oil, resin soap or neem seed kernel-extract (3%), neem oil (2%) or tobacco decoction (0.05%) is effective measures (Ravindran *et al.*, 2001). Spray of neem oil (1%, 5 litre/ha) or NKE (3%, 15 litre cow urine + 1.5 kg neem kernel extract/ha) are effective in controlling aphids (*Hyadaphis coriandri* Das) when sprayed three times at 15 days interval.

Mites (Petrobia latens)

The mites frequently attack the coriander crop and whole plant becomes whitish yellow and appears sickly. Plants get stunted at severe infestation. For control, spray of neem oil (1%) is recommended.

Diseases

Blight

The infected plants show dark brown blighty appearance on the leaves and stem resulting in withering off in affected plants. It is a serious disease in cumin cultivated areas and infection is favoured by humid conditions. For control, avoiding excessive irrigation and high humidity, using disease free seed, adopting field sanitation measures and cultivation of resistant variety are recommended.

Wilt

The infected plants exhibit drooping of leaves and epinasty later lead to death of the plant. The wilt infection may occur in patches at any stage of crop growth. After occurrence of the wilt in cumin, it becomes very difficult to reduce the damage. The disease can be managed effectively by adopting integrated approach including summer ploughing and crop rotation of at least three year with non-host crops preferably mustard or pearl millet.The occurrence of disease can be reduced by the application of oilcakes in the field with high soil pH 8.2. Use of disease free seed minimizes the disease incidence. The antagonistic fungi *Trichoderma viride/ T. harzianum* 4 g / kg seed treatment is effective in reducing the wilt incidence (Aghnoom *et al.* 1999). Extreme dry climate condition should be avoided as it favours the disease. Adoption of field sanitation measures is vital. Locate wilt resistant variety suiting to the production area (GC 3 and GC 4). Inoculation with VAM fungi reduced the incidence of wilt in cumin. Amendment of soil with a combination of mustard residues and mustard oil-cakes (2.5 + 0.5 t ha^{-1}) was effective in control of *Fusarium* population in soil and incidence of wilt on cumin.

Powdery Mildew

White powdery mass appears on the leaves and twigs of the plant in the initial stages. Later on whole plant is covered with this whitish powder. For control, the application of Sulphur dust, wettable sulphur suspension (0.6%) and sulphur spray are recommended. The restricted use of sulphur is permitted and described in guidelines from Spices Board. Foliar spray of onion extract (5%) is effective in reducing the severity. Growing resistant/ tolerant varieties is recommended. The incidence is more during February-March, thus sowing of coriander should be completed in first week of October, as late sown crop has more severe damage. Uprooting of all affected plants followed by burning or burying is necessary to reduce the inoculam load.

Control Against Frost

Cumin crop is susceptible to frost during flowering and seed formation stage. During the frost danger period to protect the crop it is necessary to make provision of smoke in the field.

Harvesting and Yield

Generally, cumin crop takes about 90–120 days to reach maturity. Harvesting should be done when plants turn yellowish brown preferably early in the morning by cutting with sickles and not by uprooting the whole plants. The harvested plants should be stacked for few days and dried in the partial sun in the threshing yard. There after the crop is beaten and trampled on a clean threshing floor. Under the scientific management condition the seed yield of 8–10 q/ha of cumin of improved varieties can be obtained mechanical thresher may be followed in large scale cultivation followed by winnowing, cleaning and grading through mechanical devices.

Post-harvest Management

Processing

The cumin seeds are dried in partial shade in the sun and the moisture content of seed should be kept to 9%. High seed moisture content may lead to chances of storage contamination by fungus. The dried, cleaned, graded produce is packed in the standard sized packs/containers and appropriately labelled. The dried seed is filled in gunny bags lined with degradable environment friendly plastic film. Waste generating packaging material should be avoided. Each bag is sealed and stored under clean, dry and ventilated place.

Processing methods of cumin should be based on mechanised, physical and biological processes. Care should be taken to maintain the vital quality of an organic ingredient throughout each step of its processing. Processing method is to be selected in such a way that it will limit the number and quantity of additives and processing aids. The mature dried seeds are distilled to obtain the essential oil. Hydro or steam distillation method is generally used for extraction of essential oil. The dried seeds are crushed or as such are steam distilled to yield 2.5–4.5% of valuable volatile oil depending upon variety and location. The oleoresin from cumin has good demand in the international market. The processed products include cumin essential oil, cumin powder, cumin oleoresin and fixed oil. Volatile oil should be kept in well sealed bottles or aluminium containers.

Quality Specifications for Export

Whole Seed

The quality specifications for cumin whole seed can be classified into three categories viz., commercial requirements, cleanliness and health specifications.

Commercial Specifications

These requirements vary from country to country and crop to crop. It depends on many factors like customer needs, acceptability and country specifications. The commercial specifications of seed spices crops are colour, appearance, taste, pungency, texture, shape, volatile oil and packaging. The general commercial and general requirements for coriander seed are:

Table 7.5. General commercial requirements

Cumin	Light brown, large seeded, non-splitting, free flowing, aromatic, high volatile oil (min. 2.2 %)

Cleanliness Specifications

The permit table cleanliness specifications as per ASTA, ESA and ISO are given in Table 7.6.

Table 7.6. Cleanliness specifications *ASTA cleanliness specifications for cumin*

Crop	*Whole insect dead (by count)*	*Excrets mammalian (mg/lb)*	*Excreta other (mg/lb)*	*Mould (% wt)*	*Insect defiled/ infected (%/wt)*	*Extraneous foreign matter (%/wt)*
Cumin	4	3	5	1	1	0.5

Other quality standards for cumin as per ESA and ISO

Crop	*Ash level % w/w (min.)*	*Acid insoluble (max.) ash % w/w*	*Moisture content % (max)*	*Volatile oil % (Min.)*
Cumin	14	3.0	13	1.5

Health Specifications

The micro contaminants such as pesticide residues, microbial counts, aflatoxins, heavy metals, etc., comprise the health specifications. These contaminants make the food unsafe. The importing countries have limited the maximum level of these contaminants and limits are: Maximum Residue Level = 0.05–0.2%, Microbial counts = 1×10^2/g–1×10^6/g (*Salmonella, E. coli*, yeast, moulds),Aflatoxin = 5 ppb–10 ppb (Max.) and Heavy metals = Arsenic (5 mg/kg), copper (20 mg/kg) lead (10 mg/kg), Zinc (50 mg/kg).

Essential Oil

The specifications for cumin oil are (Singhal *et al,* 1997): Colourless or pale yellow colour, specific gravity (25 ºC), 0.905–0.925 ; Optical rotation (20 ºC), + 3 – + 8; refractive index, 1.501–1.506; solubility (80% ethanol), 8 vol; aldehydes (cuminic aldehyde) 40–52%.

Cumin Oleoresin

The cumin oleoresin should be prepared with recommended organic solvents followed by the subsequent removal of the solvent as per specifications of importing countries.

Cumin Powder

The cumin powdered hygienically through permittable devices should at least 95% of the ground product pass through a U.S. Standard No. 30 sieve.

FENNEL

The fennel (*Foeniculum vulgare* Mill.) from family *Apiaceae,* a native of Southern Europe and Mediterranean area is an important seed spice. It is cultivated on a large scale in Romania, Russia, Germany, France, Italy, India, Argentina and USA. Besides the other growing countries are Bulgaria, China, Denmark, Egypt, Syria, Morroco and Japan. The major fennel producing states in India are Gujarat, Rajasthan and Uttar Pradesh. In many other states grown at small scale are Punjab, Tamil Nadu, Bihar, Karnataka, Maharashtra and J&K. The major importing countries of Indian fennel are Canada, Germany, Hong Kong, Indonesia, Israel, Mauritius, Netherlands, USA, Saudi Arabia, Bahrain, South Africa, Singapore, UK, etc. The leading states recently came up are Egypt and China.

Cultural Requirements

The cultural requirements including practices as per procurement management for organic fennel production are described here.

Climate

Fennel is a cool season crop, mainly grown during winter season in north India. A dry and cold weather favours higher seed production. A temperature of 15–20 °C is the optimum for growth and high temperature results premature flowering and very low seed yield. Crop is susceptible to frost injury at flowering stage. The optimum temperatures for seed germination is 20–29 °C. During maturity crop requires warm climate. High winds at the time of maturity can cause shattering and very hot winds at flowering reduce seed setting. Cloudy weather and high humidity at the time of flowering and seed formation increases the chances of aphid infestation and attack of Ramularia blight disease.

Soil

Fennel thrives well on drained loamy soil rich in plant nutrients. However, heavy soils rich in organic matter with better water holding capacity are more desirable than light soils. A neutral to slightly alkaline soil in the pH range of 6.5–8.0 is highly preferable and have moderate degree of sodic tolerance. Fennel is poor host of root knot nematodes, thus in soils having problem of root knot nematode, fennel can give reasonably good yield. Soils containing lime with proper drainage have also been considered suitable.

Cropping System

Fennel can be grown as mixed or inte-rcrop provided all the other crops are grown following organic farming methods. It is desirable to include leguminous crops or other suitable crops exhibiting lesser competition. Fennel is long duration crop and have slow initial growth thus crop is very conducive for successful inter-cropping with any of the vegetable crop such as chilli, potato, spinach, carrot, cole crops, coriander and fenugreek. As per one of report, on the basis of net return and benefit cost ratio sugarcane + fennel proved a good inter-crop combination. Inter-cropping with garlic was also reported to have least depressive effect on fennel and reduced the aphid infestation. Likewise rotation crops should also be grown following organic farming methods.

Fennel can be included in the following crop rotations under the Punjab conditions; Paddy – Fennel, Maize – Fennel, Moong – Fennel, Mash – Fennel, Kharif fodder – Fennel. Some of the other cropping systems suggested for seed spices growing area can be: Green gram/black gram – Fennel and Cluster bean/cowpea – Fennel.

Cultivated Varieties

Varieties selected should have adaptation to the soil and climatic conditions and preferably have resistance/tolerance to pests and diseases. Efforts should be made to use organic seed of recommended variety for organic farming. When certified organic seed of fennel is not available, initially conventional seeds can be used. The description of some of the important cultivated varieties recommended for different states are given in Table 7.6.

Table 7.6. Recommended varieties of fennel for different states

Variety	*Salient characters*
	Rajasthan
NRCSS-AF-1	• Suitable for growing both for early sowing and as *rabi* crop. • The seeds are attractive, bold, medium sized, fragrant with volatile oil content upto 1.6%. • It is from late maturity group and produces average yield of 19.5 q/ha during *rabi* season and 25.1 q/ha when grown as an early crop. • The variety has tolerance to Ramularia and Alternaria blight.
RF-101	• It bears large umbel with long and bold grain. • It matures in 150–160 days and gives an average yield of 15.5 q/ha.
RF-125	• The plants are early, short stature with compact umbels Long bold grains with17.3 quintal of seed yield/ hectare.
RF-143	• Very suitable for the loamy and black cotton soils. • Medium duration type, medium tall, having an average yield of 1200 kg/ha and oil content 1.87%.
	Gujarat
PF-35	• The plants are tall and spreading with medium sized hairless and green seeds. • It is moderately tolerant to sugary disease, leaf spot and leaf blight. • It matures in 215–225 days and gives an average yield of 12.89 q /ha.
Gujarat Fennel-1:	• The plants are tall, bushy type suited for *kharif, rabi* and drought. • The plants bear large sized umbel and more number of umblets. The seeds are bold, long dark green colour and contain 1.6% essential oil.
Gujarat Fennel 2 (Gf-2):	• It is tall and long duration variety. • Average yield is 18 q\ha.
Gujarat Fennel 11:	• This variety is suitable for rabi season growing under irrigated conditions. • It produces an yield of 24 q/ha
	Tamil Nadu
Co-1 :	• The plants are medium with profused branching. • It matures in 210–220 days and gives an average yield of 5.67 q/ha. • It is suited for intercropping, saline and alkaline soil and hilly area, drought prone and water logged area.
	Haryana
Hisar Swarup	• Has a yield potential of 17 q/ha. • Plant has an average height of 132 cm, spreading, late in maturity, • Resistant to lodging and shattering of grains. • Fruits are long bold, oil content 1.6%.

Cultural Practices

Preparation of Land

Field preparation should start with early rain. The first ploughing should be done by soil turning plough and afterward 2-3 ploughing should be done by *desi* plough or harrow. The ploughed field should be made fine and levelled by planking. There should be good initial moisture level in soil for better germination. In order to manage problem of soil pests summer ploughing should be done.

Sowing Time

Sowing of fennel may be done in both late *kharif* (early) and *rabi* season. For early fennel, seed is sown is nursery in the month of July and about 45 days old seedlings should be planted in the main field. Last week of September to first week of October time is ideal for direct sowing of the fennel as main season crop. Delayed sowing results in low yield levels. Sowing of fennel on 21 October in Punjab has been reported to give highest yield. The high oil content was recorded when it was harvested at full bloom stage.

Sowing Method

Fennel is propagated through seeds, seeds may be sown directly in the field or seedlings raised in nursery first and then planted in main field. About 2.5 kg seed is required for raising seedling in nursery for one hectare area but for main season crop 8–10 kg seed is required for direct seeding in one hectare land. Seed soaking for 8-12 hours before sowing helps to improve germination.

Raising Seedling in Nursery

For raising seedlings in the nursery, select fertile land having good drainage. About 2.5 kg seed is sown in a nursery area of 1000 m^2 for raising seedling for one hectare land. In the well-prepared seedbeds, seeds are broadcasted or sown in rows. The beds are covered with straw or stalk mulch to protect against high temperature and strong sunlight and are irrigated. The seeds germinate within 10–12 days. The mulch is removed after germination. Irrigation is given as per requirement. The seedlings are ready for transplanting after about 6 weeks.

Sowing/Transplanting

The transplanting of seedling should be done at line-to-line spacing of 45 to 60 cm and plant-to-plant distance of 20 cm in the evening hours otherwise plants may die in hot sunny days. Irrigation should be given immediately after transplanting. For direct seeding the sowing is done either by broadcasting method or in line. The line sowing of seed is more convenient to facilitate proper weeding and other intercultural operations.

Manurial Requirement

Manurial requirement varies from region to region due to type and fertility status of the soil. Therefore organic manure should be applied based on soil testing report. For good production of fennel about 15 t/ha well decomposed FYM should be applied at least 3-4 weeks before sowing. The biofertilizer Azospirillum applied both as seed inoculation (500 g/ha seed) and soil application (2.5 kg /ha) along with application of 10 t well-rotten farm yard manure proved best and produced highest fennel yield under organic manurial studies.

Irrigation

Fennel being long duration crop requires more irrigations than other seed spices. Depending upon temperature and soil type, 6–10 irrigations during growth period are needed. The flowering and seed formation stages are critical stages for irrigation. The higher seed yield has been reported when irrigation applied at IW/CPE ratio of 0.90. Irrigation should be given judiciously using uncontaminated water.

Inter-cultural Operation

The crop growth is slow initially and has to face severe weed crop competition. Moreover, wider row spacing and frequent use of water favours growth of weeds. To keep the field weed free, about 3 hand weeding and hoeing are required. Earthing up of plants at the time of last hoeing and weeding proves beneficial and prevents the plants from lodging.

Crop Protection

Fennel is prone to attack by some of the serious diseases and insect-pests. The plant protection measures under organic farming system should involve emphasis on crop management practices time of sowing, balanced nutrition, crop rotation, green manuring, etc., for reducing the incidence of diseases and pests, selection of resistant varieties and adoption of bio-control measures.

Pest Monitoring

Pest monitoring is an important part for organic cultivation of fennel and should be done once in seven days to work out economic threshold level (ETL). It will help to keep close watch on the appearance of insect pests, diseases and bio-control fauna (Singh and Selven, 2004).

For sucking pests: Population should be counted on three leaves, i.e., top, middle and lower per plant.

For cut worms and defoliators: Per cent damage assessment can be made by counting total number of plants and affected plants. Management strategy suggested include set up yellow pan/sticky traps for monitoring of aphids/whitefly @ 10 yellow pans/sticky traps per hectare. Locally available empty yellow palmoline tins coated with grease/Vaseline/ castor oil on outer surface may also be used. The larvae of cutworm and leaf defoliators should be collected and destroyed.

Aphids

Aphids attack fennel crop at flowering stage and cause serious damage. Both nymphs and adults suck the sap from leaves, flowers and fruits. For the control, clip off the heavily infested plant parts. The botanical insecticide, nemarin (1%) and seed extract (1%) of neem (*Azadirachta indica*) Karanj (*Pongamia sp*) have been reported to reduce aphid population by more than 50% within 7 days. Conserve predators Coccinellids, Chrysopids and Syrphids. Spray of NSKE (5%) or Azadirachtin (0.03%)is recommended. Avoid excessive irrigation under cloudy and high dew conditions

Thrips

Both nymph and adults congregate in between the leaf sheath and stem causes drying of leaves. The infestation is commonly observed during March-April. Control as above for aphids

Midge

The midge is phytophagous and female insert its ovipositor through the fruit wall to lay eggs between the pericarp and ovule in the developing fruits and field conditions. Larvae feed upon and destroy the embryo.

For the control, avoid delay in sowing (October first week sowing exhibit less infestation (4%) than (32%) in delayed sowing), use of clean packaging material and maintenance of suitable temperature and humidity during storage are recommended.

Diseases

***Alternaria blight*:** First symptoms are seen on lower buds which turn yellow or brownish.

***Ramularia blight*:** Infection causes minute angular brown necrotic areas on the lower surface of the older leaves. Infection spread on all plant parts and seed setting is adversely affected.

Control measures recommended for Alternaria and Amularia blight are avoiding excessive irrigation and high humidity, use of disease free seed, adopting field sanitation measures and cultivation of resistant variety.

Powdery Mildew

White powdery mass appears on the leaves and twigs of the plant in the initial stages. Later on whole plant is covered with this whitish powder.

For the control, spray can be repeated after 15 days if required.The application of Sulphur dust, wettable sulphur suspension (0.6%) and sulphur spray (0.1%) and the restricted use of sulphur is permitted as described in guidelines from Spices Board.Foliar spray of onion extract (5%) is effective in reducing the severity. Growing of resistant/tolerant varieties,uprooting of all affected plants followed by burning or burying and early planting manages the disease well.

For the control of nematode (*Meloidogyne javanica*), use of disease free seed, adoption of field sanitation measures, cultivation of resistant variety, application of neem cake 1000 kg per hectare and covering nursery beds with 25 ì LDPE film for solarization and rabbing with castor husk @ 7 kg per m^2 are recommended.

Control Against Frost

Fennel crop is susceptible to frost at flowering and seed formation stage. Irrigate the crop to protect it from frost. During the period of frost danger an arrangement of smoke should also be made in the field. Wind breaks should be set to protect crop from northern cool winds.

Harvesting and Yield

Time of harvesting depends upon the type of the products marketed. Mostly the crop is harvested before the fruits are fully ripe. For green *saunf* used for chewing purpose, umbels are harvested about 30–40 days after flowering when these are still green and have attained half length size. Since not all the fruits mature at a time, therefore harvesting of umbels has to be done 4-5 times as and when they become ready. Umbels are plucked when fruits have green colour and are fully mature. Harvested umbles should be dried in shade under well-aerated conditions particularly for green fennel. Umbels should never be piled as it may deteriorate the quality. The dried umbel are separated and cleaned by winnowing for removing chaffs, dust and dirt. Under scientific management of the crop, a yield of 20–25 q/ha can be harvested.

Post-Harvest Management

Processing

The fennel seeds are dried in partial shade in the sun and the moisture content of seed should be kept to 9 per cent. High seed moisture content may lead to chances of storage contamination by fungus. The dried, cleaned, graded produce is packed in the standard sized packs/containers and appropriately labelled. The dried seed is filled in gunny bags lined with degradable environment friendly plastic film. Waste generating packaging material should be avoided. Each bag is sealed and stored under clean, dry and ventilated place.

Processing methods of fennel should be based on mechanised, physical and biological processes. Care should be taken to maintain the vital quality of an organic ingredient throughout each step of its processing. Processing method is to be selected in such a way that it will limit the number and quantity of additives and processing aids. The mature dried seeds are distilled to obtain the essential oil. The mature dried seeds of fennel are distilled to obtain the essential oil. Hydro or steam distillation method is generally used for extraction of essential oil. On an average, the dried seeds yield 0.7–2.0% essential oil. The percentage of essential oil varies depending upon variety and type of fennel. The oleoresins prepared from fennel have good demand in the international market. The processed products include cumin essential oil, cumin powder, cumin oleoresin and fixed oil. The essential oil of fennel is used for scenting soap. The by-product *ark saunf* after extraction of essential oil possess good medicinal properties for curing indigestion problems at home level remedies.

Quality Specifications for Export

Whole seed

The quality specification for fennel can be classified into three categories viz. commercial requirements, cleanliness and health specifications.

Commercial Specifications

These requirements vary from country to country and crop to crop. It depends on many factors like customer needs, acceptability and country specifications. The commercial specifications of seed spices crops are colour, appearance, taste, pungency, texture, shape, volatile oil and packaging. The general commercial and general requirements for fennel seed are:

Table 7.7. General commercial requirements

Fennel	Light green and light brownish green, bitter or sweet, small or large seeds, aromatic, free flowing, high volatile oil (min. 1 % - Germany, 3% - Netherlands, 2% - UK)

Cleanliness Specifications

The permit table cleanliness specifications as per ASTA, ESA and ISO are given in Table 7.8.

Table 7.8. Cleanliness specifications
ASTA cleanliness specifications for fennel

Crop	*Whole insect dead (by count)*	*Excrets mammalian (mg/lb)*	*Excreta other (mg/lb)*	*Mould (% wt)*	*Insect defiled/ infected (%/wt)*	*Extraneous foreign matter (%/wt)*
Fennel	2	2	2	2	1	0.5

Other quality standards as per ESA and ISO

Crop	*Ash level % w/w (min.)*	*Acid insoluble ash % w/w (max.)*	*Moisture content % (max)*	*Volatile oil % (Min.)*
Fennel	9	2	12	1.5

Health Specifications

The micro contaminants such as pesticide residues, microbial counts, aflatoxins, heavy metals, etc., are the health specifications. These contaminants make the food unsafe. The importing countries have limited the maximum level of these contaminants and limits are: Maximum Residue Level = 0.05 to 0.2%, Microbial counts = 1×10^2/g to 1×10^6/g (*Salmonella, E. coli*, yeast, moulds), Aflatoxin = 5 ppb to 10 ppb (Max.) and Heavy metals = Arsenic (5 mg/kg), copper (20 mg/kg) lead (10 mg/kg), Zinc (50 mg/kg).

Essential Oil

The specifications for fennel oil are (Singhal *et al.,* 1997): colourless or pale yellow colour, specific gravity (20 °C), 0.965 to 0.977, optical rotation (20 °C), +11 to + 24, refractive index, 1.528 to 1.539,solubility (80% ethanol), 5–8 vols, congealing point, not below 5° and as high as 10° in good oils, anethol 50 to 80%.

Fennel Oleoresin

The fennel oleoresin should be prepared with recommended organic solvents followed by the subsequent removal of the solvent as per specifications of importing countries.

Fennel Powder

The fennel powdered hygienically through permittable devices, should at least 95 % of the ground product pass through a U.S. Standard No. 30 sieve.

FENUGREEK

Fenugreek is a native of south-eastern Europe and Western Asia. India is also said to be a native of fenugreek and found growing wild in Kashmir, Punjab and upper Gangetic plains. Fenugreek, the multi-use and commercially important spice crop, is extensively grown almost in every part of the country during winter season for its seeds, tender shoots, and fresh leaves. There are two cultivated species of genus *Trigonella, viz. foenum-graecum* (common fenugreek) and *corniculata* (*kasuri* type fenugreek). The major fenugreek producing countries are India, Argentina, Egypt, Southern France, Morocco, Spain, Turkey and China but India enjoys the status of largest producer in the world. In India, its production is concentrated mainly in Rajasthan, Madhya Pradesh, Maharashtra, Haryana, Punjab, Gujarat and Uttar Pradesh states. The major importing countries for Indian fenugreek are France, Germany, Canada, Israel, Japan, Jordan, Korea, Kuwait, Malaysia, Netherlands, Saudi Arabia, Singapore, South Africa, Switzerland, UAE, UK and USA.

Cultural Requirements

The cultural requirements including practices as per procurement management for organic fenugreek production are described here.

Climate

Fenugreek performs well in province of low to moderate rainfall but cannot withstand heavy rainfall. Continuous moist and cloudy weather invites insect-pests and a number of diseases. The crop requires cool climate for its better growth and is capable of tolerating 10–15° frost and freezing weather for a short period. Having wider adaptability, the crop can be grown successfully both in tropical and temperate regions up to an altitude of 2000m above mean sea level.

Soil

Fenugreek is well-adapted to a wide range of soils but grow best on well-drained loamy soils. Organic matter rich clay-loam soils may also be used provided adequate drainage facilities are available. However, the crop does not thrive best in sandy or gravely soils. In rain fed areas, black cotton soils are best suited for its successful cultivation.

Cropping System

Fenugreek can be grown as mixed or intercrop provided all the other crops are grown following organic farming methods. Being leguminous crops it fits well as component crop for most of intercropping systems involving fennel, coriander, ajowan, dill and winter vegetable crops. Likewise rotation crops should also be grown following organic farming methods.

Some of the other cropping systems suggested are: Maize/Pearl millet-Fenugreek, Sesame-Fenugreek, Sesame-Fenugreek-Summer Maize.

Cultivated Varieties

Varieties selected should have adaptation to the soil and climatic conditions and preferably have resistance/tolerance to pests and diseases. Efforts should be made to use organic seed of recommended variety for organic farming. When certified organic seed of coriander is not available, initially conventional seeds can be used. The description of some of the important cultivated varieties recommended for different states is given in Table 7.9.

Table 7.9. Recommended varieties of fenugreek for different states

Variety	*Salient characters*
	Rajasthan
RMT-1	• Moderately resistant to root rot and tolerant to powdery mildew disease. • The plants are semi-erect, tall, and moderately branched. • The bold and yellow-coloured seeds contain 21.0% protein. • An average grain yield of 15 q/ha can be obtained over crop duration of 140-150 days.
RMT-143	• Moderately resistant to powdery mildew disease. • Suitable for cultivation in heavier soils of Chittor, Bhilwara, Jhalawar and Jodhpur in Rajasthan. • Seeds are bold with typical yellow colour • Yields 16 q/ha seed in crop duration of 140–150 days.
RMT-305	• Plants are dwarf. Determinate and multipodded • Early in maturing • Resistant to powdery mildew and root knot nematode • Average seed yield 13 q/ha
NRCSS-AM 1	• The crop grown exclusively for leaf purpose gives a yield of 76 q/ha from 3 cuttings. • The seeds with less bitterness are bold in large pods of 8 mm size. • The number of seeds per pod ranges from 17 to 20 with 1000 seed weight of 17-21 g. • Average seed yield is 20.7 q/ha within a crop duration of 137 days.

(Contd.)

Variety	*Salient characters*
NRCSS-AM 2	• The crop grown exclusively for leaf purpose gives a yield of 72 q/ha from 3 cuttings. • The seeds with more bitterness are small in pods of 4 mm size. • The number of seeds per pod ranges from 16 to 18. • An average seed yield of 18.1 q/ha can be obtained in crop duration of 138 days.
	Gujarat
GM 1	• Plants are dwarf • Average yield 18.6 q/ha
	Tamil Nadu
CO-1	• A dual-purpose variety • Suitable for intercropping and relay cropping because of short-duration and dwarf growing habit. • Grain yield of 6.8 q/ha in crop duration of 85–90 days.
	Bihar
RAJENDRA KRANTI	• Moderately resistant to powdery mildew, caterpillar, and aphids • Suitable for growing as a pure crop or as an intercrop. • Plants are medium tall and bushy with medium-sized golden yellow seeds. • Average seed yield of 12-14 q/ha in total crop duration of 120 days.
	Andhra Pradesh
LAM SELECTION-1	• Tolerant to root rot, powdery mildew, caterpillars, and aphids. • Plants are bushy, green with medium sized golden yellow seeds. • Average yield 7.4 quintals of grain per hectare in 90 days crop duration.
	Haryana
HISAR SONALI	• Plants are bushy, semi-erect with bold yellow attractive grains (13-15 g/1000 grains). • Moderately resistant to leaf spot and root rot complex of diseases. • Matures in 140–150 days and produces an average seed yield of 19.0 q/ha.
HISAR SUVARNA (HM-103)	• An early high yielding variety • Resistant to *Cercospora* leaf spot and moderately resistant to powdery mildew. • Quick growing dual purpose variety. • Average seed yield of 16 q/ha can be obtained.
HISAR MUKTA (HM-346)	• High yielding, quick growing seed type variety from medium maturity group. • Seeds are bold with tan green seed coat. • Resistant to downy mildew and moderately resistant to powdery mildew. • Average seed yield of 20 q/ha.
HISAR MADHAVI (HM-350)	• A quick growing, dual purpose variety • Suitable for irrigated as well as rain-fed conditions. • Resistant to powdery mildew and moderately resistant to downy mildew. • Average seed yield is 19 q/ha under irrigated conditions.
	Uttar Pradesh and Uttaranchal
Pant Ragini	• Dual purpose variety • Tall, bushy type • Resistant to downey mildew and root rot • Maturity in 170–175 days Madhya Pradesh and Chhatisgarh

(*Contd.*)

Variety	*Salient characters*
PUSA EARLY BUNCHING	• Quick growing with upright shoots • Suitable for taking 2-3 leaf cuttings. • High yielding variety with bold seeds. • Average seed yield of 12–15 q/ha can be obtained in crop duration of 156 days.
PUSA KASURI(Other kasuri type local varieties: Kasuri methi, Nagauri Methi, Marwari methi and Champa methi	• A late flowering variety with rosette type leaves gives 5-6 cuttings. • Heavy yielder of green leaves with special fragrance. • The leaves are broad, succulent and rich in vitamin C. • Average seed yield of 6-7 q/ha and the crop grown exclusively for greens produces an average yield of about 80 q/ha.

Cultural Practices

Sowing Time

Fenugreek, being cool season crop, is sown in the month of October–November in northern plains, whereas, in hilly tracts, it is sown from March—May, depending on altitude. In southern states of India, particularly Karnataka, Andhra Pradesh and Tamil Nadu, fenugreek is sown twice, once in *rabi* (September—December) and again in *kharif* season (June–July). *Kasuri* type varieties require preferably extra cool weather for longer duration, is thus raised more successfully in northern states during winters than the southern states of India.

Seed rate

In order to sow one-hectare area, about 20–25 kg seed of *common fenugreek* and 10–12 kg of *kasuri fenugreek* type is required. The initial soil moisture should be adequate to ensure satisfactory germination. The seeds germinate in 5–7 days in common fenugreek whereas kasuri fenugreek germinates late by another 3–5 days. Germination is usually delayed, if the temperature is too low, soil moisture is inadequate, sowing depth is more than 2 cm.

Seed Treatment

Seeds should also be inoculated with *Rhizobium* culture prior to sowing, especially when the crop is sown in new field as the symbiotic nitrogen fixation takes place more effectively if an interrelationship is established between a particular legume and a specific *Rhizobium* strain.

Sowing Method

Fenugreek is sown either in lines or by broadcasting seeds in well-prepared flat seedbeds and raking the bed surface prudently, however, sowing in lines is comparatively better since it facilitates the intercultural operations. For fresh greens, seeds are usually sown sparsely in 15–20 cm apart shallow furrows and for seed crop 25–30 cm apart and later the plants are thinned to a spacing of 10–15 cm within lines. Although the depth of sowing seeds depends on soil type and soil moisture at the time of sowing, but being small size, the seeds of *common fenugreek* are usually sown at a depth of 2-3 cm and *kasuri fenugreek* at 1.0–1.5cm.

Manurial Requirement

The fenugreek crop has been reported to show very good responsive behaviour towards absorption of both macro as well as micro-nutrients. The crop removes N, P and K in the ratio of 2:1:2 from the soil, in the uptake order of 10, 3.5, 8.2 kg, respectively. Therefore, to maintain a steady state of productivity, organic manure should be added in ample quantity. Application of FYM (10 t/ha) had a beneficial effect on the enhancement of vegetative growth and resulted in higher dry matter production of fenugreek. Application of one tonne per hectare of neem cake has also proved beneficial. The microbial inoculation with *Azospirillum, Azotobactor* and *Rhizobium* have been reported to show suitable means for organic cultivation of fenugreek. The *Azospirillum* in combination with organic manure 10 t/ha yield high organic seed yield of 13.9 q/ha, which was 25% less than the yield obtained from combinations applied with chemical nitrogen with reduced incidence of root rot (8.4%).

Irrigation

Fenugreek, being primarily an irrigated crop, requires light irrigation at frequent interval for its quick growth but can also be cultivated under rain fed conditions in certain parts of the country. The irrigations are given at 7–10 days interval, depending on soil type, season, rainfall, and other temporary weather conditions. In general, frequent and light irrigations are essential for quick foliage growth and as a thumb rule, each cutting should be followed by a light irrigation. The early growth period and seed setting are the critical stages for irrigation requirement. Too much irrigation is also as harmful as the scarcity of moisture, since excessive moisture in any form and at any stage increases the incidence of root rot and powdery mildew. Application of seven irrigations at 1.0 IW/CPE ratio was found suitable as a general recommendation for getting higher production. Sprinkler irrigation with IW/CPE ratio of 0.75 was found optimum for Gujarat conditions.

Inter-cultural Operations

Fenugreek, being leguminous crop, needs proper soil aeration for normal development of root system, thus, hoeing and weeding during early stages of plant growth are very essential to make the soil loose around the roots and to control the weeds, since weeds due to slow growth of fenugreek seedlings may pose problem in initial stages, however, in later stages, when the crop canopy is fully developed, weeding is not at all required as the crop itself suppresses the weeds. Generally 2-3 hand weeding are required to keep the crop weed free and reducing the crop-weed competition for growth resources like nutrients, moisture, light and others. The first hoeing and weeding is done at the time when the plants are 5 cm tall.

Harvesting and Yield

The common fenugreek becomes ready for cutting fresh green leaves and young shoots in about 20 days after sowing while *kasuri* type fenugreek is ready in 25–30 days after sowing and subsequent cuttings may be taken at an interval of 15–20 days. The crop after harvest is bunched and marketed. Depending upon variety and season of growing the seed crop takes about 80-165 days from sowing to harvesting. The entire plant is either pulled out or cut from the base with sickle when 70% of the pods turn yellow, and made into small bundles for drying them in sun. Seeds are separated by beating with stick and winnowing. The seed is cleaned and sun dried before packing in jute bags containing polyethylene lining.

The fenugreek yield depends upon the variety used, location and season of growing. Usually, the cultivars yielding higher grain yield tend to produce low fresh green leaf yield. Under irrigated conditions, the common type fenugreek varieties normally give a fresh green leaf and seed yield of 70–80 and 15–20 q/ha, respectively, and *kasuri* type 80–100q green leaves per hectare.

Storage

Green leaves are very perishable in nature; therefore, they are marketed soon after harvesting. However, well-dried leaves can be stored for about 10–12 months. The leaves can be stored only for about 24 hours after harvesting under ambient conditions, however, in cold stores at 0 °C temperature and 90–95% relative humidity the storage period can be extended up to10 days. Seeds are stored in gunny bags lined with polythene film. Vacuum gravity separator is used for cleaning fenugreek seeds. The properly cleaned fenugreek seeds are stored with an initial moisture level of 7-8% and at an equilibrium relative humidity of 40%. Fenugreek seeds well packed is stored in ventilated dry and cool place under ordinary conditions till sowing of next season crop.

Post-Harvest Management

Keeping in view the requirement of targeted market, the post-harvest operations are advised to follow suitably for green leaves as vegetable, dehydrated leaves and seeds. In case of fresh greens, the edible portion is the tender leaves and stem, which are cut to length of about 7–10cm. After harvesting, the yellow, diseased and damaged leaves are trimmed off and thereby healthy and disease free leaves are tied into small bunches for the convenience in handling and marketing. Since the dried leaves can be stored for one year for further use in off-season, the fenugreek leaves are sun-dried or dehydrated in a suitable dehydrator. However, during dehydration, chlorophyll is oxidized and ascorbic acid is lost, so as to retain green colour of leaves, blanching in boiling water (80 °C) for 3–6 minutes can be practiced. The palatability of fenugreek leaves improve with steaming and with addition of seasoning rather than with boiling or frying. The average vitamin C value in fenugreek leaves is 43.10 mg/100g and after boiling in water or steaming and then frying, the leaves lose 10.8% and 7.4% of this vitamin, respectively.

Crop Protection

Fenugreek is prone to attack by some of the serious diseases and insect-pests. The plant protection measures under organic farming system should involve emphasis on crop management practices time of sowing, balanced nutrition, crop rotation, green manuring, etc., for reducing the incidence of diseases and pests, selection of resistant varieties and adoption of bio control measures.

Pest Monitoring

Pest monitoring is an important part for organic cultivation of Fenugreek and should be done once in seven days to work out economic threshold level (ETL). It will help to keep close watch on the appearance of insect pests, diseases and bio-control fauna.

For sucking pests: Population should be counted on three leaves, i.e., top, middle and lower per plant.

For cut worms and defoliators: Per cent damage assessment can be made by counting total number of plants and affected plants

Management strategy: Set up yellow pan/sticky traps for monitoring of aphids/whitefly @ 10 yellow pans/sticky traps per hectare. Locally available empty yellow palmoline tins coated with grease/vaseline/castor oil on outer surface may also be used. Collect the larvae of cutworm and leaf defoliators.

Insect-pests

Aphids (*Acryrthosiphon pisum,* Henis. *Myzus persicae,* Sulzer and *Aphis craccivora*, Koch.)

The most common insect that attacks the crop is aphid, both nymphs and adults suck sap from the tender leaves thus reducing marketability. For control, clip off the heavily infested plant parts and spray the crop with pressurized water. Application of fish oil, resin soap or neem seed kernel-extract (3%), *neem* oil (2%) or tobacco decoction (0.05%) is effective measures.

Leaf eating caterpillar (*Spilarcitia oblique* and *Spodoptera litura* Fabricius)

The large number of caterpillar appear and destroy the leaves. The eggs are laid in clusters and young larvae gregariously feed on leaves. For the control, destroying the egg mass and gregarious young larvae, hand picking and killing of large larvae, spray of neem seed kernel-extract (4%) in the early stage of larvae and use of nuclear polyhedrosis viruses (NPV) are effective biological controlmeasures.

Pod borer (*Helicoverpa armigera* Hubner): The pest is found to feed on leaves, flowers and pods. Eggs are laid on young leaves, which are later damaged by young larvae. Control measures are same as suggested for leaf eating caterpillar.

Mites (*Petrobia latens*): Infestation of mites can be seen occasionally. The insect mostly feeds on young leaves. Plants get stunted at severe infestation. For the control, spray of neem oil (1%) is effective.

Disease

Root rot (Rhizoctonia solani Kuhn.): Symptoms include varying degrees of rotting of the root leading to foliage yellowing in 30–45 day-old-plants. It is a serious problem in growing areas and drastically reduces the yield. For the control, use of soil amendments like neem cake @ 10q/ha and farm yard manure @10–25t/ha is beneficial. Adoption of proper crop rotation and deep summer ploughing is useful. The antagonistic fungi *Trichoderma viride* (5 kg/ha) and soil application of *neem* cake (150 kg/ha) effectively control the root rot disease. Use of bioinoculants *Azospirillum* or *Azotobactor* plays significant role in reducing incidence of root rot to 8.2%.

Powdery mildew (Erysiphe polygoni D.C. and Laveillula taurica Lev.): The symptoms include the appearance of whitish fungal growth on leaves. The disease is managed by spraying of 0.2% Sulphur or dusting of sulphur powder, 20-25 kg/ha.Growing of resistant/ tolerant varieties is also suggested.

Downy mildew (Peronospora trigonella Gaum.): Symptoms include the presence of yellow patches on the upper surface and a cottony-grayish-white mycelium at the lower surface. For the control, hot water treatment to seed before sowing and spraying of Bordeaux mixture (1%) as restricted use have proved beneficial.

Damping off (Pythium aphanidermatum): The infected seeds and seedlings appear water-soaked, discoloured and soft, emitting a bad odour. Control measures suggested include use of soil amendments like *neem* cake @ 10q/ha and farmyard manure @10-25t/ ha, adoption of proper crop rotation and deep summer ploughing and application of antagonistic fungi *Trichoderma viride* and neem cake.

Leaf spot (Cercospora traversiana SOCC): Presence of round spots beneath the leaves, which are white at the center and brown at the periphery, indicates the disease.The control can be achieved by following at least 2-year crop rotation in infested areas and by, using healthy seed.

Rust (Uromyces anthyllidis): The disease is observed in the form of minute brown pustules on both the surfaces of leaves. For the control, avoiding excessive irrigation and use resistant variety.

Fenugreek Mosaic Wilt

This is a virus disease. Symptoms include vein clearing, severe mottling, curling at the edges, leaf size reduction, stunting of the plant and ultimately wilting. For the control, remove all the infected plants and weeds and use disease free seeds.

Quality Specifications for Export

Whole seed: The quality specification for fenugreek can be classified into three categories viz. commercial requirements, cleanliness and health specifications.

Commercial Specifications

These requirements vary from country to country and crop to crop. It depends on many factors like customer needs, acceptability and country specifications. The commercial specifications of seed spices crops are colour, appearance, taste, pungency, texture, shape, volatile oil and packaging. The general commercial and general requirements for fenugreek seed are:

Table 7.10. General commercial requirements

Fenugreek	Yellow seeded or green seeded, small or large seed size, free flowing, volatile oil (min. 0.02%)

Cleanliness Specifications

The permit table cleanliness specifications as per ASTA, ESA and ISO are given in Table 7.11.

Table 7.11. Cleanliness specifications

ASTA cleanliness specifications for fenugreek

Crop	*Whole insect dead (by count)*	*Excrets mammalian (mg/lb)*	*Excreta other (mg/lb)*	*Mould (% wt)*	*Insect defiled/ infected (%/wt)*	*Extraneous foreign matter (%/wt)*
Fenugreek	4	3	10	1	1	0.5

Other quality standards of fenugreek as per ESA and ISO

Crop	*Ash level % w/w (min.)*	*Acid insoluble ash % w/w (max.)*	*Moisture content % (max)*	*Volatile oil % (Min.)*
Fenugreek	7	1.5	12	0.3

Health Specifications

The micro-contaminants such as pesticide residues, microbial counts, aflatoxins, heavy metals, etc., are the health specifications. These contaminants make the food unsafe. The importing countries have limited the maximum level of these contaminants and limits are : Maximum Residue Level = 0.05 to 0.2%,Microbial counts = 1×10^2/g to 1×10^6 /g (*Salmonella, E. coli*, yeast, moulds) ,Aflatoxin = 5 ppb to 10 ppb (Max.) and heavy metals = Arsenic (5 mg/kg), copper (20 mg/kg) lead (10 mg/kg), Zinc (50 mg/kg).

Essential Oil

Fenugreek seeds contain little quantity of essential oil of 0.02% and the specifications for fenugreek oil are not available. The seeds should contain 0.78-1.9% diosgenin.

Fenugreek Oleoresin

The fenugreek oleoresin should be prepared with recommended organic solvents followed by the subsequent removal of the solvent as per specifications of importing countries.

Fenugreek Powder

The fenugreek powdered hygienically through permittable devices, should at least 95 % of the ground product pass through a U.S. Standard No. 25 sieve.

Marketing System for Seed Spices Export

The export of seed spices mainly involves indirect exports through links with agents or export management firms which will eventually lead to the handling of own exports. The other entry options are licensing, contract production (supply), management contracting and joint ventures. The details can be sought from Spices Board at Cochin.

Two Modes for export of organic spices:

- Export via independent representative (agent)
- Establishment of overseas sales subsidiary own or with foreign party collaboratly.

Marketing channels involved are

- Direct channel: Foreign distributor, foreign retailer, state controlled trading companies and end users.
- Indirect channel: Export brokers, export agents, export management organizations and co-operative exporters.

References

Malhotra, S.K. and Vashishtha, B.B. 2004. Organic production of Seed Spices in India-Status & Strategy. Seed Spices Newsletter. **4**(1):1-4.

Anon. 2005. Organic Spices Production. Spices Board, Cochin.

Malhotra, S.K. and Vashishtha, B.B. 2008. Coriander In: Eds. Parthasarthi, V.A., Kandiannan, K. and Srinivasan, V. Organic Spices. New India Publishing Agency, New Delhi, p 523-46.

Shiva, K.N. and Malhotra, S.K. 2008. Fenugreek In: Eds. Parthasarthi, V.A., Kandiannan, K. and Srinivasan, V. Organic Spices. New India Publishing Agency, New Delhi, pp 547-76.

Malhotra, S.K. 2008. Fennel In: Eds. Parthasarthi, V.A., Kandiannan, K. and Srinivasan, V. Organic Spices. New India Publishing Agency, New Delhi, p 577-94.

Vashishtha, B.B. and Malhotra, S.K. 2008. Cumin In: Eds. Parthasarthi, V.A., Kandiannan, K. and Srinivasan, V. Organic Spices. New India Publishing Agency, New Delhi, pp 595-610.

8

Organic Farming Technologies for Arid Horticulture

T. A. More and B. D. Sharma

INTRODUCTION

In India, about 62% of cropped area is rain fed, where there is little or no use of fertilizers and other agro-chemical due to poor resources with small holding farmers. Thus promotion of organic farming in India is advocated initially in the arid and semi-arid areas of the country in some selected crops like ber, pomegranate, aonla, custard apple, tamarind, fig and other minor fruit of economically importance. Rather than promoting organic farming *en masse*, it would be appropriate to carefully delineate areas or crop where fertilizer use is nil or nominal or demarcate export oriented crops that can give reasonable yield of high quality produce without the use of chemicals. It is worthy to mention that arid fruit and vegetable crops hold great promise in this regard.

From the limited research carried out, there is a definite promise for organic farming of fruit and vegetable crops in arid region. In horticulture, it is easier to manage fruit crops organically than vegetable and flower crops due to perennial growth habit. The arid fruit crops which are considered high valued health food and being easily grown organically for years are the best suited for organic cultivation, e.g., custard apple, fig, jamun, tamarind, pomegranate, aonla, sapota, guava, citrus and ber. Therefore, in future more efforts are needed to standardize organic techniques in these crops.

Vegetables are rich and cheap source of vitamins and mineral which act as a protective food in daily diet. They are also consumed raw as a salad and are major source of fibre in diet. However, vegetable crops are vulnerable to disease and pests, therefore they need chemical sprays to protect from them. Similarly for getting optimum yield, they need chemical fertilizers. Thus in commercial vegetable production with the increase in chemical inputs, the risk of degradation of environment and residue problems has increased. Therefore, there is need to grow vegetables organically today.

Components of Organic Farming

There are large number of organic sources of nutrition and among them green manuring, composting, biofertilizer, organic cakes, vermicompost and biodynamic are important.

Among the different components of organic farming, the use of bio-fertilizers is currently gaining interest as a cheap, safe alternate to conventional chemical fertilizers. Bio-fertilizers are bacterial cultures such as *Rhizobium* from leguminous and *Azotobacter* and *Azosprillum* from non-leguminous crops and have the capacity of fixing nitrogen from the atmosphere. Similarly the phosphate solubilizing bacteria (PSB) have proven utility in making unavailable soil phosphorus in to available form. *Azosprillum* has been recommended in a number of crops for increasing the production and productivity of crops. Nitrogen fixing systems offer an economically attractive and ecologically sound means of reducing external inputs and improving internal resources. *Azotobactor* is a common soil bacterium, which is present widely in Indian soils. Soil organic matter is the important factor that decides the growth of these bacteria. Poor organic matter content in the Indian soils is a limiting factor that not only affects the proliferation of the bacteria in soils but also limits its N- fixing capacity. *Azospirillum* is a group of bacteria found in loose association with the root system of many crop plants. These bacteria grow better under reduced oxygen levels. They fix nitrogen from 10–40 kg/ha. *Azospirillum* inoculation helps the plant in better vegetative growth saving nitrogenous fertilizers by 25–30%. In the sand soil ecosystems of India, pioneer plants are subject to deficiency in major nutrients (N, P and K) and organic matter, wide fluctuation in soil moisture and temperature. It is widely accepted that arbuscular micorrhizae are important for the establishment, growth and survival of seedlings, particularly in marginal habitats, where the symbiosis improves stress tolerance as well as conserves soil structures. Mycorrhizal associations are believed to help plants grow. The potential for manipulating mycorrhizal associations to increase productivity in plantation, forestry or plant establishment during ecosystem recovery after severe disturbance are major focus of research initiatives. There is much interest in their potential utilization in agriculture and horticulture.

Prospects of Organic Farming in Horticulture

The green revolution in India though markedly enhanced the food production and brought some major food security in the recent years in horticulture sector it has ushered in the use of: (a) heavy doses of fertilizers, neglecting organic manures and soil amendments, (b) herbicides, pesticides and fungicides to control weeds, pests and diseases, and (c) growth regulators to monitor growth and yields and improved quality and helf life to an unprecedented scale (Shivshankar, 2000). These modern intensively costly practices seem to etched into the system eternally while undermining our resources and drastically changing our environment effecting our health and happiness. In this context to restore our natural resources, to safeguard our environment and to obtain pesticide residue free fruits, vegetables, spices and other commodities, organic farming is now regarded as the best solution. Production of fruits (59 MT), vegetables (113 MT), spices (3.0 MT) and other horticultural crops on 9% of the total cultivated area under all crops consumes around 24% of the pesticides used in the country. Some portion of these crops are exported. Even if 40–45% of the area under these crops is raised organically, it can redress the ecological disaster to a great extent. Besides, these crops fetch a high premium of 20–30% in vegetables and some times 100–200% in fruits.

Let us focus on the basic principles of adoption of organic farming in arid horticulture. Bio-mass and crop residues can be utilized well to make compost along with FYM and bio-gas slurry. Legumes be incorporated, green leaf manure and slashed weeds be used for mulching to reduce the soil erosion. Pests and diseases have to be controlled by a combination of intercrops, rotations, mechanical cultivation, use of botanical and biological methods, etc. other general practices depending upon the situation and crops. Solarized soil inoculated with VAM and *Trichoderma* is useful in nurseries. The bio-agent *Trichoderma* can be used for seed treatment, potting mixtures and main field.

Impact of Organic Farming in Arid Horticultural Crops

Cropping System: Crop rotation and intercropping are the key components to obtain success in organic farming. Cultivation of host plant as inte-rcrop or in rotation may lead to increase in the harmful pathogens as against this inter-cropping with leguminous crop may increase availability of nitrogen and ultimately increase the yield and improve the quality of fruits. Saia oats (*Avena strigosa*) and Sudan grass were investigated as rotation crops for strawberry and interplant as companion crop (La Mondia, 2002). Production of rotation crop suppressed the densities of the pathogens, weeds and white grub. It has been found that inter-cropping of marigold reduced the parasite as compared to sole crop of guava. Integration of two methods, i.e., use of bio-agents preferably with *Aspergillus niger* or *T. harzanium* and inter-cropping with crop like marigold or turmeric effectively control the guava wilt.

Fertilization: In addition to the common cultural practices, organic production of fruit entails proper nutrient management. Thus, the most important aspect of organic fruit production is supply of nutrients through organic sources. The concentrate organics like oil cakes, bone/fish meal will be useful in supplying major nutrients. Bio-fertilizers like *Azotobactor, Azosprillum* and PSB are of immense use in supplying unavailable nutrients and have immense importance in fruit production. Green manuring not only helps to improve soil health but also is useful in reduction of weed intensity.

Tamarind seed pelleting with *Azosprillum* in combination with seedling inoculation of VAM at 5 g per poly-pot aided the production of elite seedlings in the nursery (Parameswari *et al.,* 2001).

Studies conducted at MPKV, Rahuri showed increase in yield to the extent of 8.87 t/ha in acid lime and 7.7 t/ha in sweet orange with application of bio-fertilizers (VAM @ 500g + PSB 100 g + *Azosprillum* 100 g + *T. harzanium* 100 g per plant). Furthermore, application of organic manures viz., FYM, vermi-compost and neem cake resulted in the highest juice content (49%) with the highest TSS (15.5 °Brix) in pomegranate. Similar results were obtained in aonla. In typical arid conditions of western Rajasthan, application of vermi-compost, cattle manure and sheep manure gave the higher fruit yield and better quality parameters over inorganic production in pomegranate fruit crops. The fruit yield increased by 55, 60 and 78% on the application of cattle manure, sheep manure and vermi-compost, respectively over application of chemical fertilizers. The juice content (55%) and TSS (15.5° Brix) were estimated on the application vermi-compost were 56 % higher over inorganic fertilizer application. The application of bulky manure increased water holding capacity of sandy soils and water use efficiency by pomegranate fruit crops at CIAH, Bikaner (Sharma *et al.,* 2002). Hiwale (2004) reported the maximum plant height and fruit yield was recorded in plants treated with the combination of FYM (50%) + castor cake (25%) + urea (25%) in pomegranate cv. Ganesh in semi-arid condition of Gujarat (Table 8.1). Oil cakes were found helpful in increasing the soil health and fruit yield of pomegranate (Gaur *et al.*, 1971; Shinde, 1977 and Wavhal, 1981). Semi-hard wood cuttings of pomegranate varieties viz., Ganesh, Jyoti and RCR-1 treated with *Trichoderma harzanium* recorded maximum rooting and number of leaves (Patil *et al.,* 2004). In typical semi-arid condition of Karnataka, inoculation of VAM fungi resulted increase in fresh weight of biomass and leaf nutrient content of papaya. Treatment of alkaline soil with distillery effluent encouraged growth of aonla and also improved soil conditions (Singh and Gaur, 2004). Application of P solubilizers significantly increased the fruit weight and vitamin C content of guava over control. P solubilisers were found to have more beneficial influence on fruit yield and physico-chemical parameters of guava than that of N fixers under acid soils of Chotanagpur region (Dey *et al.*, 2004). Fruit yield and fruit quality of guava improved on the application of *Azosprillum* and vermicompost in comparison to control treatment (Ram and Nagar, 2004) (Table 8.2). Experiment on sweet orange revealed that with addition of vermicompost (20 kg/tree) + neem cake (8 kg/tree) with organic plant protection an additional yield of 8.22 t/ha was obtained.

Table 8.1. Effect of organic and inorganic N application on yield and quality of pomegranate cv. Ganesh

Treatments	*Fruit yield (kg/plant)*
FYM	8.23
Castor cake	8.20
FYM + castor cake	8.84
FYM + castor cake + urea	10.75
Urea	5.80
CD (P=0.05 %)	2.19

Source: Hiwale (2004).

Table 8.2. Effect of organic /biodynamic treatments on yield and quality parameters of guava

Treatment	*Yield (kg/tree)*	*TSS (%)*	*Acidity (%)*
Vermicompost	11.45	11.20	0.13
Vermiwash + FYM	5.90	9.67	0.23
FYM + *Azotobactor*	7.88	10.20	0.28
FYM + *Azosprillum*	8.61	9.80	0.29
Chemical fertilizers	8.09	8.20	0.24

Source: Ram and Nagar (2004).

Plant Growth Regulator: Plant growth regulators have immense importance in quality improvement of fruit crops. However, use of these chemicals is not permitted in organic cultivation. Therefore, specific technique should be evolved for quality improvement, e.g., in grape, techniques like berry thinning, stem girdling, cane girdling, paper wrapping, spreading shade net, etc. Similar techniques need to be standardized in other fruit crops.

Mulch: Data revealed that in aonla-guava cropping system under salt affected soil conditions, organic mulches (sugarcane trash) improved the plant growth, besides stabilizing soil temperature and reduced the frequency of irrigations and weed intensity (Prasad *et al.*, 2004). The decomposition of organic mulch materials in soil helps to increase the organic carbon content and availability of N (Borthaker and Bhattacharya, 1992; Chandra, 2000). Organic mulch materials (maize straw, paddy straw, grasses, subabool lopping, rice husk) encouraged plant growth, reduced moisture evaporation and also regulated soil temperature in aonla orchard under semi-arid conditions of Gujarat (Singh and Singh, 2004). Plant growth in aonla was maximized with sugarcane trash and paddy straw mulches and proved cheaper and maintains soil energy as compared to black polythene mulch (Rao, 1995; Shukla, 1996 and Mishra, 2004). Maximum plant height, spread and fruit yield were recorded in full moon terracing and subabool lopping as compared to control at semi-arid condition of Gujarat (Bagle *et al.,* 2004) (Table 8.3).

Table 8.3. Effect of *in-situ* water harvesting and mulches on yield and quality of pomegranate fruits

Treatments	*Yield (kg/tree)*	*Juice content (%)*
Maize straw	10.22	45.00
Maize straw + full moon terracing	11.00	47.00
Subabool lopping + full moon terracing	12.46	47.32
Control	8.32	40.25
CD (P=0.05%)	1.64	6.25

Biological Control Measures for Diseases and Pests

Biological control of pests and diseases has now been widely adopted in several fruit crops. Research outcome on orchard management showed high promise of some, e.g., *Verticillium lecanii* for control of mealy bug, thrips, white fly and scales in pomegranate, grape, guava and custard apple (4-6 g/l). Furthermore NSKE @ 5 % spray also provides an alternate to chemical pesticides in controlling pests of arid fruit crops. The nematode and soil borne pathogens can be effectively controlled by means of *Trichoderma viridae* + *Paecilomyces*, neem cake (Table 8.4).

Table 8.4. Evaluation of bio-agents against guava wilt

Treatments	*Average wilt (%)*	*% control*
Penicillium citrinum	47.33	52.67
Aspergillus niger	21.66	78.34
Trichoderma harzanium	24.66	75.34
Control	100.00	0.00

Source: Mishra *et al.* (2004)

Control of diseases is the most limiting factor on organic fruit production and hence, selecting resistant varieties or rootstocks is of prime importance. Growers should practice sanitation by cleaning up debris, avoiding the incorporation of plant material of same crop carrying diseased plants and removing disease vectors. In organic farming a good defense against plant disease is to maintain the crop in good health and vigour but not the excessive nutrients and moisture.

In vegetable, the application of FYM @ 20 t/ha was significantly superior in increasing the head size, and head yield as compared to recommended dose of NPK in cabbage (Table 8.5) (Chaurasia *et al.*, 2004). The highest net income with the cost benefit ratio of 6.81 was found with FYM application in moringa based cropping system in semi-arid condition of Gujarat (Raja *et al.*, 2004). Vermi-compost resulted in significant increase in plant height, number of leaves, leaf area index and gave 24% higher tuber yield in comparison to control (Sood and Lal, 2004). The experiment on organic cultivation of cabbage revealed that maximum yield was produced with inorganic fertilizers but it was at par with that produced by organic package consisting of FYM @ 20 t/ha + neem cake @ 250 kg/ha + soil treatment with *Trichoderma* + *Azosprillum* + PSG + NSKE + trap crop of mustard. The experiment on organic cultivation of bitter gourd revealed that maximum yield was produced with organic manure treatment consisting of neem cake + poultry manure + vermophos + sulphate of potash + *Azotobacter* + *Azosprillum* + PSB + FYM + NSKE that control (Table 8.6).

Table 8.5. Organic cultivation of cabbage

Treatments	*Yield (q/ha)*	*B : C ratio*
Organic package	300.00	2.60
Inorganic with trap crop	329.00	4.00
Recommended as STCR with trap crop	346.00	4.13

Table 8.6. Organic cultivation of bitter gourd

Treatments	*Yield (q/ha)*	*B : C ratio*
FYM + poultry manure + vermiphos + SOP + Package	250.00	1.83
Neem cake + poultry manure + vermiphos + SOP + Package	263.00	1.98
Control (RDF + FYM)	232.00	1.66

Studies conducted on effect of mulches on onion during kharif revealed that maximum yield was obtained in the sugarcane trash mulch with 40.70 % increase over the control and maximum B: C ratio (Table 8.7).

Table 8.7. Effect of mulches on onion during kharif season

Treatments	*Yield (q/ha)*	*B: C ratio*
Black polythene mulch	28.50	1.57
White polythene mulch	23.96	1.66
Wheat straw mulch	26.65	5.06
Sugarcane trash mulch	28.69	6.63
Control	20.39	4.21

Weaknesses of Organic Farming

Before initiating organic cultivation one shall consider following weaknesses of organic farming in fruit and vegetable crops in arid region, availability of organic materials, initial yield gap, heavy load of pests and diseases, no concrete organic means to control diseases once appeared, only prevention is way of controlling diseases.

Strengthening, required in Organic Farming on development of resistant varieties suitable rootstocks, standardization of organic mulch, effective combination of organic manures, standardization of organic practices (e.g., girdling, berry thinning, wrapping with papers, disease forecasting unit).

CONCLUSION

Organic agriculture is an alternative and appropriate management system intended to guarantee sustainable production system of safe food with minimum environmental impact. Fruits and vegetables can be considered to be highly prospective for organic farming since they are consumed fresh. Organic production of arid horticultural crops will be successful if sufficient biomass can be generated in and around the farms. Development of biogas plants and agro-forestry for providing alternate source for fuel, addition of crop residues, green manuring, recycling of on farm wastes and enhancing nutrient value of manures through proper composting, particularly vermicompost, adoption of crop rotation involving legumes, etc. are some of the strategies that will definitely help to promote organic farming of horticultural crops in arid region. In order to ascertain and guarantee the consumer or the importer that the produce is genuinely raised organically, the producer has to follow IFOAM basic organic standards. Organic standards, certification, labeling and inspection are more important for organically grown fruit and vegetable crops. Let our people, and generations to come breathe in healthy air, savour unpolluted fruit and vegetables and enjoy nature. In this context let us strive to pool all our indigenous knowledge on safe horticulture practices and evolve new organic farming technologies for which concentrated efforts are needed from the scientists, planners, personnel involved in the development of horticulture, extension agencies and the farmers themselves.

References

Bagle, B.G., Singh, A.K. and Trivedi, M.K. 2004. Effect of *in-situ* water harvesting and moisture on growth, yield and quality of pomegranate cv. Ganesh. In: Proceeding of *National Symposium on Organic Farming in Horticulture* (Eds. Pathak, R.K., Kishun, R, Khan, R.M. and Ram, R.A.) CISH, Lucknow. pp. 236-39.

Borthakur, P.K. and Bhattacharya, R.K. 1992. Organic mulches in guava orchard. *Indian Journal of Horticulture.* **37**(2): 43-44.

Chaurusia, S.N.S., De, Nirmal and Singh, K.P. 2004. Effect of organic sources of nutrients on the growth and nutrient content on the growth, yield and yield attributing characters of cabbage (*Brassica oleracia* var. Capitata). In: *Proceedings of National Symposium on Organic Farming in Horticulture* (Eds. Pathak, R.K., Kishun, R, Khan, R.M. and Ram, R.A.) CISH, Lucknow. pp. 158-61.

Chandra, S. 2000. Efficiency of drip irrigation and mulching on aonla + ber cropping system in sidic soil. Ph.D thesis. N. D. U.A.T., Faizabad, Uttar Pradesh.

Dey, P. Rai, Kumar, S., Nath, V., Das, B. and Reddy, N.N. 2004. Effect of bio-fertilizer on physico-chemical characteristics of guava fruit. In: *Proceedings of National Symposium on Organic Farming in Horticulture* (Eds. Pathak, R.K., Kishun, R, Khan, R.M. and Ram, R.A.) CISH, Lucknow. pp. 221-23.

Gaur, A.C., Neelkanthan, S. and Dargan, K.S. 1971. *Organic manure*, ICAR, New Delhi, pp. 3-10.

Hiwale, S.S. 2004. Effect of organic and inorganic fertilizers on growth, yield and quality of pomegranate cv. Ganesh. In: *Proceedings of National Symposium on Organic Farming in Horticulture* (Eds. Pathak, R.K., Kishun, R., Khan, R.M. and Ram, R.A.) CISH, Lucknow. pp. 259-61.

La Mondia, J.A., Elmer W.H., Mervosh, T.L. and Cowles, R.S. 2002. Integrated management of strawberry by rotation and intercropping. *Crop Protection*, 21(9): 837-46.

Mishra, A.K., Prakash, Om, Pandey, B.K., Prasad, B., Shukla, S.K. and Prasad, D. 2004. Biological control og guava wilt. In: *Proceedings of National Symposium on Organic Farming in Horticulture* (Eds. Pathak, R.K., Kishun, R, Khan, R.M. and Ram, R.A.) CISH, Lucknow, pp. 302-05.

Mishra, L.N., Pathak, R.A., Pandey, A.K. and Nagar, A.K. 2004. Effect of mulching on growth of juvenile plant in aonla and guava cropping system in sodic soil. In: *Proceedings of National Symposium on Organic Farming in Horticulture* (Eds. Pathak, R.K., Kishun, R, Khan, R.M. and Ram, R.A.) CISH, Lucknow, pp. 224-25.

Parmeswari, K., Srimathi, P. and Malarkodi, K. 2001. Influence of bio-fertilizer pelletization on elite seedling production of tamarind (*Tamarindus indica*) in nursery. *Seed Research*, **29**(1):58-62.

Patil, P.B., Patil, C.B., Swamy, G.S.K. and Kumar, S. 2004. Influence of micro-organism on rooting of pomegranate cuttings. In: *Proceedings of National Symposium on Organic Farming in Horticulture* (Eds. Pathak, R. K., Kishun, R, Khan, R.M. and Ram, R.A.) CISH, Lucknow. pp. 209-11.

Prasad, J, Pathak, R.A. and Pathak, R.K. 2004. Effect of mulching on improvement of soil and plant growth in aonla based cropping system under salt affected soil conditions. In: *Proceedings of National Symposium on Organic Farming in Horticulture* (Eds. Pathak, R.K., Kishun, R, Khan, R.M. and Ram, R.A.) CISH, Lucknow, pp. 151-54.

Raha, S., Apparao, V.V. and Bagle, B.G. 2004. Influence of in-situ water harvesting technique on moringa based cropping system in semi-arid region. In: *Proceedings of National Symposium on Organic Farming in Horticulture* (Eds. Pathak, R. K., Kishun, R, Khan, R. M. and Ram, R. A.) CISH, Lucknow, pp. 162-65.

Ram, R.A. and Nagar, A.K. 2004. Effect of different organic treatments on yield and quality of guava cv. Allahabad safeda. In: *Proceedings of National Symposium on Organic Farming in Horticulture* (Eds. Pathak, R.K., Kishun, R, Khan, R.M. and Ram, R.A.) CISH, Lucknow, pp. 306-10.

Rao, V.K. 1995. Effect of various mulches on soil properties and plant growth of aonla. M. Sc. (Ag.) thesis, N.D.U.A.T., Faizabad.

Sharma, B.D., Dhandar, D.G., Vashishtha, B.B and Bhargava, R. 2003. Impact of integrated nutrient management on soil fertility and yield of pomegranate under arid agro-ecosystem. In: *6th Agricultural Science Congress*. Organized by NAAS at Bhopal, February 13-15, 2003. Abstr. P-88, pp. 62.

Shinde, D. G. 1977. Effect of N, P and on growth, yield and quality of pomegranate (Punica granatum L.) M.Sc. (Ag.) Thesis, MPAU, Rahuri.

Shivsankar, K. 2000. Prospects of organic farming in horticulture. In: *National Seminar on Hi-Tech Horticulture*. Organized by NAAS, New Delhi & IIHR, Bangalore, June 26-28, 2000, Abstr. 3.7, pp 60-62.

Shukla, A.K. 1996. Efficacy of drip and mulching on young plantation of aonla + ber cropping model in sodic soil. M. Sc. (Ag.) thesis, N.D.U.A.T., Faizabad.

Singh, A.K. and Gaur, G.S. 2004. Effect of distillery effluent on soil properties and growth of aonla buddlings in alkaline soils. In: *Proceedings of National Symposium on Organic Farming in Horticulture* (Eds. Pathak, R.K., Kishun, R, Khan, R. M. and Ram, R. A.) CISH, Lucknow, pp. 205-08.

Singh, A.K. and Singh, Sanjay 2004. Effect of mulches and soil properties and growth of aonla cv. NA-7 under semi-arid ecosystem of Gujarat. In: *Proceedings of National Symposium on Organic Farming in Horticulture* (Eds. Pathak, R.K., Kishun, R, Khan, R.M. and Ram, R. A.) CISH, Lucknow, pp. 197-200.

Sood, M.C. and Lal, S.S. 2004. Evaluation of vermicompost as an organic source of plant nutrient in potato. In: *Proceedings of National Symposium on Organic Farming in Horticulture* (Eds. Pathak, R.K., Kishun, R, Khan, R.M. and Ram, R.A.) CISH, Lucknow, pp. 212-14.

Wavhal, K.N., Rane, D.A.; Kamble, K.B. and choudhary, S.M. 1985. Nutritional studies in pomegranate. *Third National Workshop on Arid Zone Fruit Research*. MPAU, Rahuri.

9

Soil Health – The Key to Success of Organic Agriculture

K.V. Ahamed Bavappa

INTRODUCTION

Soil is a highly dynamic living system which supports the growth of a large diversity of macro and micro flora and fauna. These include a variety of insects, nematodes, earthworms, enchytraeids, fungi, actinomycetes, bacteria and a host of known and unknown organisms which together support plant life through a variety of physical and bio-chemical processes. Soil being the platform for all agricultural activities, deserves careful and continued attention to keep it healthy on a long-term basis. The true indicators of soil health are soil biological activity as the driving force in the decomposition of organic matter, formation of humus, nutrient transformation, evolution and maintenance of stable soil structure, biological fixation and solubilisation of nutrients and biological control of soil-borne diseases.

Agriculture has been traditionally organic base dependent. In the most primitive form, the nutrients released through biodegradation of fallen leaves, roots and crop residues by the native biological fauna supported the crop growth. With improvements in farming, use of composted plant and animal wastes, crop rotations, use of organic by products such as cakes, cattle, poultry and fish manures and raising green manures came into being. However, these technologies failed to increase production of food and raw materials proportionate to population growth resulting in the food shortages of 1960s.This necessitated Green Revolution, ushering in the use of high doses of chemical fertilizers and poisonous plant protection chemicals adversely affecting the health of the soil.

Increasing Organic Matter Production

Since improvement of soil health depends on continued availability of organic matter for the growth of micro flora, high density planting systems research was initiated from

1908's in Sri Lanka and India (Bavappa and Jacob 1981, 1982, Bavappa *et al.*, 1986) keeping in view the pathways for enhancing the yield potentials of plantation crops (Corley, 1985; Simmonds Norman, 1985). The crops included plantation and fruit crops and tubers. One such high density multi-species model (HDMS) yielded 40 tonnes of organic matter in one year, against nine tonnes by the mono crop of coconut (Table 9.1). A similar high density model planted at CPCRI without cocoa gave 31 tonnes of organic matter.

Table 9.1. Bio mass from different crops of the HDMS at RARS Ambajipeta

Crops	*Organic matter (t/ha)*
Coconut	9.06
Cocoa	17.03
Cinnamon	1.30
Pepper	0.04
Pineapple	0.50
Banana	9.22
Turmeric	0.63
EF Yam	1.00
Colocasia	0.46
Weeds	1.08
Grant total	40.32

Nutrient Availability from Organic Matter

A study of the nutrient availability from the above sources of organic materials showed that barring for K, the available N and P were adequate to meet the nutrient needs of coconut (Table 9.2).

Table 9.2. Nutrient availability in the organic matter

System	*Biomass (t/ha/year)*	*Nutrient (kg/ha/year)*		
		N	*P*	*K*
HDMS–CPCRI	31	125	78	115
HDMS–Ambajipeta	40	269	68.6	77.8
Nutrient requirement/ha for coconut		85	54.0	204

Organic Matter from Legumes

In addition to the organic matter from the crops, additional quantities of organic matter can be generated by growing leguminous crops (Table 9.3) in coconut basins as well as in the interspaces of the crops.

Table 9.3. Leguminous cover crops as green manure in coconut basin

Legume species	*Laterite soil biomass (kg/basin)*	*Nitrogen (g/basin)*	*Sandy soil biomass (kg/basin)*	*Nitrogen (g/basin)*
Calopogonuim mucunoides	27.21	186.53	14.71	102.61
Mimosa invisa	24.97	197.55	17.00	153.19
Pueraria phaseolodies	28.45	196.19	19.43	121.29

Apart from the above, a wide range of organic materials such as green leaf, cattle and poultry manure, composts, oil cakes, etc., brought from outside also serve as good sources of organic matter and are in wide use.

Effect of Organic Matter on Soil Properties

Addition of organic materials to soil improves the soil physical and chemical properties (Table 9.4).

Table 9.4. Effect of organic matter on soil properties

Nutrients/Treatments	*Coir pith +NPK*	*Forest leaves+ NPK*	*NPK alone(Control)*
Organic carbon (%)	0.170	0.159	0.06
Av.N (ppm)	43.6	50.3	27.0
Exch.ca.(ppm)	94.7	127.3	62.1
Exch.Mg.(ppm)	8.32	10.0	5.68
Av.Mn.(ppm)	71.9	47.4	18.4
Av.Fe(ppm)	1.97	4.64	1.9
Water holding capacity (%)	33.80	27.20	23.40
Bulk density(g/cc)	1.37	1.50	1.60

There is considerable increase in available N, Mn, Mg, exch. Ca and water holding capacity.

Effect of Organic Matter on Microbial Population

The microbial load of the soil was increased by the application of organic materials considerably (Table 9.5).

Table 9.5. Microbial load in the soil

Parameters tested	*Green manures incorporated*			
	P. phaseoloides	*C. mucunoides*	*M. invisa*	*Control*
Bacteria	39.25	16.00	2.84	18.61
Fungi ($x10^4$)	8.8	11.39	10.12	2.30
Actinomycetes ($x10^4$)	18.75	14.67	5.27	6.86
Asymbiotic N-fixers ($x10^3$)	69.30	63.15	30.12	41.44
P-solubilizing fungi ($x10^3$)	7.85	4.15	6.38	1.17
P-solubilizing bacteria ($x10^4$)	7.51	6.16	4.47	1.75

Coconut in the HDMS plot which received 1/3rd the quantity of the full dose (276.6 kg NPK) yielded 22 nuts/palm more compared to the monocrop which received 317 kg NPK (Table 9.6). The extra yield in spite of a lower level of applied nutrients presents interesting discussion (Bavappa, 2002).

Table 9.6. Yield of coconut and component crops (HDMSCS) under different manurial doses

Manurial dose	HDMS			Monocrop
	Full	2/3rd	1/3rd	
Total nutrients applied to the System-N,P,K (kg/ha)	792.7	534.2	276.6	317.1
Crop yield				
Coconut (mean of 14 years) (nuts/tree)	151	149	151	129
Clove (g/tree)*	419	614	327	–
Banana (kg/bunch)*	5.54	5.1	4.54	–
Pineapple (kg/plant)*	–	0.82	0.79	0.7

*Mean of 5 years

Preventing soil erosion coupled with programmes for water harvesting are the primary activities to be undertaken in all efforts for soil health improvement. It is estimated that one million ton of fertile top soil is lost annually from Kerala alone. The underground water table has gone down from about 60 meters to over 200 meters. All-round efforts are called for in this sector for conservation of valuable resources.

The role of organic matter in improving soil health is amply borne out by the results obtained from cropping and farming system studies undertaken with different crops. By choosing appropriate crop combinations and planting in high densities, the organic matter production capability of the system can be substantially increased. Recycling the organic matter so generated improves the physical, bio-chemical and nutrient status of the soil in addition to providing a soil condition that favours the growth and activity of the micro flora and fauna giving yield increase of 22 nuts as a 'System Effect'. The microbial growth and interactions accelerated by the rhizosphere biomass of different crops giving diverse types of root exudates and enhancing the production of a wide range of plant growth promoting substances appear to have contributed for this yield increase in spite of a lower fertilizer input. Intensive research in this area is urgently called for to discern this highly cost effective mechanism.

The process of soil health improvement can be further accelerated by improving the quality of the applied organic matter through composting and fortifying them with bio-agents capable of fixing nitrogen, solubilizing phosphate, mobilizing potash and controlling soil borne diseases. Reducing the tillage operations and mulching with organic materials provide favourable environment for the growth and activity of earth worms and microflora. Avoiding the use of fertilizers and plant protection chemicals needs no emphasis. Soil health improvement and its maintenance on a sustainable basis, thus necessitate a multi-pronged well-planned approach which in the ultimate analysis would not only pave the foundation for a successful Organic Agriculture movement but also an efficient low cost and cost effective farming system which will have far reaching implications on the socio-economic fabric of the farming community.

References

Bavappa, K.V.A and Jacob, V.J. 1981. A model for mixed cropping, Ceres No. 82, 14(3), FAO Rome, pp. 44-46.

Bavappa K.V.A. 2002. Plantation Crops based Cropping and Farming System options for Asia and South East Asian countries. Plantation Crops Research and Development in the new Millennium, 23-29.

Bavappa K.V.A. and Jacob V.J. 1982. World Crops, 34(2): 47-50.

Bavappa K.V.A. 1986. Coconut and arecanut based HDMS Cropping Systems, Journal of Plantation Crops, 14(2): 74-87

Corley, C.H.V. 1985. Yield potentials of Plantation Crops International Potash Institute, Bern, 35-53.

Simmons Norman, W.C. 1985. Farming System Research – A review, Paper No. 43 The World Bank, Washington DC, 80, p.

10

Advances and Approaches in Ecofriendly Pest Management in Organic Agriculture

C.P.R. Nair

Organic agriculture aims to maintain the farming to be a live production system with sustainable nature. It supports us to live in harmony with nature. Intensive agriculture, on the other hand with a goal of high productivity, makes use of inputs like high yielding varieties, fertilizers, chemical pesticides and farm machineries. Heavy dependence of modern agriculture on these inputs causes several disruptions to the ecological balance. Modern agriculture has definitely helped us to satisfy the nutritive demand of an ever increasing population to a certain extent. At the same time it has been damaging the wealth of this planet largely. Man as a social being has the moral responsibility to sustain nature for future generations. When farming is viewed in this direction, the concept of organic farming gains practical significance as it is a production system respecting the sustainability of nature.

Pests and diseases along with weeds are known to cause an estimated crop loss of about 35% of the potential food production worldwide (Sharma, 2002). In spite of our heavy dependence on insecticides for providing protection to the varieties of crops grown, there has been increasing incidence of pests and diseases causing heavy crop loss mainly due to elimination of natural enemies. In our attempt to save the crops, pest control strategy in modern agriculture has a heavy dependency on pesticides. Excessive use of pesticides has resulted in pesticide residue in food commodities and environment, outbreak of new pests changing the conventional pest scenario of agricultural crops, resurgence of pests and pesticidal resistance among target pest species.

Pest Management Concept in Organic Farming

Pest management requires an overall rescheduling of various components of IPM in organic farming to maintain ecosystem sustainability. In organic agriculture, pest control strategies that compromise with the nature are suitably blended and employed to realize desirable results. Pest control strategies of a preventive rather than reactive nature are advisable in organic farming. The non-chemical and bio-rational methods of pest management that constitute the major components of IPM approach are ideal, ecofriendly and feasible in a sustainable crop production system like organic farming.

Biocontrol

Among the various components of IPM, the most effective tool for organic agriculture is the biological pest suppression that utilizes the natural enemies of pests *viz.*, entomophaga and entomopathogens. One of the major steps involved in the biological control of pests is conservation of natural enemies which helps in regulating a particular species of pest in the ecosystem. The biodiversity of the ecosystem is conserved, in general, in biological pest suppression. A comparatively low level of pest incidence that occurs due to natural disturbances in the ecosystem can be corrected more easily and feasibly by intervention of an appropriate biocontrol agent. When a situation arises where natural enemies present in the ecosystem exert a low level of pest suppression, the need for augmentation through release of recommended natural enemies becomes more imperative. Biocontrol method is an advanced approach in the management of some of the major pests of coconut, oil palm, coffee and black pepper besides many annual crops. When an exotic pest is to be tackled, the need for deliberate introduction and colonization of its natural enemy to suppress the introduced pest becomes essential.

In our attempts to enrich the biodiversity by organic farming practices, several methods of conservation of natural enemies like providing nesting places, perching sites, water pan, nectar and pollen rich plants can be practiced (Sharma, 2002). Attempts to restore natural control mechanisms especially the natural enemies by selectivity of biodiversity within the ecosystem become more relevant in organic farming. Manipulation of the ecosystem inducing added activity of the natural enemies will be beneficial to reduce the pest population resulting in increased yield.

Entomophaga

Entomophaga constitutes one of the major components in the natural control of crop pests. Many species of spiders, coccinellids, reduviids (Fig. 10.1), predatory ants, mites and thrips are well documented as effective predators. Spiders are important predators of insect pests. They suppress pest species to low densities at all growth stages. They are generalist predators and polyphagous in habit. Spider fauna contributes a major group of natural enemies in cropping systems like paddy, tea, etc. A detailed literature on spiders in tea ecosystem is given by Hazarika and Chakraborti (2007) and they stated that the spiders occupy a major share among the natural enemy complex of tea pests. Spiders suppress rice pests like BPH, GLH, WBPH, etc. (Kamal *et al.*, 1992). Habitat manipulations like conserving ground litter, using safer pesticides *etc.* contribute to their conservation.

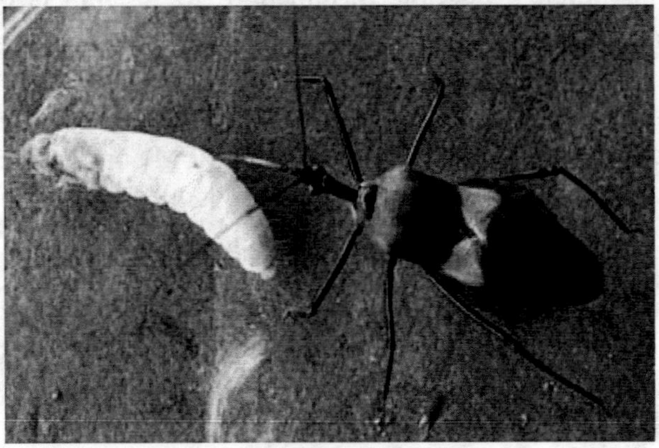

Fig. 10.1. Predator (Reduviid bug)

Among the natural enemies used for biological control of insect pests, the parasitic hymenoptera has been the most successful group comprising families like Ichneumonidae, Braconidae, Encyrtidae, Trichogrammatidae, Pteromalidae, Eurytomidae, Bethylidae, Elasmidae, etc., with more than 279 successful cases reported world over (DeBack and Rosen, 1991; Narendran, 2001). *Bracon brevicornis, Goniozus nephantidis* (Fig. 10.2) (Braconidae), *Brachymeria nosatoi* (Chalcididae), *Elasmus nephantidis* (Elasmidae), *Xanthopimpla nana nana* (Ichneumonidae) are effective in biosuppression of black headed caterpillar, *Opisina arenosella* Wlk. of coconut palm in the field (Pillai and Nair, 1983). Members of the family Trichogrammatidae are well known egg parasitoids in the field. Various species of Trichogrammatids like *T. japonicum* and *T. evanescence* are used against stem borers of sugarcane in India (David and Easwaramoorthy, 1985).

Fig. 10.2. Larval parasitoid (*Goniozus nephantidis*)

Entomopathogens

Microbial agents are important components in a biological control programme and in organic farming. Use of these agents has wider scope in organic agriculture. Various groups of micro-organisms like bacteria, virus, fungi and nematodes are employed in the bio suppression of many major pests of crop plants. In recent years, the market for microbial pesticides is growing worldwide though it contributes less than 1 % of the total crop protection products (Rabindra *et al.*, 2001).

Of over 750 species of fungi known to be pathogenic to insects, only six have been commercialized (Ambethgar, 2001). Pathogens such as *Beauveria bassiana, Metarhizium anisopliae, Nomuraea rileyi, Verticillium lecanii* and *Hirsutella thompsonii* (Fig. 10.3) are the best known of these fungi. Soil treatment with conidia of *Nomuraea rileyi* along with a solid substrate has been recommended for management of *Spodoptera litura* in ground nut which is a close canopy crop (Vimala Devi, 1995). Fungi require relatively high humid microclimate (>70% RH).

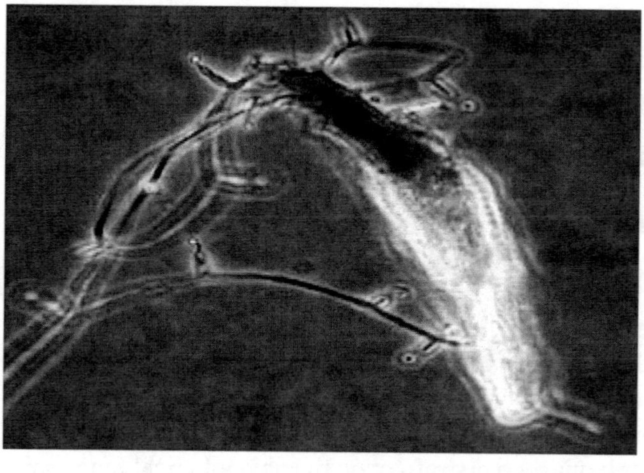

Fig. 10.3. Acaropathogen (*Hirsutella thompsonii*)

Bacillus popilliae and *Bacillus thuringiensis* are two major species of bacteria employed in the management of many species of crop pests particularly the caterpillar pests affecting many vegetable and fruit crops. In India, *Bacillus thuringiensis* (B.t) is used in the management of pests like diamond back moth on cruciferous vegetables and pod borers on pulses and okra.

As plantation crops are cultivated in contiguous areas, use of microbial agents is more feasible to achieve effective management of certain pests. In coconut and oil palm, two classical examples of microbial agents viz., the green muscardine fungus *Metarhizium anisopliae* (Fig. 10.4) and the *Oryctes rhinoceros* virus (OrV) have been proved to be highly effective for managing the black beetle, *Oryctes rhinoceros* L.

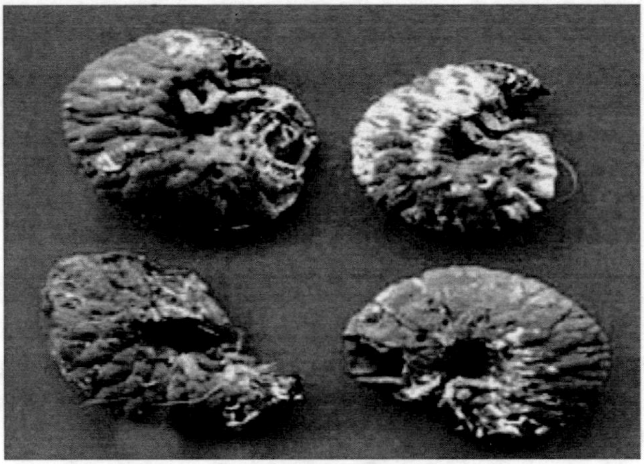

Fig. 10.4. *Oryctes* grubs infected by *Metarhizium*

The entomopathogenic fungus, *Metarhizium anisopliae* (Metch.) Sorokin infects grubs of rhinoceros beetle especially during wet season. Mass culture of this fungus in cheaper substrates like coconut water or cassava chips is standardized with low cost technologies and the fungus can be applied at the recommended dose in the breeding sites of the beetle. The OrV is a potential viral pathogen of rhinoceros beetle in many countries. The successful biocontrol attempts carried out by Central Plantation Crops Research Institute in many coconut growing tracts in India with this virus have established the potential of this pathogen.

Botanical Pesticides

Botanical pesticides are having a prominent position in pest management under organic agriculture. As botanical pesticides are safe for the non-target organisms, they are preferred in organic pest control strategies. Botanical pesticides are used either as repellents or as primary insecticides having adverse effect on the physiological functioning of insect pests.

Derivatives from various species of plants have been in use for plant protection from ancient days onwards in agriculture. More than 600 plant species are reported to have pesticidal properties. Tulsi, Mahuva, Lantana, Tobacco, Marigold, *Clerodendron infortunatum* (Fig. 10.5), etc., are some important plants used for pest management. Insecticidal properties of neem were proved beyond doubt in ancient times and in modern agriculture neem based pesticides are getting importance because of their versatile ability to control many pests of agricultural importance. *Azadirachtin* is the active principle having insecticidal property present in neem tree. Decoction of neem leaf, neem oil cake, powdered neem seed and seed oil are used in various dilutions mixed with washing soap to effect satisfactory control of various groups of foliage pests in many field crops.

Trials conducted by Indian Institute of Horticultural Research have proved that sprays

Fig. 10.5. *Clerodendron infortunatum* (IGR)

of neem seed powder or neem soap at 1% were effective in the management of tomato fruit borer. Pulverised neem seed powder extract at 4% is recommended for management of foliage pests of cruciferous plants (Amerika Singh *et al.*, 2006). In coconut mite management, use of neem pesticides is recommended either as spray or as root feeding (Nair *et al.*, 2005).

Semiochemicals

Semiochemicals which are considered to be a modern tool in IPM concept has a pivotal role in eco-friendly pest management programmes. Semiochemicals are compounds released or emitted by plants and animals for specific communication purposes. If the communication is between the same species (Instraspecific), they are called pheromones, and on the other hand if the communication is between two different species (Interspecific), they are called kairomones. Insect pheromones fall into several categories usually related to functions viz., sex pheromones, aggregation pheromones, alarm pheromones, trail pheromones, etc., Pheromones that have been used most successfully in pest monitoring/control are sex pheromones of Lepidoptera and aggregation pheromones of Coleoptera. The application of pheromones in crop protection may be indirect as population monitoring agents or direct as tools in mass trapping, lure and kill and mating disruption techniques (Yadav *et al.*, 2004).

In India successful attempts have been recorded for the control of sweet potato weevil, *Cylas formicarius*, sugarcane pests like early shoot borer, *Chilo infuscatellus*, inter node borer, *Chilo sacharophagus indicus*, tobacco leaf eating caterpillar, *Spodoptera litura* and rice yellow stem borer, *Scirpophaga incertulus* (Ganeshwara Rao *et al.*, 1994) and coconut

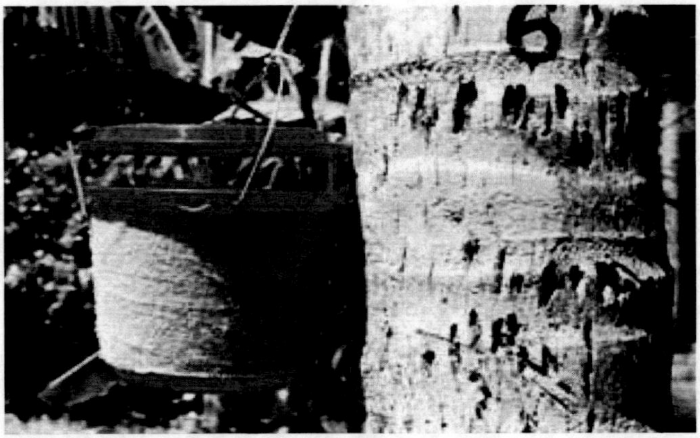

Fig. 10.6. RPW Pheromone trap

red palm weevil, *Rhynchophorus ferrugineus* (Fig 10.6) (Nair and Saritha, 2003).

Kairomones are volatile compounds capable of evoking behavioural response favourable to the receiver. Mostly they are released by the host plant and can be used to trap the insects. Intensive efforts are required to develop strategies exploring the possibilities of semiochemicals in organic agriculture.

Light Traps or Attractant Traps

Some of the pests like diamond back moth of cabbage, fruit flies damaging mango and cucurbitaceous vegetables, red palm weevil of coconut, etc., are attracted to food lure traps employing the natural food substances. The synthetic attractant like methyl eugenol is employed to trap fruit flies. White grubs are major pests of coconut, arecanut, tuber crops, sugarcane and ground nut and they are successfully managed using light traps. Trapping of adult beetles during their emergence period coinciding with onset of monsoon is found to be one of the effective strategies in IPM of white grubs. Light traps have been successfully used in monitoring of yellow stem borer of rice (Chandramohan and Chelliah, 1984).

Agronomic Practices

Historically, agronomic practices or cultural manipulations of plants were the most important methods for controlling and preventing crop losses. Farmers continuously refined them with experience. Among the oldest techniques for managing pests are sanitation, destruction of alternate hosts, tillage, avoidance and planting of crop cultivars that resist pest attacks. The worldwide awareness of safe environment provided impetus to foster non-chemical pest management strategies. Of these strategies, cultural control constitutes the most farmer-oriented approach, where weak points in the biology and behaviour of insects are exploited and pressure is exerted on the population by manipulating the environment. An understanding of natural defense mechanisms present in the host against the pest is useful to prepare a pest control programme where agronomic methods are to be utilized. Also, a thorough knowledge of pest, host, host-plant interaction and optimum environment for both are necessary in this method. In an agricultural country like India, agronomic management practices are of immense importance because of low cost of inputs, easy in their adoption and acceptance by farmers. The initial build up of a pest in crops is many times influenced by the cultural and crop management methods followed in farming. Nutritional management of the plant is important in managing pest problems. Adjustment of sowing/planting date keeping in mind the peak activity of the damaging stage of the pest helps in avoiding the critical stage of the crop from pest attack. Thus synchronization of most susceptible stage of the crop with the inactive period of insect reduces the infestation and chemical intervention.

Sanitation

Removal and destruction of breeding materials, manipulating over-wintering sites, pruning of infested plant parts etc. are sanitation methods that can be employed in organic agriculture. Sanitation is important to prevent introduction of insects, pathogens, nematodes and weeds into pest free fields and to reduce losses in infested fields. Sanitation includes all measures aimed at eliminating or reducing the amount of inoculums present in a plant or field and at preventing the spread of pests to other healthy plants and plant products. Sanitation practices such as use of healthy seed material, burning and destruction of crop refuse, clean storage, etc., reduce pest population, discourage breeding and hibernating sites and prevent carry over to the next crop season. These methods are economical, effective and easy to adopt. Depending upon the crops and types of pests that affect the crops, sanitation methods can be practiced by farmers. Destruction of stubble soon after harvesting in cereal crops is advisable to remove the surviving stages of pests like stem borer and gall midges in rice.

Trashing in cardamom reduces the thrips damage on cardamom capsules. Crown cleaning in coconut reduces the coreid bug damage and also removes the nesting sites of rodents. Phyto-sanitation is the key factor in the management of major pests of plantation crops like rhinoceros beetle and red palm weevil of coconut and stem borers affecting crops like cashew, coffee, cocoa, etc. Avoiding breeding sites of black beetle in the immediate vicinity controls its attack in coconut plantations. Soil solarization for six weeks during peak summer and fortification with *Trichoderma harzianum* reduces the foot rot disease in black pepper nursery, besides helping in control of nematodes.

Tillage

Tillage is an integral factor in destroying food sources and habitat of the pests within the field. In addition tillage plays a major role in rhizosphere microflora and microfauna. Tillage operation changes the texture, nutritional composition and pH of the soil which becomes less favourable to weeds and soil inhabiting insects. Ploughing exposes the grubs and pupae of coconut white grubs to predation by birds. Deep ploughing (Fig 10.7) during the pre-monsoon and post-monsoon seasons helps in control of soil insects like white grubs, termites and army worms. Many times tillage operations are also essential in managing nematodes and rodents. Deep ploughing of the field after the harvest reduces the activity of fruit fly, red pumpkin beetle and cut worm as these insects remain in the soil in earthen cocoon to complete the dormant stage of their life cycle.

Fig 10.7. Predatory birds in ploughed field

Resistant Cultivars

Genetic resistance is one of the oldest methods of pest control. Growing resistant cultivars is the most effective and economic means of controlling plant pests. Resistant cultivars are also the first line of defense against pests. Resistant varieties should be used in concert with other pest suppression or pest control measures. Plant resistance provides a built-in ability to allow fewer pests and cut off the extra load of insecticides in the crop. Keeping in mind the diversity and intensity of pests in a particular place, selection of resistant/less susceptible varieties holds well in pest management. Plant resistance has been well-utilized in the management of pod borers of peas and cotton. The Bt Cotton has become highly popular among farmers for its high level of resistance to bollworms.

Tall varieties of coconut are found to be tolerant to red palm weevil damage and Malabar varieties of cardamom are tolerant to shoot and capsule borer damage. Orange dwarfs are tolerant to mite damage in coconut. In paddy cultivation, use of resistant varieties especially to tackle the brown plant hopper and green leaf hopper problems has been successfully implemented in more than 20 million hectares in Asia (Panda and Khush, 1995).

Trap Crops

Trap crops or catch crops are species of plants which are planted to attract and retain a pest species or to provide a more favourable habitat to increase natural enemies. Trap crop

(Fig. 10.8) must be definitely and distinctly attractive to the pest than the main crop. Trap crop provides protection either by preventing the pest from reaching the main crop or by concentrating them in certain parts of the field, where they can be economically destroyed. Trap crops have been utilized in control of nematodes. Marigold as a trap crop for nematodes in vegetable fields and black pepper garden is common at many places. Use of mustard and marigold as trap crops have become highly popular in management of pests affecting vegetable crops, particularly tomato and cabbage.

Fig. 10.8. Trap crop (Marigold) in cardamom

Removal/Regulation of Alternate Hosts of Pests

Many economically important pests are known to survive on collateral hosts which constitute mainly the weeds grown in the vicinity of the agricultural land or a subsidiary crop grown in the field. The coreid bug of coconut not only infests coconut but also infests collateral hosts like guava, cashew, cocoa and tamarind. Knowledge on the feeding behaviour and biology of the pests of cultivated crops would enable the farmer in utilizing the technology effectively.

Crop Rotation

Crop rotation and fallowing were reported to be of common usage in Roman times as methods of increasing crop production and avoiding losses due to adversity. Continuous production of one crop over an area virtually assures an increase of those pest species adapted to attack that particular one. Substitution of a crop belonging to a different plant family will eliminate or greatly reduce population of those pest species associated with the earlier crop. Crop rotations are very effective in breaking insect build-up, especially for insects of limited host range. Crop rotation with a view to avoid continuous supply of food plants for pests particularly the oligophagous pests is an advisable step in eco-friendly pest management. Similarly to tide over the colonization of pests, adjustment of sowing time not to synchronize with the crop phenology and pest stages is another method practiced by farmers. Crop rotation practice helps in reducing population build up of soil insects and nematodes. It also enhances the biodiversity which in turn is beneficial to the natural enemies.

Pest Surveillance

Pest surveillance and monitoring system play a major role in pest management in an organic farming system. The farming community has to be aware about the pest problems of the crop, the behavior and population ecology of the pests and their natural enemies for employing appropriate technology at a time of urgency. This is possible only through an effective transfer of technology programme by an interactive involvement of research-extension agencies and farming community.

Future Thrusts

The increasing awareness on the ecological aspects of pest management and the inclination to conserve ecosystem without disruption for maintenance of sustainability among various stake holders are appealing in recent years. Utilization of frontier areas like information technology can make tangible effects in safer pest management systems. Proper planning on the adoption of apt control measures to save the crop is possible through reliable prediction of pest out break based on weather forecast. Employment of modern tools like remote sensing and geographical information system (GIS) will be useful in assessing crops loss and pest incidence and using the information from such sources, advance action plan can be taken up to schedule effective plant protection strategies over larger areas within short period. Exploring the potential of farmer's knowledge and finding out scientific basis for such indigenous technical knowledge of farmers are to be priority areas in our attempts to mange the pest problems in organic agriculture. Bio informatics is a recent field by which speedy dissemination of information is achieved. This helps in proper understanding and timely action to tackle situations like accidental introduction or sudden out breaks of pest species.

References

Ambethgar, V. 2001. Exploitation of entomopathogenic fungi in biological control of crop pests, pp. 39-57. In*: Biocontrol Potential and its Utilization in Sustainable Agriculture* Vol 2 Insect pests. (Eds. R.K. Upadhay, K.G., Mukerji, K.G. and Chamola, B.P), Kluwer Academic / Plenum publishers, New York.

Amerika Singh, Sharma, O.P. and Garg, D.K. 2006. *Integrated Pest Management-Principles and Applications.* Vol II, CBS Publishers and Distributors, New Delhi, p. 656.

Chandramohan, N. and Chelliah, S. 1984. Light trap studies on rice yellow stem borer, *Scirpophaga incertulus. National Seminar on Behavioural and physiological approaches in pest mananagemmt in crops.* TNAU, Coimbatore, p. 230.

David, H. and Easwaramoorthy, S. 1985. Recent advances in biological control of sugarcane pests. *In Advances in Biological Control research In India* (Eds. K.J. Joseph and U.C. Abdurahiman) Proc. of National Seminar on Entomophagous Insects, IISR, Calicut, India, pp. 200-213.

DeBach, P. and Rosen, D. 1991. *Biological Control by Natural Enemies*. 2nd Edition, Cambridge University Press, New York, pp. 255-85.

Ganeshwara Rao, A., Krishnaiah, K., Cork, A. and Hall, D.R. 1994. Disruption of sex pheromone mediated communication in yellow stem borer, *Scirpophaga incertulus* Walker with controlled release pheromone formulation. *J. Entomol. Res.* **18**: 287-88.

Hazarika, L.K. and Chakraborti, S. 2007. *Spiders of the Tea Ecosystem in Assam- A hand book.* Assam agricultural University, Assam, p. 38.

Kamal, N.Q., Begam, A., and Biswas, V. 1992. Studies on the abundance of spiders in the rice ecosystem. *J. Insect Sci.* **5**(1): 30-32

Nair, C.P.R. and Saritha S. Nair. 2003. Role of pheromones in the integrated management of coconut red weevil. *Indian Coconut Journal.* **33**(12): 14, 22-23.

Nair, C.P.R., Rajan, P. and Chandrika Mohan. 2005. Coconut eriophyid mite, *Aceria guerreronis* Keifer- An overview. *Indian J. Plant Protec.* **33**(1): 1-10.

Narendran, T.C. 2001. Parasitic hymenoptera and biological control p 1-12. *In: Biocontrol Potential and its Utilization in Sustainable Agriculture* Vol 2 Insect pests. (Eds. R.K. Upadhay , K.G. Mukerji and B.P. Chamola) Kluwer Academic/ Plenum publishers, New York, p. 421.

Panda, N. and Khush, G.S. 1995. *Host Plant Resistance to Insects.* CAB International, Oxford, p.431.

Pillai, G.B. and Nair, C.P.R. 1983. *Xanthopimpla nana nana* Schultz (Hymenoptera : Ichneumonidae) a new pupal parasitoid of *Opisina arenosella* Wlk. on coconut. *J. Plantn. Crops* **11**: 65-69.

Rabindra, R.J., Kennedy, J,S., Sathiah, N., Rajasekaran, B. and Srinivasan, M.R. .2001. *Microbial Control of Crop Pests.* Proc. of Winter School on Emerging Trends in Microbial Control of Crop Pests, TNAU, Coimbatore, p. 321.

Sharma, A.K. 2002. *A Hand Book of Organic Farming.* Agrobios India, Jodhpur, p. 627.

Vimala Devi, P. S. 1995. Soil treatment with *Nomuraea rileyi,* a promising technique for the control of *Spodoptera litura* in ground nut. *Biocontrol Sci. Technol* **5**:361-64.

Yadav, J.S., Prasad, A.R., Prasuna, A.N. and Jothi, K.N. 2004. Semiochemicals as potential tools in pest management. *In: Frontier Areas of Entomological Research* (Eds. Subrahmanyam, B., Ramamurthy, V.V. and Singh, V.S.), Entomological Society of India, IARI, pp. 411-22.

11

Achieving Eco-friendly Management of Pests and Diseases Using Bioresources in Organic Horticulture

R. Samiyappan

INTRODUCTION

India has witnessed significant increase in production of a number of horticultural crops. Economic importance of fruits, vegetables, plantation crops, flowers, etc., has been increasing due to high domestic and international demand. All over the world, including India, appreciation for safe food and healthy environment have necessitated and resulted in increase in awareness of organic horticulture. Organic farming is one of the several methods found to meet the objectives of sustainable fruit, vegetable and flower production. Major techniques that include are mulching, intercropping, and integration of cropping system with livestock farming. Adverse effects of modern horticultural practices are seen not only on the farm but also on all living organisms and hence, have been well documented. In conventional farming, use of chemical fertilizers and pesticides has been given much importance; hence, the situations persuaded us to think differently. Negative effects due to modern farming are manifested through soil erosion, water shortages, salination, water and soil pollution and genetic erosion. Organic way of doing horticulture is one of the widely used methods which is considered as the best alternative to reduce or prevent the ill effects of chemicals based conventional farming.

Organic farming is defined as a system that is designed and maintained to produce agricultural products by employing methods and substances which maintain the integrity of organic agricultural products until they reach the consumer. This is carried out by using substances to fulfill any specific fluctuation within the system so as to maintain long-term soil biological activity. Organic farming is an approach to sustainable production of horticulture crops, which mainly emphasizes protection of environment, food quality and health, sustainable natural resources management and other social justice objectives. By organic farming, the activity of major pests, plant pathogens and plant nematodes occurring in vegetables, fruits, flowers and plantation crops are managed by ensuring sustainable agro-ecological situations.

Management of Ecosystem Services

Soil structure and soil health play a major role in determining where different kinds of farming take place and the quantity and quality of agricultural and horticultural output are decided. Soils with high fertility and high levels of organic matter seem to enhance natural biocontrol of plant pathogens and nematodes as suggested by the fact that in some soils, pathogens cause little or no disease, despite prevailing of an apparently favourable environment. It is well-known that the activity of the soil microbial community can change very rapidly with the available fresh nutrients, changes in environmental variables like temperature and moisture and with changes in pH or metal concentrations. Less is known on how these effects on activity and the subsequent new situation in competitive advantage of different parts of the microbial community which reveal changes in the microbial community itself. This will be seen as different activities within different groups of micro-organisms, eventually leading to an altered community composition. This can be detected using different techniques, e.g., isolation and determination of species/strains, microscopical counts coupled with immunological/probe detection of specific species/strains, phospholipid fatty acid (PLFA) patterns, DNA based techniques, etc.

Organic amendments are recognized as initiators of two important disease control processes: increase in dormancy of propagules and their digestion by soil micro-organisms. Addition of organic matter increases the general levels of microbial activity and the more microbes that are active, the greater the chances that many of them will be antagonistic to pathogens. Use of chemical pesticides in the management of plant diseases leads to greater disturbance in the ecological niche. Soil fumigation with general biocides has short or long-term impacts on soil community composition and abundance. Biocides may either suppress the microbes or fauna directly with toxic effects or indirectly through reduced vegetation or other non target effects. Methyl bromide and benomyl have direct toxic effects on bioagents, nematodes and earthworms (Edwards and Bohlen, 1996). Phenoxy acetic acid herbicides do not depress soil fauna directly with toxic effects, but indirectly through reduced vegetation and smaller additions of organic matter to soil.

Further, plantations in horticultural cropping systems have several advantages. Root growth is more extensive and less ephemeral than with annual crops and supports a rich soil community with many omnivores and predators. Soil community composition with perennial crops resembles that of natural ecosystem soils more closely than is the case with annual crops. Fertilizer addition affects soil microflora and indirectly influences soil mesofauna by changing their food resources. Addition of nitrogen may acidify soil and consequently inhibit microbial growth and activity. Nitrogen may also affect the quality of microbes as a food source for mesofauna (Darbyshire *et al.*, 1994).

When pathogens are not inhibited by naturally occurring antagonists it is possible to enhance biocontrol by adding more effective ones. *Agrobacterium tumefaciens* var. *radiobacter* strain 84 and *Peniophora gigantean* have been successfully introduced and used against crown gall in fruit trees and *Fomes annosus* in conifers, respectively. Many other micro-organisms have produced beneficial plant responses when introduced into the soil or plant rhizosphere. Examples include *Trichoderma* spp, *Pseudomonas* spp, *Alcaligens* spp, *Agrobacterium tumefaciens* and others. Directed biocontrol involves mass introduction of antagonistic micro-organisms in soil with or without a food base to inactivate pathogen propagules thereby reducing their numbers and adversely affecting infection. Clearly, soil microbes and fauna play important roles in ecosystem function.

Biocontrol Agents for Eco-friendly Management of Pests and Diseases

Various soil micro-organisms which are capable of exerting suppressive effects on plant pests and diseases either in culture or protected environment have potential for use in both agriculture and horticulture crops besides showing enhanced yield. The modes of action are nitrogen fixation, phosphate solubilization and mobilization, plant growth regulation, siderophores production, lytic enzymes production and antibiotics production and induced systemic resistance. The microbial system can siphon out appreciable amounts of nitrogen from the atmospheric reservoir and enrich the soil with this important but scarce nutrient. Rhizosphere bacteria that favourably affect plant growth and yield of commercially important crops are denominated by PGPR include bacteria belonging to the genera *Azotobacter, Azospirillum, Arthrobacter, Bacillus, Burkholderia, Enterobacter, Klebsiella, Psuedomonas, Xanthomonas* and *Serratia*. The Plant Growth Promoting Rhizobacteria (PGPR) are those that affect plant growth directly by producing and secreting plant growth promoting substances such as auxins, gibberllins and cytokinins, by eliciting root metabolic activities with bacterial surface components and/or by supplying biologically fixed nitrogen. Other PGPR affect plant growth by indirect mechanisms such as suppression of bacterial, fungal and nematode pathogens (biocontrol) by production of various metabolites, induced systemic resistance and/or by competing with the pathogen for nutrients or for colonization of space. Recently, Saravanakumar *et al.* (2007) reported that foliar application of *Pseudomonas fluorescens* strain TNAU Pf1 significantly reduced the incidence of blister blight disease (*Exobasidium vexans*) besides recording increased yield in tea plantations. Further, it was also observed that application of *P. fluorescens* TNAU Pf1 on tea plants attracted the natural enemy population against mite incidence. Radjacommare *et al.* (2007) demonstrated that application of PGPR and *Trichoderma* spp. reduced the incidence of major diseases of vanilla plants. *Trichoderma* spp. are common saprophytic fungi in the rhizosphere, which have received considerable attention as potential biocontrol agents. The beneficial effect of *Trichoderma* is due to direct mycoparasitism on pathogenic fungi and to effects on plants, such as enhanced plant development and induced resistance (Harman, 2000).

Interestingly, the concept of endophytes in the management of plant pests and diseases is emerging as an effective eco-friendly approach. Endophytic bacteria colonize an ecological niche similar to that of plant pathogen, especially vascular wilt pathogens, which might favour them as potential antagonistic agents. Endophytic strains of *P. fluorescens* 89B-27 and *Serratia marcescens* 90-166 were observed to induce resistance in cucumber to *P. syringae* pv. *lachrymans* as well as to the fungal pathogen, *F. oxysporum* f. sp. *cucumerinum* and *C. orbiculare* (Liu *et al.*, 1995). Similarly, the endophytic nature of *Bacillus pumilus*, Gottheil SE34 and *P. fluorescens* 63-28, *P. fluorescens* CHAO and *P. fluorescens* WES417r has been demonstrated on different agricultural and horticultural crops. These endophytic bacteria may be in a particularly advantageous ecological position in that they may be able to grow and compete on the root surface, but also may be capable of developing within the root, relatively protected from the competitive and high-stress environment (Struz *et al.*, 1999). Recently, several strains of endophytic growth promoting bacteria have been isolated from coconut plants and they showed higher inhibition to the growth of *Ganoderma* spp. which cause basal stem rot disease in coconuts. Interestingly, the application of bioagents enhanced FYM (plant growth promoting endophytic bacteria combined with plant growth promoting fungus, *Trichoderma viride*) reduced the incidence of basal stem rot disease in coconuts palms (Rajendran *et al.*, 2007).

It is also interesting to note that mycopesticides based on the entomopathogens, *Beauveria bassiana* and *Metarhizium anisopliae* are valuable tools for integrated pest management strategies in organic horticulture. The *Beauveria* infects the insects by breaching the host cuticle and this is advantageous from the point of view of pest control, because propagules do not have to be ingested, and thus are active against the nonfeeding stages of insects. Recently, it is reported that *M. anisopliae* controlled the breeding of insect pest, *Oryctes rhinoceros* L. in coconut leaf vermicomposting sites (Gopal *et al.*, 2006). This indicated the possibility of formulating the bioproduct based on *M. anisopliae* against coconut rhinoceros beetle. The same researchers have reported the possibility of controlling the coconut rhinoceros beetle with microbial agents viz., *Oryctes virus* and opportunistic bacteria pathogen, *Pseudomonas alcaligenes*. Similarly, endophytic *B. bassiana* isolated from coffee plants has been inoculated onto cocoa seedlings against coco pod borer (*Conopomorpha cramerella*). The study demonstrated the possibility of establishing a cocoa endophyte when the radicles of seedlings are inoculated. These researches raised the hope for exploiting endophytic bacterial and fungus strains in the eco-friendly management of pests and diseases in horticultural crops.

The consequence of using bio-inoculants with these effects is to enable reduced application of synthetic chemicals, thereby satisfying environmental requirements with a possible cost benefit. Currently, there is a range of microbial inoculant products successfully marketed for agriculture. With the improvements in modern science and technology, it is likely that the future will see a greater use of such products in the management of pest and diseases in organic horticulture.

Earthworms

Earthworms are key regulators of soil structure and soil organic matter dynamics in several horticulture ecosystems. They are greatly impacted by agricultural management and only little is known about how these factors interact in maintaining soil health. The vermi-compost made from decomposition of agro wastes by earthworms besides improving the soil health enhances the survival and activities of many biocontrol microbes. Vermi-compost is a very important low external input component in organic horticulture as it not only increases the soil fertility through the addition of plant growth hormones and increased levels of soil enzymes, but is also responsible for the dissemination of important micro-organisms as they are rich in microbial diversity, population, and activity. Production of vermi-compost from coconut leaf waste with the help of earthworms is being extensively used for the cultivation of high value, export oriented crops like pepper, cardamom, vanilla, tea, coffee etc. in South India (Gopal *et al.*, 2006). Recently, it has been reported that production of vermi-compost from coconut leaves and cow manure mixture amplified plant beneficial microbial communities viz., spore forming *Bacillus*, high population of antagonistic fungi, free living nitrogen fixers, phosphate solubilizers, cellulose degraders, fluorescent pseudomonads and silicate solubilizers, actinomycetes and *Trichoderma* spp. (Gopal *et al.*, 2009). The presence of these plant beneficial microbes greatly aid in the dissipation of available phosphorus and other nutrients and plant growth promoting substances in vermi-composts. However, the potential impact of the feeding habits of the earthworms on the microbiota of the ingested material during gut passage and consequently on the microbial potential and activities of soil is less known.

Conservation of Bioagents for Eco-friendly Approach

In organic horticulture, floral and faunal biodiversity fulfils important ecosystem services: it (i) reduces soil erosion; (ii) improves pollination of crops; and (iii) is an important component of pest and disease management (i.e., conservation biocontrol). Therefore, conservation biocontrol is incorporated in a phased approach for arthropod pest management in organic horticulture. Together with suitable cultural practices, the enhancement of soil quality, the choice of resistant cultivars and the implementation of functional biodiversity (e.g., incorporation of non-crop habitats such as hedgerows and wild flower strips) help to prevent pest and disease outbreaks.

The goal of biological control is to use naturally occurring organisms to maintain the pathogen population at a level that does not cause any disease or yield loss. Conservation of biological control aims to conserve existing populations of naturally occurring organisms by adopting farming practices that favour populations of these beneficial microbes. Examples of such practices include the use of less toxic herbicides and/or the timing of farming practices to minimize their disruption on beneficial species. However, little is known about the conservation of beneficial microbes in the sustainable horticulture and it is possible to maximize our ability to tailor farming practices to optimize plant productivity while encouraging the activity of beneficial soil organisms and the functions they perform.

CONCLUSION

As organic farming is a method of production that combines traditional agriculture with innovation and science, it is set to sustain or enhance soil quality, enhance biological diversity and preserve natural resources. These goals are accomplished by using crop rotation, biological pest control and natural inputs and by strictly limiting synthetic fertilizers and pesticides. In order to harness potential benefits of bioagents in commercial farming, the consistency of their performance must be improved to make the horticulture production more sustainable. This requires research in many diverse areas as these biological systems involve complex interactions among the host, rhizosphere, phyllosphere, microflora and fauna and the environment. Further, the survival and competitive ability of the microbial strains to be introduced must be improved with the use of modern scientific technologies.

References

Darbyshire, J.F., Davidson, M.S., Chapman, S.J. and Ritchie, S. 1994. Excretion of nitrogen and phosphorus by the soil ciliate *Colpoda steinii* when fed the soil bacterium *Arthrobacter* sp. *Soil Biology and Biochemitry*, **26**:1193-99.

Edwards, C.A. and Bohlen, P.J. 1996. *Biology and ecology of earthworms*, 3rd ed. Chapman and Hall, London.

Gopal, M., Gupta, A. and Thomas, G.V. 2006. Prospects of using *Metarhizium anisopliae* to check the breeding of insect pest, *Oryctes rhinoceros* L. in coconut leaf vermicomposting sites. *Bioresource Technology*, **97**:1801-06.

Gopal, M., Gupta, A., Sunil, E. and Thomas, G.V. 2009. Amplification of plant beneficial microbial communities during conversion of coconut leaf substrate to vermicompost by *Eudrilus* sp. *Current Microbiology* (In Press).

Harman, G.E. 2000. Myths and dogmas of biocontrol. *Plant Disease*, **84**:377-93.

Liu, L., Kloepper, J.W. and Tuzun, S. 1995. Induction of systemic resistance in cucumber against *Fusarium* wilt by plant growth promoting Rhizobacteria. *Phytopathology*, **85**:695-98.

Posada, F. and Vega, F.E. 2005. Establishment of the fungal entomopathogen *Beauveria bassiana* (Ascomycota: Hypocreales) as an endophyte in cocoa seedlings (*Theobroma cacao*). *Mycologia* **97**:1195-1200.

Radjacommare R., Usharani R. and Samiyappan R. 2007. Genotyping antibiotic producing fluorescent pseudomonads to select effective rhizobacteria for the management of major vanilla diseases. *Annals of microbiology* **57**:163-70.

Rajendran L., Karthikeyan G., Raguchander T. and Samiyappan R. 2007. *In vitro* evaluation of bacterial endophytes influence on *Ganoderma lucidum* (leys) karst. mycelial growth. *Journal of Plant Protection Research* **47**:425-36.

Saravanakumar, D.C., Vijayakumar, N. Kumar and R. Samiyappan. 2007. PGPR Induced Defense Responses in Tea Plants against Blister Blight Disease. *Crop Protection*, **26**:556-65.

Struz, A.V., Christie, B.R., Matheson, B.G., Arsenault, W.J. and Buchanan, N.A. 1999. Endophytic bacterial communities in the periderm of potato tubers and their potential to improve resistance to soil-borne plant pathogens. *Plant Pathology*, **48**:360-69.

SECTION – III
INPUT MANAGEMENT

12

Organic Amendment for Production of Black Pepper (*Piper nigrum*)

Hemant G. Hegde and M.S. Lokesh

INTRODUCTION

The major producing countries of black pepper are India, Brazil, Indonesia, Malaysia, Thailand, Sri Lanka and Vietnam. In India, this crop occupies an area of 2.2 lakh ha with a production of 70,000 tonnes. India earned Rs. 306.2 crore by exporting 28,750 tonnes of black pepper during 2005-06. Black pepper is being cultivated in Kerala (96%), Karnataka (3%) and to a lesser extent in Maharashtra, Andhra Pradesh, Tamil Nadu and North Eastern regions (Anonymous, 2005). In Karnataka it is cultivated in an area of 10,690 ha mainly in Kodagu, Uttara Kannada, Dakshina Kannada, Shimoga and Chickmagalore districts as mixed crop in plantations such as arecanut, cardamom, coffee and rubber with a production of 2360 tonnes during 2005-06 (Anonymous, 2005).

In Uttara Kannada black pepper is mainly cultivated in moist valleys as mixed multistoried crop in arecanut plantations using arecanut palms as standards for training black pepper. This system of cultivation offers considerable scope for organic production of black pepper.

The study was done in farmer's arecanut garden in Sirsi, Uttara Kannada district, Karnataka for five years from 2000–04 by using organic amendments viz., burnt soil, vermicompost, *Azospirillum* and FYM, as well as their combinations. The garden selected was continuously under organic cultivation where no chemicals except prophylactic use of Bordeaux mixture sprays against kole roga of arecanut and foot rot of black pepper were adopted. Burnt soil was prepared by piling litter and woody waste material on the ground over which a thin layer of soil was covered and fire was lit from one side and allowed to burn slowly. This mixture of ash and burnt soil was used as organic manure.

Application of burnt soil along with FYM to vines produced significantly the highest fresh berry yield of 6.74 kg/vine during first year itself as compared to the other treatments. The remaining treatments were at par. Similar trend was recorded where burnt soil with FYM application to vines resulted in significantly higher yield (6.52, 5.94, 6.20 and 6.75 kg/vine) during each year of the study (Table 12.1). The pooled data of five years revealed that fresh berry yield obtained in treatment consisting of FYM 10 kg and burnt soil 10 kg / vine was significantly the highest (6.43 kg/vine) over all the treatments. The results clearly showed that FYM along with burnt soil recorded the highest fresh berry yield. This may be attributed to the readily available nutrient in burnt soil which has led to significant increase in black pepper yields. The pooled data also showed that FYM + leaf manure along with 10 kg soil as practiced by the farmers yielded significantly lower than FYM alone which

Table 12.1. Effect of organic farming in black pepper (2000-01 to 04-05)

Treatments	*Mean fresh berry Yield (kg/ vine)*					*Mean*	*B:C ratio*
	00-01	01-02	02-03	03-04	04-05		
FYM 10 kg/vine + 10 kg soil	5.36	5.84	5.13	4.98	5.73	5.41	3.85
Vermicompost (10 kg) + 10 kg soil	5.06	5.40	5.02	4.19	4.20	4.77	3.77
FYM 10 kg + Burnt soil 10 kg	6.74	6.52	5.94	6.20	6.75	6.43	4.17
FYM 10 kg+ 10 kg soil +*Azospirillum* 50g/vine	5.05	5062	4.72	4.70	4.75	4.97	3.87
FYM 10 kg + 10 kg soil + Leaf manure 10 kg	5.61	5.82	4.09	3.62	4.15	4.66	3.33
Sem +	0.27	0.16	0.18	0.06	0.28	0.19	
CD (P = 0.05)	0.85	0.49	0.55	0.17	0.88	0.56	

might be attributed to the sudden nutrition depletion caused by nutrient competition by decomposing micro-organisms at the time which coincided with the spiking of black pepper. The use of burnt soil is a common practice among the farmers to supplement nutrient as well as to avoid soil borne pathogens and applied for various crops like coconut, arecanut, ginger, and sugar cane in this region. The ash and burnt soil mixture might have contributed to higher yields due to the ready availability of nutrients from burnt soil. The preparation of burnt soil is a simple practice which could be efficiently utilized by the farmers to burn unwanted litter and woody waste materials which otherwise is being burnt for sanitation purposes. The B: C ratio also revealed that benefit obtained was 4.17, which was higher than any of the other treatments. Hence, the practice of application of burnt soil could form a profitable method of manuring of black pepper to obtain higher and sustainable yields organically. Krishnakumar *et al.* (1999) opined that black pepper can be grown quite successfully using organic methods. In Malaysia, forest soil after clearing forest trees for planting of black pepper was used for preparation of burnt soil. A portion of top soil, stripped off from the deforested land was covered on a pile of wood and slow roasting was resorted to for a period of one to three weeks, followed by thorough mixing of sterile soil. The wood ash and the soil provided the burnt soil which had a pH of 7-8 as compared with pH of 4-5 for fresh soil (Sadanandan, 2000). The burnt soil altered the physical characteristic of the rooting medium including reducing the bulk density of soil, increasing the soil pH and release of nutrients (De Waard, 1969). Thus, application of burnt soil provided the vines with readily available nutrients for immediate uptake and resulted in improved yields.

References

Anonymous, 2005. Area, production and productivity of black pepper. *www.ipcnet.org.*

Devasahayam, S., Anandaraja, M., Thankamani, C.K., Saji, K.V. And Jayashree, E., 2006, Black pepper *In : Major Spices-Production and Process* (Eds. Parthasarthy V.A. and Rajeev, P). Indian Institute of Spices Research, Calicut -673012, Kerala.

De Waard, P.W.F. 1969. *Yield Stability and Foliar Diagnosis of Black pepper in Sarawak Malaysia*, Wageningen Univ. Ph.D. Thesis.

Krishnakumar, V., George, C.K. and Potty, S.N. 1999. Organic farming in pepper. *Planters'-Chronicle*, 95(8): 357-61.

Sadanandan, A.K. 2000. Agronomy and nutrition of black pepper. In: Black pepper *Piper nigrum* (Ed. Ravindran P. N.) Harwood Academic Publishers, Amsterdam, pp.166-23.

13

Organic Cultivation of Medicinal Plants as Inter/ Mixed Crops in Coconut Gardens

H.P. Maheswarappa, D.V. Srinivasa Reddy, V. Krishnakumar
P. Subramanian and R. Dhanapal

INTRODUCTION

Coconut (*Cocos nucifera* L.), widely known as 'Kalpavriksha' or " tree of heaven" is an important perennial oil yielding crop of humid tropics and is a versatile crop providing food, medicine, health drink, shelter, fuel, timber and fibre. It is grown in 92 countries and India, Indonesia, Philippines and Sri Lanka are the major growing countries. In India, it is mainly grown in the southern states like Kerala, Karnataka, Tamil Nadu and Andhra Pradesh. Coconut palms planted at the recommended spacing of 7.5m × 7.5m does not utilize fully the available resources such as land space, aerial space, water and nutrients. Active root zone of coconut is confined to 25% of the available land area and the remaining area could be profitable exploited for raising subsidiary crops. In adult plantation of more than 25 years old, about 45–50% of sunlight is wasted without being intercepted by coconut.

Plant-based drugs are available throughout the world in one or the other form. The use of medicinal plants in the health care management programme has not remained under the domain of the developing world, but their use has also been ever increasing in the developed industrialized countries. The interest in the plant medicines in developed countries has been dramatically surged up to 60% since 1989. Even now the market continues to grow at the rate of 15% annually. The increasing demand of medicinal plant based products have created problems of supply of raw materials and one of the major difficulties being faced by the industries and entrepreneurs is that of obtaining sufficient quantities of raw materials of desired quality for the internal consumption as well as for export. To address this situation, measures are needed to promote the cultivation and domestication of medicinal plants.

Performance of different medicinal plants was tried as inter/mixed crops at Central Plantation Crops Research Institute (CPCRI), Kasaragod during 2003–06 in 30 years old coconut garden spaced at 7.5m × 7.5m. The crops were cultivated with only organic manure in the form of vermi-compost. Details of package of practices followed for different crops are given in the Table 13.1. During summer months irrigation was provided with perfo system at IW/CPE ratio of 1.0. As crops vary in their growth habit, the harvesting period also varied. The economic products were harvested as and when ready.

Table 13.1. Details of package of practices followed for different medicinal plants

Crop	*Planting material*	*Duration*	*Spacing*	*Organic manure*
Orila – *Desmodium gangeticum* DC	Seed	8 months	30 x 30 cm flat bed	Vermi-compost : 10t/ha (At the time of sowing)
Moovila – *Pseudarthria viscida*	Seed	8 months	30 x 30 cm flat bed	Vermi-compost : 10t/ha (At the time of sowing)
Chittadalodakam – *Adhatoda beddomei*	Cutting	1½ to 2 years	60 x 60 cm flat bed	Vermi-compost : 15t/ha (two splits)
Karimkurinji – *Nilgiranthus ciliatus*	Cutting	1½ to 2 years	60 x 60 cm flat bed	Vermi-compost : 15t/ha (two splits)
Nagadanthi – *Baliospermum montanum*	Cutting	1½ to 2 years	60 x 60 cm flat bed	Vermi-compost : 15t/ha (two splits)
Panikkorkka – *Coleus aromaticus* Benth	Cutting	Perennial	30 x 30 cm flat bed	Vermi-compost : 10t/ha (two splits)

Among the annuals tried, *coleus aromaticus* could not establish well in the coconut garden. Among the annuals, orila had higher height and number of branches compared to moovila (Table 13.2). Among the biennials, the plant height was more in Nagadanthi and number of branches produced were more in Nilgirianthus.

Table 13.2. Growth characters of medicinal plants (mean of 2 seasons)

Crop	*Plant height (cm)*	*No. of branches/plant*
Orila	86.6	6.4
Moovila	64.3	5.3
Nilgirianthus	107.7	22.1
Chittadalodakam	128.5	5.7
Nagadanthi	206.7	4.5

The mean dry root yield of orila was 764.3 kg/ha, whereas, moovila produced dry root yield of 590.8 kg/ha (Table 13.3). Among the biennials, the dry root and stem yield of Nilgirianthus was 2.45 and 17.34 t/ha, respectively. The dry root and stem yield in the case of Chittdalodakam was 1.73 t/ha and 9.04 t/ha, respectively. The Nagadanthi crop produced 9.26 t/ha of dry roots.

Table 13.3. Economic yield of medicinal plants (kg/ha)

Crop	2003-04		2005-06		Mean	
	Root (DW)	Stem (DW)	Root (DW)	Stem (DW)	Root (DW)	Stem(DW)
Orila	664.3	–	864.3	–	764.3	–
Moovila	570.0	–	611.7	–	590.8	–
Nilgirianthus	2530.0	16903.3	2436.7	17773.3	2483.3	17338.3
Chittadalodakam	1685.7	9007.0	1765.3	9068.7	1725.5	9037.8
Nagadanthi	8868.3	–	9660.7	–	9264.5	–

DW = Dry weight

The cost of cultivation was higher during the first season due to high cost of planting materials. Planting materials were produced in the farm itself in the subsequent years and hence cost towards this item was less. Cost of production was higher for Nilgirianthus crop and was lower for annuals (Table13.4). Among the annuals, orila recorded the highest net return of Rs. 14,929/ha compared to moovila in 8 months duration. Among the biennials, the highest net return obtained was with Nilgirianthus (Rs.1,93,049/ha) in 18 months duration. This was followed by Rs. 1,65,290/ha in the case of Nagadanthi crop. Sujatha *et al.* (2006) also reported higher return from Nilgirianthus when grown as inter-crop in arecanut garden.

Table 13.4. Economics of medicinal plants (Mean of two seasons) (Rs./ha)

Crop	Cost of production	Gross return	Net return
Orila	8,000	22,928	14,929
Moovila	8,000	17,724	9,724
Nilgirianthus	30,000	2,23,049	1,93,049
Chittadalodakam	20,000	1,24,888	1,04,888
Nagadanthi	20,000	1,85,290	1,65,290

The coconut equivalent (copra) of intercrops (Table 13.5) was significantly higher under Nilgirianthus crop and the lowest was with Moovila crop. Inter-cropping of medicinal plants did not affect the yield of coconut rather there was increase in the yield of coconut. The pre-experimental (2002-03) yield of coconut was 103.2 nuts/palm/year, whereas, the yield obtained during 2005-06 was 133.0 nuts/palm/year. From the above study it can be concluded that, growing medicinal crops organically as mixed crops in coconut garden is economically beneficial depending upon the market facility for selling economic produce of the medicinal plants.

Table 13.5. Coconut equivalent (copra) of intercrops yield (kg/ha)

Crop	*2003-04*	*2005-06*	*Mean*
Orila	442.9	576.2	509.5
Moovila	380.0	407.8	393.9
Nilgirianthus	4880.8	5032.6	956.7
Chittadalodakam	2750.7	2799.8	774.1
Nagadanthi	3941.7	4293.2	117.5
CD (P=0.05)	372.56	332.38	67.42

Reference

Sujatha, S., Ravi Bhat, Balasimha, D. and Kannan, C. 2006. Crop diversification in arecanut plantation through intercropping of medicinal and aromatic plants. *J. Plantation Crops* **34**(3): 318-22.

14

On Farm Recycling of Biomass for Sustainable Farming: An Analysis of Farmers' Experiences in the Production and Use of Vermi-compost

C. Thamban, C. Palaniswami and K. Muralidharan

INTRODUCTION

There is a growing awareness among farmers about organic farming. However, lack of availability of quality organic manure is a major difficulty experienced by farmers. Adoption of vermi-composting of organic wastes available in the farm would be a potential way to overcome this difficulty. Technology has been standardised at Central Plantation Crops Research Institute (CPCRI) for the production of vermi-compost as quality organic manure using coconut leaves. Many farmers have started vermi-compost production using on farm wastes including coconut leaves.

Farmers' practices and experiences pertaining to the on farm recycling of biomass through vermi-composting were studied in two gramapanchayats in Kannur District of Kerala State by randomly selecting forty farmers adopting vermi-composting practices. Data on various aspects of vermi-compost production and use were collected from farmers using a pre-tested interview schedule. Besides, samples of vermi-compost were collected from four units representing different substrate combination used for vermicomposting and the nutrient contents were analysed.

The profile of the farmers revealed that majority of them was between 35–55 years of age. All of them were literate. 43% of them were having high school level education while 13% were degree holders. Three-fourth of the farmers was depending on farming alone as their occupation. The profile features indicate that educated, comparatively young and full time farmers are taking up vermicomposting technology as an integral part of sustainable farming practices. It was found that more than 50% of the farmers started the practice of vermi-compost production after the year 2000.

Factors Influencing the Adoption

The perceived benefit of vermi-composting to enhance soil and plant health was the most important factor, which influenced the farmers to adopt vermi-composting technology. Lack of availability of green leaves for manuring crops, motivation by extension personnel of local Krishibhavan, scope to keep the farm clean by removing the biomass for recycling were some other factors, which prompted the farmers to go for vermi-composting. Earlier studies have shown that lack of availability of organic manures and their high cost is a major problem adversely affecting the sustainability of farming and many farmers have started showing keen interest in vermi-composting technology (Thamban *et al*, 2006).

80% of the farmers attended training on vermi-compost production technology. CPCRI and Krishibhavan were the important sources of information about vermicompost production technology. Fellow farmers, farmer organisations like INFARM and private agencies were the other sources of information. 50% of the farmers procured earthworms from fellow farmers while 33% of the farmers obtained the earthworms from CPCRI. Private agencies and NGOs also distribute earthworms to farmers and organise awareness programmes on organic farming practices and vermi-compost production.

Majority (67%) of the farmers used a mixture of coconut leaves, banana leaves and wastes, arecanut leaves, leaf litters, etc., in varying proportions as the substrate for vermi-composting (Table 14.1). Only 33% of the farmers used coconut leaves alone as the substratum. A sizeable proportion of farmers (32%) used less than the recommended quantity (100kg for one tonne substratum) of cow dung for vermicomposting.

Table 14.1. Substrate used for vermi-composting

Substrate	*Number of units*	*Percentage*
Leaf litters + Coconut leaves + Arecanut leaves	9	22
Coconut leaves +Banana leaves and pseudostem	10	25
Coconut leaves + Leaf litter +Arecanut leaves + Banana leaves and pseudostem	8	20
Coconut leaves alone	13	33
Total	40	100

68% of the farmers obtained 200–400 kg vermi-compost per production cycle. In majority (60%) of the cases, 60–75 days were taken for vermi-composting while in 18% cases it took more than 90 days. Majority (65%) of the farmers used the entire quantity of vermi-compost in their own farms while the remaining farmers used part of the vermi-compost in their farms and part sold to other farmers at Rs. 5-6 per kg. Studies have shown that a substantial number of farmers produce vermicompost more than required for their farms and sell surplus to other farmers (Ninawe, 2008). Farmers are benefited not only by the improvement of soil fertility through vermicompost application but also by earning income through the sales of surplus vermicopmpost and earthworms.

The nutrient content of vermicompost produced by farmers in terms of N, P, K and organic carbon were higher for the units in which a mixture of coconut leaves, banana leaves and wastes, arecanut leaves, leaf litters, etc., were used in varying proportions as the substrate for vermicomposting (Table 14.2) when compared to units in which coconut leaves alone were used as substrate. Vermi-compost prepared from Leaf litters + Arecanut leaves + Coconut leaves had the highest N content of 2.19 % with lowest C:N ratio of 8.26. Wherever leaf litter was added as a component substrate, the nutrient contents were high. In general vermi-compost prepared by the farm unit had the favourable C:N ratio for facilitating liberal availability of nitrogen to the crops. Application of 15 kg vermi-compost per palm per annum contributes about 200–328 g N, 15-41 g P_2O_5 and 61–75 g K_2O to the coconut palm.

Table 14.2. Nutrient content of vermicompost prepared from different substrates

Substrates	*N (%)*	*P_2O_5 (%)*	*K_2O (%)*	*Organic carbon (%)*	*C: N ratio*
Leaf litters + Arecanut leaves + Coconut leaves	2.19	0.28	0.49	18.08	8.26
Coconut leaves + Banana leaves and pseudostem	1.33	0.11	0.45	17.76	13.36
Coconut leaves + Leaf litter + Arecanut leaves + Banana leaves and pseudostem	2.03	0.25	0.51	20.47	10.07
Coconut leaves alone	1.34	0.22	0.41	17.15	12.80

Farmers experience many constraints in vermi-compost preparation and use (Table 14.3). As per farmers' perception, the most important constraint experienced in the production of vermi-compost was damage to earthworms caused by ants and rats (77%). High cost for constructing cement tank, lack of development schemes to provide incentives for constructing cement tank and difficulty in marketing the surplus vermi-compost were the other important constraints pointed out by farmers. Lack of knowledge about vermicomposting technology was also cited as a constraint by 20% of the farmers, which indicate the necessity for organizing appropriate training programmes on vermi-composting technology for farmers.

Table 14.3. Constraints experienced by farmers in vemi-composting

Sl. No.	*Constraints*	*Number of farmers*	*Percentage*
1.	Damage to earthworms caused by ants and rats	31	77.00
2.	High cost for constructing cement tank	28	70.00
3.	Lack of incentives for construction of tank	24	60.00
4.	Difficulty in marketing the surplus vermi-compost	14	35.00
5.	Transportation of biomass to the vermi-composting tank	13	33.00
6.	Lack of knowledge about vermi-composting technology	8	20.00
7.	Difficulty in regulating the moisture content in the unit	6	15.00
8.	Damage to earthworms caused by flatworms (*Bipalium sp*)	5	12.00

The perceived benefits of vermi-composting to enhance soil and plant health were the most important factor, which influenced the farmers to adopt vermicomposting technology. The results of the study indicate the necessity for implementing suitable development schemes to promote vermicomposting technology among farmers and also for organising training programmes to educate the farmers about the scientific methods of vermi-compost preparation. There is also a need for standardising the methods of vermicompost preparation using different combination of substratum including coconut leaves and other farm wastes.

References

Ninawe, A.S. 2008. Vermicomposting for building soil fertility. *LEISA INDIA*. **10**(2): 21.

Thamban, C., Sairam, C.V., Muralidharan, K., Krishnakumar, V. and Arulraj, S. 2006. Multidimensional analysis of coconut based farming systems adopted by farmers. *Journal of Plantation Crops*. **34**(3): 715-19.

15

Effect of Different Nutrient Enriched Compost on Cashew Root Stock

Mini Poduval and Tapas Biswas

INTRODUCTION

Successful graft production in horticultural crops largely depends upon healthy and vigorous rootstocks besides good scion material, season and grafting skills. Any attempt made to improve the quality of rootstocks not only increases the percentage graft success but also reduces the cost of nursery management. The common practice for raising cashew rootstock is the use of potting mixture consisting of soil, FYM and sand in a ratio 1:1:1. Inorganic fertilizers are also used as nutrient sources. However, nutrient content of FYM is very meager (N –0.05%, P_2O_5 –0.25%, K_2O –0.25%) (Reddy and Reddy, 1999). On the other hand, chemical fertilizers impart many undesirable effects. Cashew rootstocks raised on conventional rooting medium attain maturity for grafting in about 60–75 days or sometimes even 90 days. In red and laterite zone of West Bengal, grafting period is limited only for four months (June – September) and the delay in rootstock maturity further reduces the grafting season. Application of nutrient enriched organic manure is known to improve soil physical and chemical parameters such as texture, structure, fertility etc. and there by improve seed germination, root proliferation and ultimately rootstocks raised on such medium attain early optimum maturity for grafting. Micro-organisms are of significance in the intricate biological processes occurring in soils. The fertility of soil depends not only on its chemical composition but also on the qualitative and quantitative nature of micro-organisms inhabiting it (Rao, 2001).

The study was conducted at the Regional Research Station, Bidhan Chandra Krishi Viswavidyalaya, Jhargram, West Bengal during 2004–06 Kharif season. Three different composts produced from cashew leaf litter, groundnut waste and general farm waste were utilized as a source of nutrient enriched phospho- compost for rising of cashew rootstocks. The nutrient content of various composts and FYM is given in Table 15.1.

Table 15.1. C:N ratio and nutrient content of different organic manures

Organic manure	*C:N ratio*	*Nitrogen (%)*	*Phosphorus (%)*	*Potassium (%)*
Ground nut waste compost	12.0	5.4	2.5	1.40
Cashew leaf compost	15.0	2.34	4.0	0.50
General farm waste compost	17.0	2.1	0.50	1.80
FYM	18.0	0.5	0.25	0.5

Polythene bags of 22 cm × 15 cm size of 300-gauge thickness with 30 holes per bag were used for raising the rootstocks. The poly bags were filled with the pot mixture containing equal quantities of sand, soil and organic manures. Four different sources of organic manures were used for preparing different potting mixtures. Pre-soaked seeds of Red Hazari variety were selected and sown in the center of the polybags at about 2–3 cm below the soil surface during kharif season. Regular weeding and watering of polybags were attended and prophylactic plant protection measures were taken at an interval of 15 days. No chemical fertilizers were applied during the period of study.

Maximum germination was noticed under Sand + Soil + General Farm Waste Compost. Different compost materials irrespective of type increased the germination percentage over control (Table 15.2). Higher germination percentage is very important for attending early maturity for grafting. The increased germination percentage due to different compost materials could be attributed to production of biologically active substances by micro-organisms present in the compost (Rajaswami and Bagyaraj, 1993; Verma, 1993). Significant increases were noticed in plant height, stem girth, number of leaves /plant and root length over control. The increase could be attributed to the improved soil structure and texture supporting easy root penetration thereby providing easy access to nutrients resulting in increased nutrient uptake. Micro-organisms present in compost produced biologically active metabolites like B Vitamins, indole acetic acid, gibberellins and plant growth in general. Nutrient enriched compost also supplied sufficient nutrients for the cashew seedlings to grow better than FYM alone.

Potting mixture with cashew leaf compost had maximum positive effect on shoot length. The plants provided with nutrient enriched composts attained the required stem girth of 2–2.5cm within 73 days after sowing. But still the plants grown on FYM did not attain the required stem girth. Increased stem girth due to application of composts has significance in success rate of grafts. Treatments using various composts are almost at par in terms of their response on leaf production per plant. In the case of leaf area, a peculiar trend was noticed. During the initial period, the response of groundnut waste compost on leaf area was not

Table 15.2. Growth parameters of cashew root stock under different organic manures

Treat-ments	*Germin-ation (%)*	*Shoot length(Cm)*				*Stem girth(cm)*				*Number of leaves/plant*				*Leaf area(cm²)*			
		10 DAS	31 DAS	52 DAS	73 DAS	10 DAS	31 DAS	52 DAS	73 DAS	10 DAS	31 DAS	52 DAS	73 DAS	10 DAS	31 DAS	52 DAS	73 DAS
T1	87	8.66	15.89	20.17	37.85	1.19	1.45	1.82	2.29	4.9	8.5	11.2	18.6	20.80	33.60	42.22	77.73
T2	90	10.18	18.72	22.75	37.40	1.41	1.48	1.82	2.17	4.6	9.2	11.8	17.1	25.15	42.88	52.16	75.46
T3	89	7.39	19.37	24.90	37.95	1.26	1.44	1.64	2.10	4.6	8.9	11.4	16.4	22.76	43.96	51.82	79.73
T4	84	8.60	13.87	16.30	22.95	1.26	1.44	1.49	1.92	4.6	7.0	9.9	14.3	19.08	28.49	36.06	60.79
S. Em+		1.053	0.0192		0.5831		3.8332										
CD(P=0.05)		2.104	0.0384		1.164		7.657										

T1 = Sand + Soil + Groundnut waste compost; T2 = Sand + Soil + General Farm Waste Compost T3 = Sand + Soil + Cashew Leaf Compost; T4 = Sand + Soil + FYM

Table 15.3. Effect of organic manures on root length, shoot and root dry weight of cashew rootstock

Treatments	*Root length (cm)*				*Shoot dry weight (g)*			*Root dry weight (g)*		
	10 DAS	*31 DAS*	*52 DAS*	*73 DAS*	*31 DAS*	*52 DAS*	*73 DAS*	*31 DAS*	*52 DAS*	*73 DAS*
Sand + Soil + Groundnut waste compost	7.76	14.44	29.00	35.92	0.918	1.485	2.433	0.266	0.49	0.552
Sand + Soil + General Farm Waste Compost	5.74	14.56	27.50	32.22	0.758	1.695	2.534	0.188	0.331	0.485
Sand + Soil + Cashew Leaf Compost	9.00	16.16	28.40	35.84	0.889	1.283	2.433	0.335	0.445	0.856
Sand + Soil + FYM	11.13	15.22	21.88	26.10	0.838	1.505	1.876	0.17	0.383	0.626
S. Em+	2.268	0.138	0.098							
CD(P=0.05)	4.53	0.277	0.198							

comparable with the response of general farm waste compost and cashew leaf compost. But at 73 days after sowing, all the treatments were at par in terms of their response on leaf area. This might be due to the delay in some nutrient availability in the groundnut compost, which indirectly delayed the increase in leaf area. Root growth was far better in the cashew rootstocks raised on compost medium than the FYM pot mixture. Shoot dry weight at 73 days after sowing was better in treatments using various composts than FYM (Table 15.3).

Among the treatments, the rate of CO_2 evolution after seven days of incubation was the lowest in the case of ground nut waste compost (177.78 mg carbon kg $^{-1}$soil (Table 15.4). The rate of CO_2 evolution from general farm waste compost was initially much faster than that from other treatments. Over 50% of the carbon added as different manurial source had been respired by 14 days. The increased CO_2 evolution during initial period of incubation may be attributed to the increased accessibility of moderately available carbon to microbial attack (e.g., Cellulose, hemicelluloses). The accessibility of plant carbon to microbial attack is increased by a greater surface area and by reduction in the protection of available carbon compounds by recalcitrant plant constituents such as cutin and lignin.

Table 15.4. Carbon dioxide evolution by microorganisms within different rooting media

Treatments (Organic manure)	*Carbon dioxide evolution (mg kg $^{-1}$ soil)*			
	7 Days	*14 Days*	*21 Days*	*28 Days*
Sand + Soil + Groundnut waste compost	177.78	341.28	485.84	589.63
Sand + Soil + Cashew leaf compost	184.44	353.94	546.97	604.21
Sand + Soil + General farm waste compost	189.88	368.45	529.51	669.63
Sand + Soil + FYM	186.43	358.03	530.62	665.68
S. Em +	0.988	2.682	10.898	16.513
CD(P = 0.05)	2.418	6.562	26.667	40.408

Table 15.5. Effect of different organic manures on rate of increase of growth parameters over a period of time

Treatments	*Increase in shoot length (cm)*			*Increase in stem girth (cm)*			*Increase in number of leaves/plant*			*Increase in leaf area (cm²)*			*Increase in root length (cm)*		
	10-31 days	*32-52 days*	*53-73 days*	*10-31 days*	*32-52 days*	*53–73 days*	*10-31 days*	*32 -52 days*	*53–73 days*	*10-31 days*	*32–52 days*	*53–73 days*	*10-31 days*	*32–52 days*	*53-73 days*
T1	7.23	4.28	17.66	0.084	0.118	0.15	3.6	2.7	7.40	14.994	8.596	35.512	6.68	14.56	6.92
T2	8.54	4.03	14.62	0.052	0.108	0.11	4.6	2.6	5.30	17.728	9.282	23.294	8.82	12.94	10.12
T3	10.37	5.53	12.43	0.06	0.062	0.148	4.3	2.5	5.70	23.294	21.22	7.862	7.16	12.24	7.44
T4	5.27	2.33	5.45	0.06	0.024	0.116	2.4	2.9	4.40	9.41	7.442	24.738	4.086	6.662	4.22
S. Em+	0.677	0.026	0.544	5.332	2.529										
CD(P=0.05)	1.367	0.053	1.099	10.776	5.11										

Rooting medium using Cashew leaf compost has been graded as the best material amongst the four as soil organic matter builder followed by FYM and General farm waste compost; Groundnut waste compost being least effective in this respect. Therefore, it can be derived that during the initial period, i.e., within first 28 days, rooting medium with cashew leaf compost is the best to impart positive effect on cashew rootstock.

For all the treatments, it was noticed that the rate of increase in all the growth parameters except root length was more during the initial 21 days and also during 53–73 days (Table 15.5). The rate of increase was low during 32–52 days. There were significant differences among the treatments in terms of response on increasing the rate of growth. During the initial period, maximum rate of increase in shoot length (10.37cm) was observed with Cashew leaf compost. Later on, the rate of increase in shoot length was maximum (17.66 cm) with Groundnut waste compost. Rate of increase in girth and number of leaves per cashew rootstock was the highest with Groundnut waste compost (23.294 cm^2). During 53–73 days after germination, the rate of increase in leaf area was maximum (35.512 cm^2) with groundnut waste. Rate of increase in root length was maximum during 32–52 days after germination and groundnut waste compost had the maximum positive effect (14.56cm) on rate of increase of root length.

Therefore, study of the parameters recognized that nutrient enriched composts are the better substitute for FYM for raising cashew rootstock. It can be concluded that the rooting medium with Groundnut waste compost is the best for increasing the growth rate of rootstocks at the optimum period for grafting. Rooting media comprising Cashew leaf litter compost as well as Groundnut waste compost were found to be the best for getting robust, optimum tall, healthy root stocks at an early date.

References

Dubey, R.C. and Maheswari, D.K. 2002. *Practical Microbiology*, Pub: S. Chand & Company Ltd., Ramnagar, New Delhi, p. 192.

Panse, V.G. and Sukhatme, P.V. 1989. *Statistical Methods for Agricultural workers*, ICAR, New Delhi, p. 97-128.

Rao, Subba 2001. *Soil Microbiology*, Pub: Oxford & IBH publishing So. Pvt. Ltd., New Delhi, 26p.

Reddy, T.Y. and Reddy, Sankara G.H. 1999. *Principles of Agronomy.* Pub: Kalyani publishers, New Delhi, p. 219.

SECTION – IV
ORGANICS *VERSUS* INORGANICS

16

Evaluation of Organic Management System for Sustainable Production of Turmeric

V. Srinivasan, K.P. Sangeeth, M. Shamsudeen, C.K. Thankamani, S. Hamza
T.J. Zachariah and A. Kumar

INTRODUCTION

In India, the total production of spices is around 3.94 lakh tonnes among which, about 0.3 million tonnes (8–10%) is exported. The Indian share of the world trade in spices is 45–50% by volume and 25% in value. Turmeric is cultivated in more than 20 states with an area of 1.86 lakh ha with a production of 8.37 lakh tonnes during 2006-07 (DASD, 2008). Andhra Pradesh, Tamil Nadu, Orissa, Karnataka, West Bengal, Gujarat, Meghalaya, Maharashtra, Assam are some of the important states cultivating turmeric, where, Andhra Pradesh alone occupies 35% of area and 47% of production. The national productivity of the crop is 4501 kg ha^{-1} and wide variations in productivity are observed across the states. India exports only 6.48 % of its production of turmeric to more than 50 countries mainly as dry produce (63%) and powder (37%). The share of organic turmeric is only 11% compared to conventional turmeric. There is a great demand for organic turmeric in USA, Germany, France and Japan (ITC, 2004). Growing demand for natural colours in industry, fast food chains and pharmaceuticals offer a potential scope for organic production of turmeric.

Field experiments on turmeric cv. Alleppy Supreme were conducted in the experimental farm of Indian Institute of Spices Research at Peruvannamuzhi during 2004–06. Different management systems viz., fully organic (MP1) (with 20t FYM + 2t Neem cake + 1t Ash + 4t Vermicompost/ha, *Azospirillum* and Phosphate Solubilising bacteria, *Trichoderma* and *Pseudomonas* sp. as seed treatment and also as bimonthly soil application and spray of Bordeaux Mixture (1%) and neem oil (5 ml/l) for disease and pest control), fully inorganic

[MP2 with 100% inorganic recommended dose of fertilizers, seed treatment with Dithane M-45 at recommended dosage, spraying and drenching of mancozeb (0.3%) and spraying of malathion (0.1%) for control of pest and diseases] and Integrated system (MP3 with 20t ha^{-1} FYM + ½ N, P, K at the recommended levels of 60:50:120 kg/ha + P- Solubilising bacteria and spray with mancozeb and neem oil and malathion for disease and pest control). The experiments were laid out in blocks with an area of 75 m^2 each. Live fencing of the organic area with *Glyricidia* was maintained to avoid drift of chemicals between systems. Turmeric was planted in 3 × 1m^2 beds having 40 plants each. Full dose of organics were applied as basal and the application of bio-fertilizers and biocontrol agents were done at bimonthly intervals. The N (as urea) and K (as muriate of potash) were applied in two equal doses on 45 and 90 days after planting (DAP) in inorganic and integrated systems. The prophylactic spray of plant protection materials were done on 45 DAP and at monthly intervals after from June—September.

Results indicated that soil pH and organic carbon status did not differ significantly among the management systems at 120 DAP during 2004-05. The nitrogen content was the highest in integrated system and at par to organic management with the lowest under inorganic system. Potassium availability was significantly high under inorganic system. Significantly higher availability of phosphorus, calcium and magnesium was noticed in organic management. Among the micronutrients, the highest manganese and zinc availability was noticed in integrated system, whereas, iron and copper did not vary significantly among the treatments. In 2006-07, soil pH and organic carbon status did not differ significantly among the management systems at 120 DAP. The nitrogen content was the highest in integrated system at par to organic management with the lowest under inorganic system. Potassium availability was significantly high under inorganic system. Significantly higher availability of phosphorus, calcium and magnesium was noticed in organic management. Among the micro-nutrients highest manganese and zinc availability was noticed in integrated system where as iron and copper did not vary significantly among the treatments.

The soil nutrient availability after harvest of the crops showed increased soil pH to 5.4 under organic system (2004-05) compared to inorganic and integrated systems. Higher phosphorus, potassium, calcium and magnesium contents were noticed in organic and inorganic systems as compared to integrated system. Micro-nutrient availability got increased under all managements over the period. During 2005-06 also, the highest pH of 4.97 was recorded in the organic system which was on par with that of integrated system. Maximum phosphorus and potassium was recorded in the integrated system and the calcium availability was more in the organic system with the lowest in the inorganic system. Organic system recorded the maximum magnesium and copper availability. Similarly, during 2006-07 also the highest pH of 5.06 was recorded in the integrated system which was on par with that recorded in organic and inorganic system. Nitrogen content was more in the organic system and lowest in the inorganic system. Organic system recorded the maximum magnesium content. The iron content was higher in organic whereas integrated system showed highest copper content.

The overall comparison over years also showed that soil pH status did not differ significantly among the management systems at 120 DAP. But an increased soil pH under organic system compared to inorganic and integrated systems at harvest was noticed. Similar higher soil pH under organic farms over conventional management was observed in South Wales by Derrick and Dumaresq (2000). The organic carbon, nitrogen, phosphorus and potassium content were the highest in integrated system and on par to organic management at 120 DAP. Phosphorus and potassium followed the same trend during harvest also. Nitrogen did not vary significantly among the treatments. Significantly higher availability of calcium magnesium and zinc was noticed in organic management both during 120 DAP and harvest. Iron content did not vary significantly among the treatments (Tables 16.1 and 16.2).

Table 16.1. Overall comparison of soil status under different systems of turmeric (2004–07) at 120 DAP

Treatments	*pH (1:2)*	*OC%*	*N*	*P*	*K*	*Ca*	*Mg*	*Fe*	*Mn*	*Zn*	*Cu*
							ppm				
Organic	5.09	2.01	150.4	8.65	198	493.6	160.8	48.4	6.32	0.78	1.19
Inorganic	4.97	2.09	152.0	7.66	265	396.3	123.0	43.9	8.46	0.64	1.54
Integrated	4.82	2.15	168.7	10.3	273	426.7	140.8	49.1	8.96	0.78	1.35
CD(P = 0.05)	NS	0.11	11.3	1.8	33	62.5	21.7	NS	1.3	0.08	0.17

NS = Not Significant

Table 16.2. Overall comparison of soil status under different systems (2004–07) at harvest

Treatments	*pH(1:2)*	*OC%*	*N*	*P*	*K*	*Ca*	*Mg*	*Fe*	*Mn*	*Zn*	*Cu*
							ppm				
Organic	5.16	1.83	154.3	2.49	212	597.3	96.0	48.7	9.87	1.0	1.13
Inorganic	4.65	1.75	149.7	2.02	213	397.9	73.3	51.5	9.74	0.75	1.22
Integrated	4.80	1.93	146.5	2.68	231	431.6	83.3	48.6	9.77	0.85	1.17
CD(P = 0.05)	0.17	0.07	NS	0.53	NS	67.8	12.3	NS	NS	0.12	NS

NS = Not Significant

The overall data on microbial population showed that the total bacteria, *Azospirillum*, Pseudomonas, actinomycetes and trichodermal population was the highest in the organic system. The fungal population was lower in inorganic system, followed by organic system. The population of Phosphobacteria was the highest in integrated system and was comparable with that of organic system (Table 16.3). The presence of vermi-compost, neem cake and farmyard manure enhances the microbial population of organic system. The overall population dynamics under organic system was comparable with that of integrated system which contributes significantly to yield, nutrient and quality parameters. Application of coir pith compost, *Azospirillum*, Phosphobacteria and VAM induced maximum IAA oxidase and peroxidase activity in turmeric (Velmurugan *et al*., 2002). Mohan *et al*. (2004) recorded a linear response of turmeric growth, yield and quality of turmeric with inoculation of *Azospirillum* in combination with N levels compared to that of *Azotobacter* under Karnataka conditions. No incidence of rhizome rot was observed in all the treatments over the years. The organic and integrated system showed only up to 2.5 % shoot borer attack as compared to chemical control.

During the first year (2004-05), the fresh rhizome yield was significantly highest in the integrated management system followed by inorganic system with the lowest yield in organic system. Yield reduction of 13.6 and 25.5% was observed under organic management as compared to inorganic and integrated systems, respectively. During 2005-06, significantly high fresh rhizome yield of 33.38 t/ha was noticed in the treatment where integrated management practices are followed (Table 16.3). This was followed by 23.68 t/ha in the organic treatment which was on par with inorganic treatment (22.44 t/ha). In 2006-07 also, significantly high fresh rhizome yield of 29.98 t/ha was noticed in the treatment where integrated management practices are followed. This was followed by 26.60 t/ha in the organic treatment on par with the yield recorded by inorganic treatment (26.22 t/ha) when repeated at the first years location after raising a legume. Living mulches of legumes as cover crops can fix atmospheric nitrogen and improve soil quality (Sjoerd *et al*., 2004) and thereby yield. Application of FYM increases the organic carbon, available P and effective cation exchange capacity content in the soil thereby improving the fertility which inturn improves the fresh rhizome yield. Similar increase in plant height, number and weight of mother, primary and secondary rhizomes and fresh rhizome yield and increased nitrogen, phosphorus and potassium contents in leaves and rhizome with farmyard manure application was reported by Gill *et al*. (2004). The overall soil fertility build-up under organic system was comparable with that of integrated system. In turmeric, combined application of different organic sources like FYM + pongamia oil cake + Neem oil cake + Stera meal + Rock phosphate + wood ash has yielded on par to the conventional practice in addition with high quality (AICRPS, 2006). Application of vermi-compost at 10t ha^{-1} increased the rhizome yield from 6.7–25.5 % in small cardamom (Vadiraj *et al*., 1998).

Table 16.3. Different management systems on yield of turmeric

Treatment	*Yield (t ha^{-1})*		
	2004-05	*2005-06*	*2006-07*
Organic	31.0	23.68	26.60
Inorganic	35.9	22.44	26.22
Integrated	41.7	33.38	29.98
CD (P=0.05)	3.09	3.99	2.02

Among the quality aspects, significantly highest oil content was recorded in organic system on par with integrated system in 2004-05, but during next years oil content did not vary significantly between treatments. Inorganic system recorded the lowest oil. Oleoresin content also followed the same pattern with the highest content in organic management. During 2005–07 also the oleoresin content was the high under organic management on par with that of integrated management. Curcumin content was highest under the integrated system (5.49%), on par with organic system during 2004-05 and in the following years also significantly highest curcumin levels were observed under organic system on par with that of integrated system. Significantly highest starch content of 43.9% was noticed in 2004-05 under the organic system and it was on par with that of integrated system (39.7%) during 2005-06 (Table 16.4). The application of groundnut cake (1.1 t/ha) significantly increased the dry yield and the highest curcumin production on par to neem cake application (2.5 t/ha) (Sadanandan and Hamza, 1998). The differences in the management of soil fertility also affects soil dynamics and plant metabolism which results in quality.

Table 16.4. Quality parameters as affected by different management systems over the years

Management	*2004-05*		*2005-06*		*2006-07*	
Quality	*Oleoresin (%)*	*Curcumin (%)*	*Oleoresin (%)*	*Curcumin (%)*	*Oleoresin (%)*	*Curcumin (%)*
Organic	19.46	5.39	15.26	5.40	12.67	6.15
Inorganic	16.50	4.56	12.79	5.09	11.65	5.55
Integrated	18.71	5.49	14.99	4.81	12.47	5.85
CD (P=0.05)	0.87	0.51	0.89	0.34	0.94	0.50

The economic evaluation of the various management systems revealed that the integrated and inorganic systems recorded the highest B:C ratio of 1.83 and 1.87 followed by the organic system recording 1.50, with the prevailing price of Rs. 50 per kg of dry produce. If a 20% premium price is considered for the organically grown produce, the B:C ratio of organic system will be 1.78. If the organic inputs like FYM, vermi-compost and coir compost are internally produced within the farm itself, the B:C ratio will become 1.79 in the case of organic system. A case study on the sustainability of three apple production systems showed price premiums of 12% for the organic system to match the break-even point of the conventional system (John *et al.*, 2001).

Organic farming is an alternative agriculture to conventional methods of crop production for sustaining soil productivity and ecological balance. The overall results showed the positive influence of organic cultivation on the quality aspects of turmeric and also on the soil fertility build-up, even though the yield returns were higher under integrated management. An increased yield from third year of conversion and quality and microbial abundance reflects better nutrition of the crop under organic mode. Our results showed that organic and integrated systems are not only better for soil and the environment than their conventional counterpart but have comparable yields and quality. Premium price is the major constraints for marketing of organic turmeric. Currently, growers of more sustainable systems may be unable to maintain profitable enterprises without economic incentives such as price premiums for organic products. The organic turmeric cultivation will be a sustainable venture when a premium price for the product is assured.

References

DASD. 2008. Arecanut and Spices Database. Directorate of Arecanut and Spices Development, Calicut. 110p.

Derrick, J.W and Dumaresq, D.C. 2000. Soil properties under organic and conventional management in south wales. *Australian Journal of Soil Research.* **37**(6):1047-56.

Gill, B.S., Simerjit Kaur, Saini, S.S. 2004. Influence of planting methods, spacing and farmyard manure on growth, yield and nutrient content of turmeric (*Curcuma longa* L.). *Journal of Spices and Aromatic Crops,* **13**(2):117-120.

ITC. 2004. Marketing Manual and Web Directory for Organic Spices, Herbs and Essential Oils. International Trade Centre (UNCTAD/WTO), Genenva.

John, P.R., Jerry D.G., Preston K.A and Herbert R.H. 2001. Sustainability of three apple production systems. *Nature* **410**:926-29.

Mohan, E., Melanta, K.R., Guruprasad, T.R., Herle, P.S., Gowda, N.A.J and Naik, C.M. 2004. Effects of graded levels of nitrogen and biofertilizers on growth, yield and quality in turmeric (Curcuma domestica Val.) cv. D.K. Local. Environment and Ecology **22**(3):715-19.

Sadanandan, A.K., Kandiannan, K. and Hamza, S. 1998. Soil nutrients and water management for sustainable spices production. Proc. Water and nutrient management for sustainable production and quality of spices. (Eds. Sadanandan, A.K., Krishnamurthy, K.S., Kandiannan, K. and Korkanthimath, V.S.). Indian Society of Spices, Calicut, India. pp. 12-20.

Sjoerd, W.D. and Nathan, L.H. 2004. Living mulches of legumes in imidazolinoline resistant corn. *Agronomy Journal.* **96**:1021-28.

Vadiraj, B.A., Siddagangaiah and Potti, S. N. 1998. Effect of vermicompost on the growth and yield of turmeric. *South Indian Horticulture.* **46**(3/6):176-79.

Velmurugan, M., Chezhiyan, N and Jawharlal, M. 2002. Effect of biofertilizers on physiological and biochemical parameters of turmeric cv. BSR 2. *Proceedings of the National seminar on Strategies for increasing production and export of spices,* 24-26 Oct, Indian Society for Spices, Calicut, pp.138-45.

17

Agronomic, Nutritional and Economic Implications of Organic Elephant Foot Yam Production

G. Suja and S. Sundaresan

INTRODUCTION

Presently alternative agricultural systems that are less exploitative, less dependent on non-renewable fossil fuels like fertilizers, pesticides, etc. and that can conserve the precious soil and water resources are gaining popularity (Gliessman, 1998). This interest has resulted in part from studies documenting negative impacts of conventional agriculture on long-term profitability and resource use, including decline in soil organic matter levels, poor surface water quality, reduced water infiltration rates and finally unfavourable soil tilth (National Research Council, 1989). Such system entails large chemical inputs especially fertilizers but few carbon additions to the soil and often contribute to decline in soil quality (Susan Andrews *et al.*, 2002). Organic farming is an alternative farming strategy that focuses on soil health, environmental protection and human health by largely excluding the use of synthetic chemicals and with maximum use of on-farm generated resources. Though the use of chemical inputs cannot be altogether avoided, their use in agriculture has to be rationalized. Reduced energy use and CO_2 emissions are the other benefits of organic farming (Reganold *et al*, 2001; Stockdale *et al.*, 2001). It offers opportunities for employment generation, waste recycling and export promotion. The high quality, nutritious, clean and safe organic foods fetch a higher premium price in world markets.

Tropical tuber crops, including cassava, sweet potato, yams and aroids form the most important staple or subsidiary food to about 500 million global population. The tropical tuber crops are well adapted to low input agriculture (Nayar and Suja, 2004). Tuber crops in general and aroids in particular, like elephant foot yam, respond well to organic manures and there is ample scope for organic production in these crops (Suja and Nayar, 2006; Suja *et al.,* 2006a; Suja *et al.*, 2006b). In addition there is a great demand for organically produced vegetables, particularly aroids and yams, among affluent Asians and Africans living in Europe, United States of America and Middle East. Research and development on organic farming of tropical tuber crops have been less focused and is at a slow pace.

Studies were conducted at Central Tuber Crops Research Institute, Sreekariyam, Thiruvananthapuram, Kerala from 2004–08 for the organic production of high quality tuberous vegetables, mainly aroids (elephant foot yam, taro and tannia), and yams (greater yam, lesser yam and white yam) and eco-friendly technologies were evolved by evaluating various production systems viz., conventional, traditional, organic farming and using biofertilizers. Two crops of green manure cowpea was raised and incorporated during 2003-2004 prior to the initiation of the studies and the land was not subjected to any chemical inputs for two years prior to the study. The fertility status of the soil during the pre-experimental period is indicated in Table 17.1.

Table 17.1. Soil fertility status prior to experimentation

Stage of soil sampling	*Nutrient status of the soil*			
	Organic C (%)	*Available N (kg ha^{-1})*	*Available P_2O_5 (kg ha^{-1})*	*Available K_2O (kg ha^{-1})*
Prior to green manuring (2003)	1.026	329.38	79.64	253.49
After green manuring (pre-experiment nutrient status – 2004)	1.319	255.61	142.06	527.80

In "conventional plots" the package of practices recommendations (FYM @ 25t ha^{-1} + NPK @ 100:50:150 hg ha^{-1}) was advocated. Farmers' practice of using FYM (35t ha^{-1}) and ash (3t ha^{-}) was followed in "traditional plots". In "organic farming plots", FYM @ 35t ha^{-1}, green manure cow pea (to yield 20–25 t ha^{-1} of green matter), ash @ 3t ha^{-1} and neem cake @ 1t ha^{-1} were applied to substitute chemical fertilizers. Proper care was taken to avoid the use of any chemical inputs in organic plots. FYM @ 25t ha^{-1}, mycorrhiza, @ 5 kg ha^{-1} *Azospirillum* @ 3kg ha^{-1} and phosphobacterium @ 2.5kg ha^{-1} were applied in plots assigned for evaluating the effect of biofertilizers. Organically produced corms of a local variety of elephant foot yam, with good market preference and excellent cooking quality, procured from Peermade Development Society, Pothupara, Idukki district, Kerala, was used for the study. Field culture as per the package of practices recommendations was followed (KAU, 2002). Throughout the growth period, the crop was monitored for the occurrence of pest and disease incidence, if any.

Canopy growth characters viz., plant height, leaf spread and girth of pseudostem, observed at 3, 6 and 9 MAP in both the years were not significantly influenced by the various production systems. However, organic plants showed luxuriant growth, which may be due to better moisture retention and improved soil properties under organic manuring (Table 17.2). Mahapatra *et al.* (2006a) also reported that growth parameters of rice improved under organic management.

Table 17.2. Impact of production systems on growth characters of elephant foot yam

Production systems	*Plant height (cm)*			*Leaf spread (cm)*			*Pseudostem girth (cm)*			*Collar rot incidence (%)*
	2004-05	2005-06	2006-07	2004-05	2005-06	2006-07	2004-05	2005-06	2006-07	
Conventional	71.47	76.33	69.60	109.60	119.53	108.67	11.98	12.80	13.13	20.73 (27.08)*
Traditional	73.73	73.80	71.87	108.40	119.4	97.40	11.83	12.53	10.90	11.88 (20.16)
Organic	78.20	82.00	67.67	125.00	134.47	99.73	12.86	12.67	11.37	11.33 (19.68)
Using biofertilizers	74.33	73.40	64.47	105.80	117.07	100.27	11.67	12.66	11.87	15.27 (23.00)
CD (P=0.05)	NS	NS	NS	13.15	12.61	7.967	NS	NS	1.057	NS

NS = Not Significant

*Figures in parentheses are transformed means in angles

During the first year, the incidence of collar rot was almost the same under the different production systems. But organic and traditional plots showed lower incidence (Table 17.2). During the subsequent years of study, inoculation of farmyard manure using *Trichoderma* culture could further lower the disease incidence.

After three years of experimentation, organic plots showed significantly higher available P status and slightly higher pH, organic C, available N and K status (Table 17.3). Increases in soil organic matter, soil pH, available P and K have been measured in some organic systems (Scow *et al.,* 1994: Clark et al., 1998).Similar results have been reported in other crops under Indian conditions as well (Srivastava *et al.*, 2006 and Mahapatra *et al.*, 2006 b).

The physical properties of the soil viz., bulk density, particle density and water holding capacity remained unaltered under the influence of the various production systems after three years. However, bulk density was slightly lower and water holding capacity and porosity slightly higher in the organic plots (Table 17.4). Radhakrishnan *et al.* (2006) also observed similar results in tea gardens under the influence of organic farming. As reported by Stockdale *et al.* (2001) changes in organic matter drive underpin many of the other changes in soil biological and physical properties. The increased organic matter content of soil as evidenced from higher organic C status in organic plots might have resulted in the formation of stable soil aggregates with the resultant slight decrease in bulk density and increase in water holding capacity. The soil microbial activity remained almost the same in the various treatments.

Table 17.3. Effect of production systems on soil chemical properties at the end of three years

Production systems	*pH*	*Organic C*	*Available N (kg ha^{-1})*	*Available P_2O_5 (kg ha^{-1})*	*Available K_2O (kg ha^{-1})*
Conventional	5.137	1.269	184.491	55.207	378.480
Traditional	5.505	1.322	192.801	80.628	372.736
Organic	5.668	1.343	222.774	103.325	491.008
Using Biofertilizers	5.106	1.266	194.463	75.181	347.200
CD (P = 0.05)	0.389	NS	NS	29.155	NS

NS = Not Significaant

Table 17.4. Effect of production systems on soil physical properties at the end of 3 years

Production systems	*Dry bulk density(g cm^{-3})*	*Wet bulk density (g cm^{-3})*	*Particle density (g cm^{-3})*	*Water holding capacity (%)*	*Porosity (%)*
Conventional	1.799	1.626	2.458	13.250	33.292
Traditional	1.777	1.642	2.515	15.646	34.668
Organic	1.770	1.609	2.490	15.030	35.412
Using Biofertilizers	1.768	1.644	2.450	15.062	32.924
CD (P = 0.05)	NS	NS	NS	NS	NS

NS = Not Significant

Cooking quality of corms from the different production systems did not vary appreciably and the corms tasted equally good. Analyses of biochemical constituents of corms viz., dry matter, starch, crude protein, total sugar, reducing sugar, total phenols and oxalates revealed that production systems could not significantly influence these attributes. However, dry matter, starch and crude protein contents were slightly higher and oxalate content slightly lower in organic corms (Table 17.5). There was not much variation in the mineral composition of corms viz., P, K, Ca, Mg, Cu, Zn, Mn and Fe contents due to the various practices. This is in accordance with the report of Radhakrishnan *et al.* (2006) that quality parameters of made tea manufactured from different farming systems, including organic system, did not vary significantly. Moreover Chhonkar (2008) stated that regardless of whether the nutrients are from organic or inorganic source, plants absorb the same as inorganic ions and once absorbed the nutrients are re-synthesized into compounds that determine the quality of the produce, which is predominantly the function of genetic make up of the plants.

Table 17.5. Comparative yield, quality and economic advantages of organic farming over conventional and other production systems in elephant foot yam (mean of 4 years)

Production system	*Corm yield(t ha^{-1})*	*Dry matter (%)*	*Starch (% FW basis)*	*Oxalate (% DW basis)*	*Crude protein (% FW basis)*	*Net income (Rs. ha^{-1})*	*B:C ratio*
Conventional	52.567	19.69	14.41	0.268	1.740	2,188,49	2.09
Traditional	49.991	20.82	16.82	0.228	1.876	1,93,428	1.94
Organic	62.668	21.50	16.60	0.191	2.160	2,77,194	2.24
Using biofertilizers	47.028	21.62	16.44	0.210	2.026	1,77,574	1.89
CD (P = 0.05)	3.834						

Pooled analysis of corm yield of 4 years indicated that organic farming resulted in significantly higher yield (62.67 t ha^{-1}) in elephant foot yam (Table 17.5). Similar results were reported by Mahapatra *et al.* (2006 a) in Basmati rice and Mahapatra *et al.* (2006 b) in lentil, chick pea and wheat. During 2004-05, organic farming resulted in higher corm yield (65.867t ha^{-1}) on par with conventional practice (58.560t ha^{-1}). Traditional practice resulted in significantly the lowest yield (50.126 t ha^{-1}). The mean corm weight from organic farming, conventional and biofertilizer practice was 5.489, 4.880 and 4.530 kg, respectively. Organic farming continued its superiority (70.625 t ha^{-1}) in the subsequent year also. Traditional practice produced 54.540t ha^{-1} of corms. The conventional practice and the treatment with biofertilizers were on par (47.494 and 45.509t ha^{-1}). In the third year organic farming was on par with conventional practice (56.948 and 57.812t ha^{-1}, respectively). Traditional practice and plots that received biofertilizers produced almost similar yields (50.479 and 47.772 t ha^{-1} respectively). During the fourth consecutive year also organic farming produced higher corm yield (57.23t ha^{-1}), on par with conventional practice (46.40t ha^{-1}). Traditional practice and plots that received bio-fertilizers produced almost similar yields (44.82 and 40.47 t ha^{-1}

respectively) to that of conventional practice. The higher yield in organic farming plots (20%) may be attributed to the overall improvement in soil physico-chemical properties due to addition of various organic sources of nutrients.

Of the various production systems tested, organic farming generated the highest net income of Rs. 2.77 lakh ha^{-1} and benefit cost ratio of 2.24 as against Rs. 2.19 lakh ha^{-1} and 2.09, respectively under the conventional system (Table 17.5). Taking into account the mean yield of 4 years, organic farming (62.67t ha^{-1}) resulted in 20% yield increase over conventional practice (52.57t ha^{-1}) (yield increase of 10t ha^{-1}) and generated additional income of Rs. 0.58 lakh ha^{-1} over conventional practice (Table 17.5). The organic farming technology package standardized for elephant foot yam (farmyard manure @ 35t ha^{-1}, green manuring with cow pea to generate 20–25 t ha^{-1} of green matter in 45–60 days, neem cake @ 1t ha^{-1} and ash @ 3t ha^{-1}) is being validated and popularized through on farm trials in Kollam and Pathanamthitta districts of Kerala under National Horticulture Mission (State Horticulture Mission, Kerala) funded programme.

CONCLUSION

Organic farming is an alternative and appropriate management strategy in elephant foot yam for getting high yield of quality tubers. Moreover organic farming resulted in additional returns in elephant foot yam and helped in maintaining soil fertility.

References

Chhonkar, P.K. 2008. Organic farming and its relevance in India. pp.5-33. In: *Organic Agriculture* (Eds). Tarafdar, J.C., Tripathi, K.P. and Mahesh Kumar, Indian Society of Soil Science, Scientific Publishers (India), Jodhpur, p.5-33.

Clark, M.S., Horwath, W.R., Shennan, C. and Scow, K.M. 1998. Changes in soil chemical properties resulting from organic and low-input farming systems. *Agron.J.* **90**:662-71.

Cochran, W.G. and Cox, G.M. 1965. *Experimental designs.* Asia Pub. House. Bombay.

Gliessman, S.R. 1998. Agroecology and sustainability: The ecological framework. In: *Atti Agroecology and Sustainability,* Intercool Symposium, 20-23 July 1998, Florence, Italy, 8p.

Gupta, R.P. and Dakshinamoorthy, C. 1980. *Procedures for physical analysis of soil and collection of agrometeorological data.* 116pp. Indian Agricultural Research Institute, New Delhi.

Jackson, M.L. 1973. *Soil chemical analysis.* 2nd edn.. Prentice Hall of India Private Ltd., New Delhi, p. 498.

KAU. 2002. Kerala Agricultural University. *Package of Practices Recommendations: Crops,* Directorate of Extension, Mannuthy,Thrissur, pp. 49-53.

Mahapatra, B.S., Misra, A. Chilana, K., Ajay Kumar and Singh, S.P. 2006a. Growth and yield of Basmati rice under organic mode during initial years of conversion in relation to nutrient management practices. *J. Ecofriendly Agric.* **1**(1):1-5.

Mahapatra, B.S., Singh, S.P. Rajesh, A., Vishwakarma, V.K., Bhushan, C., Anil Kumar, Singh, R.K. Ajay Kumar, Chilana, K. and Pandey, C.S. 2006b. Performance of lentil, chickpea and wheat under organic mode during initial years of conversion in relation to nutrient management practices. *J. Ecofriendly Agric.* **1**(2):105-116.

National Research Council. 1989. Alternative agriculture. Committee on Agriculture. National Acadedmy Press, Washington, DC.

Nayar, T.V.R. and Suja, G. 2004. Organic production potential of tropical tuber crops. In: *Abstracts of XIV Swadeshi Science Congress,* 5-7 November 2004, Swadeshi Science Movement, Kerala, p. 20-21.

Radhakrishnan, B., Ranjit Kumar, Q., Ganapathy, M. N. K. and Hudson, J. B. 2006. Effect of conventional, organic and biodynamic farming systems in tea. *J. Plantation Crops* **34**:330-33.

Reganold, J.P., Glover, J.D., Andrews, P.K. and Hinman, H.R. 2001. Sustainability of three apple production systems. *Nature* **410**:926-29.

Scow, K.M., Somasco, O., Gunapala, N., Lau, S., Venette, R., Ferris, H., Miller, R. and Shenman, C. 1994. Transition from conventional to low-input agriculture changes soil fertility and biology. *Calif. Agric.* **48**:20-26.

Srivastava, A.K., Singh, S., Shivankar, V. J., Das, A.K. and Rao, C.N. 2006. *Organic Citrus Status.* 175 pp. National Research Centre for Citrus, Nagpur, Maharashtra.

Stockdale, E.A., Lampkin, N.H., Hovi,M., Keating, R., Lennartsson, E.K.M., Macdonald, D.W., Padel, S., Tattersall, F.H., Wolfe, M.S. and Watson, C.A. 2001. Agronomic and environmental implications of organic farming systems. *Adv. Agron.* **70**: 261-27.

Subbiah, B.V. and Asija, L.L. 1956. A rapid procedure for estimation of available nitrogen in soils. *Curr. Sci.* **25**:259-260.

Suja,G. and Nayar,T.V.R. 2006. Organic tuber production: Issues, prospects and future strategies. *Kisan World* **33**(11):58-59.

Suja, G., Nayar, T.V.R., Potty, V.P. and Sundaresan, S. 2006a. Organic farming: An alternative option for tuber crop production. *CTCRI News* **23**(1): 4-5.

Suja, G., Nayar, T.V.R., Potty,V.P. and Sundaresan, S. 2006b. Organic farming: a viable strategy for high yield and quality tuber crop production. *Ind. Hort.* **51**(6): 4-5.

Susan Andrews, S., Mitchell, J.P., Mancinelli, R., Karlen, D.L., Hartz, T.K., Horwath, W.R., Pettygrove, G.S., Scow, K.M. and Munk,D.S. 2002. On-farm assessment of soil quality in California's central valley. *Agron J.* **94**:12-23.

18

Effect of Organic and Integrated Nutrient Management on Coconut

T.B. Basavaraju and M. Hanumanthappa

INTRODUCTION

Organic manures are important in sustaining soil productivity especially for perennial crop like coconut, which requires continuous supply of nutrients. Any organic manure supplemented with the required quantity of inorganic fertilizers is the best manurial combination for an adult coconut palm (Thampan, 1972). Coconut produces huge amount of waste biomass such as dried leaves, sheathes, spadices, inflorescence, husk, etc. However, the husk is used mainly for extracting coir fibre and coir pith is obtained as a by-product in the coir processing industry. With high C:N ratio of above 100:1 and higher contents of lignin and polyphenols, raw coir pith is not an ideal organic input in crop production. Coir pith can be made suitable for use in agriculture/horticulture after stabilization by proper composting process by employing suitable organisms capable of degrading lignin and polyphenols and bringing down C:N ratio (Savithri and Khan, 1994). An experiment was conducted at the Agricultural Research Station, Arsikere during 1997-98–2004-05 in a 25 years old coconut garden of local cultivar-Tiptur Tall to supplement chemical fertilizers with organics by recycling coconut waste such as coir pith as a source of nutrients.

The fertilizers were applied in the form of urea, single super phosphate and muriate of potash as per the treatments. The quantity of organic manures required was calculated on nitrogen equivalent basis and applied during May-June. The coir pith was composted using the fungus *Pleurotus sajorcaju* along with amendment of urea and rock phosphate.

The integrated nutrient management practices of supplying 50 or 100% N through composted coir pith (CCP) along with the balance of NPK through fertilizers and also of providing entire NPK through neem cake + bone meal + ash produced significantly higher number of functional leaves, bunches and female flowers on the palm compared to control of no manure or fertilizer application (Table 18.1). The nut setting percentage was also significantly increased with the application of CCP and neem cake + bone meal + ash. Similar results of increase in the number of functional leaves, bunches and female flowers were reported with the use of FYM (Marimuthu *et al.*, 2001) and coir pith compost (Venkitaswamy, 2003) along with fertilizers.

Table 18.1. Effect of nutrient management practices on growth and yield attributes in coconut (Pooled data of 8 years)

Sl. No.	*Treatments*	*No. of functional leaves/palm*	*No. of bunches/palm*	*No. of female flowers/palm*	*Nut setting per cent*
1.	Control (No manure or fertilizers)	24.6	10.4	168.9	26.0
2.	Recommended dose of fertilizers (500:320:1200 g N, P_2O_5, K_2O/palm/year)	29.2	11.9	225.5	27.8
3.	CCP on equivalent N basis + balance quantity of P & K as fertilizers	30.8	12.2	251.8	28.8
4.	50% N as CCP + balance NPK as fertilizers	31.6	12.5	244.4	29.0
5.	Neem cake + Bone meal + Ash (on nutrient equivalent basis)	31.2	12.1	235.0	30.8
	S. Em ±	0.83	0.54	4.83	0.78
	CD(P = 0.05)	2.57	1.66	14.90	2.39

The mean data of nut yield of coconut, copra content and copra yield over eight years is presented in Table 18.2. The nut yield of coconut significantly increased with the application of manures and fertilizers over no-manure control. The integrated nutrient management practices of supplying 50 or 100 % N through CCP along with the balance of NPK through fertilizers and also of supplying of entire NPK through neem cake + bone meal + ash produced significantly higher nut yield compared to application of recommended dose of chemical fertilizers. The yield of nuts in the treatments of supplying of 50 or 100% N through CCP along with balance of NPK through fertilizers and application of neem cake + bone meal + ash to supply 100% NPK were on par indicating that full or 50% N could be supplied through CCP along with the balance of NPK through fertilizers or all the nutrients could be supplied through the application of neem cake, bone meal and ash

The increase in nut yield with the integration of organics with fertilizers could be attributed to increased number of bunches, female flowers and nut setting per cent due to improved availability of nutrients to coconut. Similarly copra content per nut and copra yield per palm were also significantly increased with the integration of organics with fertilizers. Similar results of higher nut yield were reported with the use FYM (Marimuthu *et al.*, 2001) and composted coir pith (Venkitaswamy, 2003) integrated with the fertilizers.

Table 18.2. Effect of nutrient management practices on nut and copra yield of coconut (Pooled data of 8 years)

Sl. No.	*Treatments*	*Nut yield/ palm/yr (Nos.)*	*Nut yield/ ha/yr (Nos.)*	*Copra content (g/nut)*	*Copra yield (kg/palm/yr)*	*Copra yield (kg/ha/yr)*
1.	Control (No manure or fertilizers)	58.15	5815	116.0	6.75	675
2.	Recommended dose of fertilizers (500:320:1200 g N, P_2O_5, K_2O/palm/ year)	73.31	7331	125.0	9.24	924
3.	CCP on equivalent N basis + balance quantity of P & K as fertilizers	84.54	8454	126.2	10.69	1069
4.	50% N as CCP + balance NPK as fertilizers	82.19	8219	131.0	10.78	1078
5.	Neem cake + Bone meal + Ash (on nutrient equivalent basis)	85.01	8501	129.8	11.03	1103
	S. Em ±	1.99	199	2.63	0.34	34
	CD @ 5%	6.15	615	8.12	1.04	104

The nutrient status of the soil in the coconut basin was significantly increased with the application of manures and fertilizers (Table 18.3).

Table 18.3. Effect of nutrient management practices on nutrient status of soil in coconut basins

a) Organic carbon (%) and available N (kg/ha)

Sl. No.	*Treatment*	*Organic carbon (%)*			*Available N (kg/ha)*		
		0–25 cm	*25–50 cm*	*50–100 cm*	*0–25 cm*	*25–50 cm*	*50–100 cm*
1.	Control (No manure or fertilizers)	0.41	0.35	0.30	262.5	244.3	223.3
2.	Recommended dose of fertilizers (500:320:1200 g N P_2O_5 K_2O/palm/year)	0.51	0.39	0.32	318.3	288.3	250.5
3.	CCP on equivalent N basis + balance quantity	0.58	0.46	0.36	366.5	300.0	260.5
4.	50% N as CCP + balance NPK as fertilizers	0.58	0.45	0.35	360.3	300.5	262.0
5.	Neem cake + Bone meal + Ash (on nutrient equivalent basis)	0.57	0.46	0.38	358.8	297.0	262.0
	S. Em ±	0.02	0.03	0.03	14.0	10.3	8.8
	CD(P = 0.05)	0.08	0.09	NS	30.8	31.8	27.1

b) Available P_2O_5 and K_2O (kg/ha)

Sl. No.	*Treatment*	*Available P_2O_5 (kg/ha)*			*Available K_2O (kg/ha)*		
		0-25 cm	*25-50 cm*	*50-100 cm*	*0-25 cm*	*25-50 cm*	*50-100 cm*
1.	Control (No manure or fertilizers)	18.68	16.93	16.25	268.8	258.0	223.8
2.	Recommended dose of fertilizers (500:320:1200 g N P_2O_5 K_2O/palm/year)	23.03	20.95	18.80	332.5	303.3	247.0

(Contd.)

Sl. No.	*Treatment*	*Available P_2O_5 (kg/ha)*			*Available K_2O (kg/ha)*		
		0-25 cm	*25-50 cm*	*50-100 cm*	*0-25 cm*	*25-50 cm*	*50-100 cm*
3.	CCP on equivalent N basis + balance quantity of P & K as fertilizers	25.50	21.03	19.93	359.5	330.0	268.8
4.	50% N as CCP + balance NPK as fertilizers	26.43	21.80	19.05	362.8	329.8	265.5
5.	Neem cake + Bone meal + Ash (on nutrient equivalent basis)	26.05	22.93	20.08	371.5	335.5	262.5
	S. Em ±	0.91	0.93	0.89	8.81	10.70	8.23
	CD (P=0.05)	2.80	2.85	2.74	27.16	32.96	25.35

The increase in organic carbon and available NPK in soil in coconut basin was higher with the integration of composted coir pith with fertilizers or application of neem cake + bone meal + ash to supply recommended nutrients. The NPK content in the reference 14th leaf also increased with the application of manures and fertilizers to coconut (Table 18.4). These findings indicated the beneficial effect of integrated nutrient management in improving soil physical, chemical and biological environment which favoured the higher nutrient uptake from the soil.

Table 18.4. Effect of nutrient management practices on leaf nutrient status of coconut

Sl. No.	*Treatments*	*Leaf nutrient content (%)*		
		N	*P*	*K*
1	Control (No manure or fertilizers)	1.75	0.120	0.85
2	Recommended dose of fertilizers (500:320:1200 g N P_2O_5 K_2O/palm/year)	1.93	0.138	0.97
3	CCP on equivalent N basis + balance quantity of P & K as fertilizers	2.07	0.148	1.03
4	50% N as CCP + balance NPK as fertilizers	2.07	0.158	1.05
5	Neem cake + Bone meal + Ash (on nutrient equivalent basis)	2.06	0.160	1.06
	S.Em +	0.024	0.005	0.016
	CD (P = 0.05)	0.074	0.014	0.050

The economic analysis of supplying NPK to coconut through different sources indicated that application of CCP to supply 50 or 100% N along with balance of NPK through fertilizers recorded higher net income and B : C ratio compared to application of neem cake + bone meal + ash on equivalent nutrient basis and also with the application of only chemical fertilizer (Table 18.5). Though application of neem cake + bone meal + ash on equivalent nutrient basis produced on par yield with the application of 50 or 100% N through CCP along with balance of NPK as fertilizer, the increased cost of production due to high cost of neem cake and bone meal reduced the net income and B : C ratio of this treatment. Hence, composted coir pith is recommended to supply 50 or 100% of the recommended N to coconut in areas where raw coir pith is available freely.

Table 18.5. Economics of coconut production in different nutrient treatments (Based on the mean data of 8 years)

Sl. No.	Treatments	Gross Return (Rs./ha)	Cost of Production (Rs./ha	Net Return (Rs./ha)	B:C Ratio
1.	Control (No manure or fertilizers)	29075	11025	18050	2.64
2.	Recommended dose of fertilizers (500:320:1200g N P_2O_5 K_2O/palm/year)	36655	14335	22320	2.56
3.	CCP on equivalent N basis + balance quantity of P & K as fertilizers	42270	13725	28545	3.08
4.	50% N as CCP + balance NPK as fertilizers	41095	14025	27070	2.93
5.	Neem cake + Bone meal + Ash (on nutrient equivalent basis)	42505	18525	23980	2.29

Note: The cost and returns were calculated based on 2004-05 market prices.

Conclusion

Raw coir pith can be made into compost of good manurial value by using ligno-cellulolytic fungi, *Pleurotus sajorcaju* and used as a source of nutrients to coconut. The composted coir pith (CCP) can be applied to coconut palm to supplement 50–100 % of the recommended dose of N (25–50 kg CCP).

References

Gomez, K.A. and Gomez, A.A., 1984. Statistical Procedures for Agricultural Research, 2nd Edition, John Wiley and Sons, New York.

Jackson, M.L., 1967. Soil Chemical Analysis, Prentice Hall of India (P) Ltd., New Delhi.

Marimuthu, R., Athmanathan, U. and Mohan, S., 2001. Integrated nutrient management for coconut. *South Indian Hort.*, **49** (Special):145-48.

Savithri, P. and Khan, H.H., 1994. Characteristics of coconut coir pith and its utilization in agriculture. *J. Plantation Crops* **22** (1):1-18.

Thampan, P.K., 1972. Coconut culture in India, Cochin, India, pp.39-48.

Venkitaswamy, R., 2003, Integrated nutrient management in coconut with composted coir pith. *Madras Agric. J.* **90**(1-3): 54-56.

19

Yield and Nutrient Status of Coconut as Influenced by Different Sources of Nutrients in Sandy Soil of Konkan Region of Maharashtra

D.D. Nagwekar, V.S. Desai, V.S. Sawant, S.C. Talashikar, P.M. Haldandar
A.D. Rangwala, S. Arulraj and B.B. Jadhav

INTRODUCTION

In India, coconut is grown in an area of 1.95 million ha with the production of 14,111 million nuts. Further it is distributed in 159 districts, 17 States and 3 Union territories. Coconut is perennial in nature and hence, nutrient management is very important in this crop. Continuous use of chemical fertilizers alone will result in imbalance in soil fertility; they are costly and also affect the environment. For achieving sustainable production in coconut and improving the soil fertility, application of organic manure is of utmost importance.

The effect of various organic material and chemical fertilizer on yield as well as soil and leaf nutrient status of WCT variety of coconut palm was studied at the Regional Coconut Research Station, Bhatye under the All India Co-ordinated Research Project on Palm during 1998-99–2005-06. Chemical fertilizers and organic manures such as vermi-compost, neem cake and bone meal were used. The vermi-compost prepared of coconut leaves by using *Eudrilus* spp. of earthworm was applied in the first week of June and chemical fertilizers were applied in three splits, i.e., June, October and February.

No significant variation in nut yield was observed among the treatments (Table 19.1) during the transition period of three years (1998-99–2000-01). However, from 2001-02–2005-06, different treatments started showing statistically significant difference in nut yield.

Table19.1. Yield of coconut (No. of nuts/palm/year) as influenced by organic and inorganic fertilizers in coconut

Sl. No.	*Treatments*	*Yield (No. of nuts/palm/year)*								*Mean cumulative yield/palm/ year 01-02 to 05-06*
		Transit period				*Post-treatment period*				
		98-99	*99-00*	*00-01*	*01-02*	*02-03*	*03-04*	*04-05*	*05-06*	
1.	Control	87	105	70	58	58	57	54	62	58
2.	Recommended Dose of NPK	100	114	91	97	101	96	96	95	96
3.	100 % Vermi-compost	96	107	77	94	89	97	92	87	91
4.	Recommended Dose of NPK (50%) + Vermi-compost (50%)	89	102	88	91	77	70	92	90	83
5.	Neem cake + Bone meal + Ash	96	103	89	68	68	74	79	70	72
	SE +	8.34	10.53	6.99	7.63	7.95	6.53	8.29	7.21	5.77
	CD(P=0.05)	N.S.	N.S.	N.S.	23.51	24.51	20.06	25.57	22.21	17.79

NS = Not Significant The treatments are: T1-Control with no fertilizers; T2- 100% recommended dose of chemical fertilizers (1000:500:1000g N, P_2O_5, K_2O/palm/ year); T3- 100% nutrient supply on N basis through vermi-compost prepared from coconut leaves; T4 - 50% of nutrient supply on N basis through vermi-compost + 50% through chemical fertilizer and T5 - Neem cake (10 kg) + Bone meal (1 kg) + ash (20 kg) per palm per year

Application of recommended dose of chemical fertilizers-T_2 (1000:500:1000 N,P_2O_5 & K_2O/palm/year) recorded the highest mean yield of 96 nuts/palm/year followed by application of 100% vermi-compost-T_3 (91 nuts/palm/year) and application of 50% recommended dose of chemical fertilizers + 50% vermi-compost-T_4 (83 nuts/palm/year) and these treatments were statistically on par and significantly superior over T_1 and T_5. The soil of experimental site is sandy loam and continuous application of chemical fertilizer resulted in normal yield. Application of organic manures to palms which were under regular application of chemical fertilizers required longer time to make use of nutrients and give high yield compared with palms under continuous application of chemical fertilizers. Venkitaswamy (2002) reported that application of 100% composted coir pith/palm/year on N basis or application of 50% composted coir pith + 50% of recommended dose N, P_2O_5 chemical fertilizers was optimum for getting maximum nut yield.

Application of recommended dose of chemical fertilizers (T_2) resulted in higher net income of Rs. 41,690 with a B:C ratio of 1.70 followed by application of 100% nutrient supply on N basis through vermi-compost prepared from coconut leaves (T_3) with net profit of Rs.32,150 and B:C ratio of 1.52 (Table 19.2). However, the lowest net income (Rs.6,850/-) and B:C ratio (1.13) were obtained from the Control (T_1). The cost of production was found to be high in T_3 treatment.

Table 19.2. Yield and economics of coconut as influenced by organic and inorganic fertilizers

Treatment	*Yield (No. of nuts/ha)*	*Total return (Rs)*	*Total cost (Rs)*	*Net profit (Rs)*	*B:C ratio*
Control	10150	60900	54050	6850	1.13
Recommended Dose of NPK	16800	100800	59110	41690	1.70
100 % Vermi-compost	15925	95550	62800	32150	1.52
Recommended Dose of NPK (50%) + Vermi-compost (50 %)	14525	87150	60954	26196	1.42
Neem cake + Bone meal + Ash	12600	75600	67350	8250	1.12

Build up organic carbon in the soil at 0–25 cm and 50–100 cm depth was noticed with the application of vermi-compost. Application of neem cake + bone meal + ash also resulted in increase of organic carbon of soil. Nitrogen build up in the soil was higher in T_3 followed by T_4 which were statistically on par. Potassium build up of soil was high with T_4 followed by T_5 and T_3 which were statistically on par at 0–25 and 50–100 cm depth. Kalpana *et al.* (2006) found that integrated nutrient management and cropping segment improve the soil physico–chemical properties through building up organic carbon and NPK content in the soil.

Table 19.3. Organic carbon, available, N, P_2O_5 and K_2O of soil as influenced by organic and inorganic fertilizers at different depths

Treatment	*Organic carbon (%)*			*Available N (kg/ha)*			*Available P_2O_5 (kg/ha)*			*Available K_2O(kg/ha)*		
	*0-25**	*25-50*	*50-100*	*0-25*	*25-50*	*50-100*	*0-25*	*25-50*	*50-100*	*0-25*	*25-50*	*50-00*
Control	0.46	0.33	0.26	162.65	97.55	84.23	20.6	9.6	7.6	308.5	216.5	174.7
Recommended Dose of NPK	0.46	0.41	0.40	183.05	125.12	94.78	27.5	13.7	9.2	391.4	268.9	205.3
100 % Vermicompost	0.55	0.43	0.42	205.45	169.15	106.64	29.1	14.7	10.5	481.6	325.9	294.4
Recommended Dose of NPK (50%) + Vermicompost(50 %)	0.47	0.38	0.34	200.18	156.75	106.05	25.6	15.8	12.1	519.1	375.9	302.4
Neem cake + Bone meal + Ash	0.54	0.45	0.36	194.28	137.62	100.06	30.9	14.9	9.2	502.1	349.1	294.6
SE (m) ±	0.016	0.013	0.008	4.40	4.32	2.02	0.79	0.52	0.14	12.42	9.20	7.22
C.D. (0 = 0.0 5%)	0.052	0.042	0.026	13.57	13.32	6.24	2.45	1.61	0.45	38.27	28.35	22.25

Depth of soil in cm

The leaf nutrient status indicated that 100% vermi-compost application recorded the highest value of N (1.98 %) followed by T_4 (1.95%) (Table 19.4). As regards to P status, it was the highest in T_2 (0.107%) followed by T_5 (0.106%) and T_3 (0.101%) which were statistically on par. The highest K was recorded in T_4 (1.0 %) followed by T_5 (0.95%). Higher leaf N and K status in organic treatments would have been due to the better uptake of N and K in application of organic treatment due to the increased availability. Venkitaswamy (2002) reported that composted coir pith treatment recorded more leaf N and K content at 2.07 and 1.55% respectively in the 14th leaf.

Table 19.4. N, P, K, content of leaf as influenced by organic and inorganic fertilizers

Treatment	*Total N (%)*	*Total P (%)*	*Total K (%)*
Control	1.42	0.096	0.50
Recommended Dose of NPK (R.D.)	1.63	0.107	0.66
100 % Vermi-compost	1.98	0.101	0.71
Recommended Dose of NPK (50%) + Vermi-compost (50 %)	1.95	0.098	1.00
Neem cake + Bone meal + Ash	1.77	0.106	0.95
SE (m) ±	0.012	0.0021	0.032
CD (P = 0.0 5)	0.037	0.0065	0.099

References

Kalpana, M., Rao, D.V.R. and Srinivasulu, B. 2006. Integrated nutrient management in coconut based cropping system under coastal ecosystem of Andhra Pradesh, *Journal of Plantation Crops,* 2006, **34** (3):258-262.

Venkitaswamy, R. and Khan, H.H. 2002. Integrated nutrient management in coconut. In: *PLACROSYM XV*, (Eds) Sridharan K and Vinod kumar P.K. Mysore, Karnataka, India 2002, pp. 73-74.

20

Performance of Coconut Palms During Pre and Post- Cyclonic Period in Response to Nutrient Management in Littoral Sand of Coastal Orissa

D.K. Dash, D.P Ray and S.C. Sahu

INTRODUCTION

Coconut is the major small holder's plantation crop cultivated in Orissa. It has high nutrient requirement for its growth and development for their long growing period. In coconut farming, only a limited percentage of farmers use chemical fertilizers and as a result vast majority of palms remain either unfertilized or under fertilized conditions. To overcome this situation, regular inclusion of nutrient in organic form is required. Coir pith is abundantly available at coir industries and a lignocelluloses waste biomass can be utilized in coconut garden. Though coir pith has this number of beneficial properties, its direct utilization as manure is not advisable as it contains large amount of lignin and phytotoxic polyphenols and less nitrogen. It can be easily converted to compost using *Pleurotus sajor caju* (Nagarajan *et al.,* 1988; Sheeba *et al.*, 1999).

Attempts have been made at the AICRP on Palms, OUAT, Bhubaneswar, Orissa to generate information on the effect of integrated use of manures and fertilizers on coconut nutrient management. The trial laid out on Sakhigopal local Tall (SKGT) variety of coconut of 18 years old during 1996 at Konark was damaged by the super cyclone that hit the coastal district of Orissa with heavy rain and wind velocity more than 250 kmph during October, 1999. However, the experiment was continued with the same treatments and data were collected on extent of damage caused by the super cyclone to the treated palms and its performances in pre and post-cyclonic period. The average yield of palms was low

(18–40 nuts per year) due to poor soil type. The composted coir pith (CCP) was prepared with lignin degrading fungus *Pleurotus sajor caju.* For preparation of one tonne of coir dust 5 bottles of *Pleurotus* spawn and 5kg of urea are used.

The effect of treatments, in general, was observed after one year of amending the treatments by retention of more number of leaves on the crown and gradual increase in leaf production (Table 20.1). Maximum number of leaves (29.9) was retained in 50% CCP + 50% of recommended dose of fertilizers treated palms followed by full dose of fertilizers during the third year of experiment. Maximum number of leaves (11.7) was produced by palms receiving 50% N through CCP (25kg / palm/ year) + 50% of recommended dose of fertilizers followed by palms (11.23) receiving full dose of fertilizers during third year. Palms under these treatments produced maximum number of leaves after second year of treatment and were found statistically on par and superior to the other treatments. This may be due to enhancement of nutrient content in soil by amendment of CCP and recommended dose of fertilizers. Organic manures influence plant growth through its effect on the physical, chemical and biological properties of soil besides providing nutrients to the plant (Thampan, 1995). Talashilkar *et al.* (2008) reported that application of vermi-compost either alone or in combination with inorganic fertilizer enhanced the available nutrient content in soil.

Table 20.1. Vegetative characters influenced by the nutrient management during pre and post cyclonic period *(Number of leaves produced per year)* number of leaves on the crown

Treatments	*96-97*	*97-98*	*98-99*	*99-2000*	*2000-01*	*2001-02*	*2002-03*	*2003-04*
a) Pre cyclomic period								
Control: Farmer practice	7.23	7.48	7.70	3.93	5.25	6.88	8.75	8.0
Recommended dose of fertilizers NPK	7.08	10.78	11.23	6.70	8.70	11.03	10.83	10.8
100% N through composted coir pith	7.80	9.45	9.58	5.33	8.08	9.75	9.88	9.9
50% N through composted coir pith	7.45	11.08	11.70	6.70	9.15	10.95	11.33	11.1
Neem cake + Bone meal + wood ash	7.20	8.48	9.08	4.13	7.65	9.38	9.38	9.5
CD(P = 0.05)	NS	1.019	0.867	0.938	1.072	0.733	2.859	0.709
c) Post cyclonic period								
Control: Farmer practice	22.8	21.5	22.5	11.7	15.4	18.5	18.6	21.3
Recommended dose of fertilizers NPK	22.5	26.0	28.7	15.4	17.6	25.9	27.3	28.4
100% N through composted coir pith	21.9	24.2	25.8	13.3	18.1	23.5	23.6	25.9
50% N through composted coir pith	21.4	27.7	29.9	17.6	19.7	26.5	28.7	29.4
Neem cake + Bone meal + wood ash	23.8	24.9	26.2	13.7	17.6	23.3	24.9	25.1
CD(P = 0.05)	NS	1.879	1.167	1.274	1.950	1.108	1.388	0.742

Spadices ranged between 6.3–6.6 in all the treatments in the initial year (Table 20.2). The impact of nutrient management on spadices and female flower production was noticed after two years. The palms responded very well to application of 50% N through CCP

(25 kg / palm/ year) + 50% of recommended dose of fertilizers for increased number of spadices (8.3) and maximum number of female flowers (167.5) in comparison to the other treatments. The increased number of female flowers (153.2) in this treatment was noticed during the 3rd year of study followed by palms applied with recommended dose of fertilizers (7.7 and 144, respectively for spadices and female flowers). The nut yield increased in the pre-cyclonic period in palms applied with Neem cake @ 4kg, Bone meal @ 1kg and Wood ash @ 5 kg/palm /year (44.3/palm/year) during the 3rd year followed palms under full dose of fertilizers (40/palm/ year). Kalpana *et al.*(2006) and Raghaba Rao *et al.*(2007) also obtained the highest nut yield when nutrients were supplied through 50% organics and 50% inorganic fertilizers.

The increased in yield may be due to increase in spadices and female flower production, increase in setting and decrease in premature nut fall due to integrated nutrient management through CCP and RDF. Adequate nutrient and moisture supply can regulate coconut productivity to an extent of 80% (Nair *et al.*, 1996). Thomas *et al.* (2001) reported the effectiveness of organic manuring to enhance nutrient content of soil and soil physical conditions. The incorporation organic manures significantly enhances organic carbon, available N, Fe, Mn and exchangeable Ca and Mg in littoral sandy soil. It is also effective in increasing nutrient content and water holding capacity and decreasing bulk density of soil.

The palms applied with recommended dose of fertilizers as well as 50% N through CCP (25kg/palm/year) + 50% of RDF retained more number of leaves during super cyclonic year as compared to other treatment and control. The former treatment retained 17.6 leaves on the crown and produced 6.7 leaves during the year 1999. During the cyclone year, palms under the latter treatment produced 8.0 spadices. The number of female flowers reduced by the cyclone damage being 82.8 in palms in the latter treatment and 81.3 in palms in the former case, wherein they recorded maximum values among the treatments. It clearly brings out that palms with proper nutrient management were less affected by the cyclone, which resulted in less abortion of flower primodia.

Table 20.2. Reproductive characters and nut yield as influenced by the nutrient management during pre and post cyclonic period

	Number of spadices produced per year							
	96-97	*97-98*	*98-99*	*99-2000*	*2000-01*	*2001-02*	*2002-03*	*2003-04*
Control: Farmer practice	6.5	6.5	6.4	1.5	5.0	5.8	7.1	6.7
Recommended dose of fertilizers NPK	6.3	7.9	7.7	3.2	7.1	10.2	9.7	10.37
100% N through composted coir pith	6.8	7.0	7.0	2.2	6.0	7.7	7.5	7.75
50% N through composted coir pith	6.4	9.0	8.3	4.5	8.0	10.5	10.8	10.9
Neem cake + Bone meal + wood ash	6.6	7.5	7.6	2.2	6.7	7.7	8.7	7.9
CD (P = 0.05)	NS	1.14	1.18	1.14	1.59	1.06	0.79	1.31

(Contd.)

	Number of female flowers produced per year							
	96-97	*97-98*	*98-99*	*99-2000*	*2000-01*	*2001-02*	*2002-03*	*2003-04*
Control: Farmer practice	101.5	101.6	96.7	40.2	66.8	82.4	84.9	83.9
Recommended dose of fertilizers NPK	102.0	131.7	144.0	81.3	106.8	131.1	152.0	163.3
100% N through composted coir pith	94.8	107.2	114.3	65.0	63.2	92.7	99.1	121.8
50% N through composted coir pith	101.3	153.2	167.5	82.8	138.8	143.1	175.1	175.9
Neem cake + Bone meal + wood ash	97.0	117.6	104.8	58.9	68.6	99.7	92.8	117.6
CD(P = 0.05)	NS	29.17	30.71	7.53	13.82	5.78	16.04	9.54
	Number of nuts produced per year							
Control: Farmer practice	19.9	20.7	21.9	4.7	4.3	14.5	14..7	17.6
Recommended dose of fertilizers NPK	19.4	24.3	40.0	18.9	21.7	31.3	39.9	41.1
100% N through composted coir pith	21.7	26.7	29.5	12.2	14.3	24.2	29.8	29.4
50% N through composted coir pith	20.4	28.4	44.3	22.6	28.6	38.7	42.1	48.6
Neem cake + Bone meal + wood ash	21.7	23.7	26.3	14.4	15.5	18.7	24.2	26.7
CD(P = 0.05)	NS	6.37	5.89	4.41	4.44	5.86	5.78	7.20

NS = Not Significant

The cyclonic wind damaged the leaves on the crown and caused breaking of bunches due to high wind velocity. The damaged palms under application of neem cake, bone meal and wood ash recorded maximum nut yield (22.6/palm) during the year followed by palms applied with full dose of fertilizers (18.9/palm). Yields under these treatments were on par and found to be superior to other treatments adopted.

The damaged palms under no manuring took four years to recover whereas palms under 50% N through CCP (25kg/palm/year) + 50% RDF as well as under full dose of fertilizers recovered by two years as regard to production and retention of leaves, production of spadices, production of female flowers and nut yield.

In the first year succeeding cyclone, more leaves was recorded in palms applied with 50% N through CCP + 50% RDF (9.15) followed by palms (8.70) under RDF. In the second year, palms under RDF recorded maximum (11.03) leaf production followed by palms (10.95) under 50% N through CCP + 50% of RDF. The number of leaves on the crown increased as the year progress. The palms under these two treatments took three years to have similar number of leaves on the crown as it was during the pre-cyclone year. It was observed that palms applied with nutrients recorded more spadices in the succeeding year of cyclone as compared to the palms without manuring.

In palms applied with 50% N through CCP + 50% of RDF and full RDF, the rate of production of spadices was high than that of the other treatments. The maximum number of spadices was produced during the succeeding year of cyclone, which was at par with the spadices production in the pre- cyclonic year. Palms under 50% N through CCP + 50% of

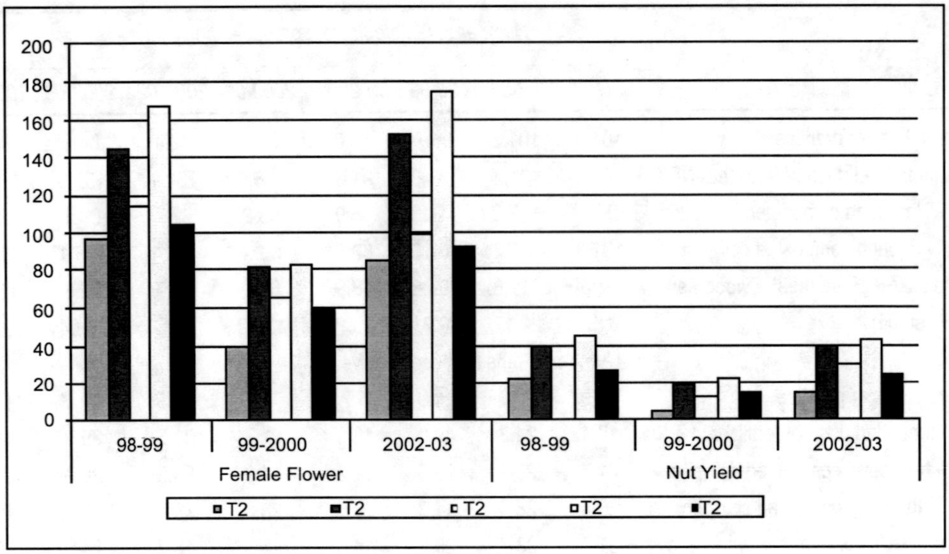

Fig. 20.1. Female flower and nut yield in pre and post-cyclone period as influenced by integrated nutrient management

RDF were found to be statistically superior in production of female flowers (138.8) in the preceding year of cyclone followed by palms (106.8) under RDF. All the other treatments including control recorded low female flower production which was found statistically at par. The number of female flower production increased in each year at a slower rate except in the case of 50% N through CCP + 50% of RDF and full RDF. Palms under RDF recorded maximum female flowers (175.1) on the third year of cyclonic damage followed by palms (152) under 50% N through CCP + 50% of RDF.

The increase in yield was faster in each year of advancement in the post-cyclonic period in palms supplied with nutrients, whereas, the increase in nut number was in slower pace in the control. The palms supplied with nutrients recorded increase in number of nuts in the succeeding years of cyclone as compared to the control. The palms under 50% N through CCP + 50% of RDF and full RDF yielded similar number of nuts of pre-cyclonic period, in the third year of damage. Both the treatments were statistically at par and found superior to the other treatments. The less damage to the palms and quick recovery of cyclonic damage in vegetative, and yield attributing character due to application of 50% N through CCP + 50% RDF as well as RDF may be due to strong internal nutrient status of palms which were made possible by enhancement of nutrient content of soil and soil physical condition.

It is, therefore, concluded that application of 50% N through CCP in combination with 50% inorganic fertilizers enhances the vegetative and reproductive characters of coconut and gives strength to overcome the adverse conditions of cyclone.

References

Kalpana, M. Rao, D.V.R. and Srinivasulu, B. 2006. Integrated nutrient management in coconut based cropping system under coastal ecosystem of Andhra Pradesh. *J. Plantn Crops.* **34(3)**: 258-262.

Nagarajan, R., Manickam, T.S., and Kothandaraman,G.V. 1988. *J. Plantn. Crops.* **16**:59-62.

Nair, K.K., Biddappa, C.C., Palaniswami, C. and Upadhayay, A.K. 1996. A critical review of plant supply, needs efficiency and policy issues for the plantation crops for the year 2000 and 2005 A.D In : Proc. Of Symposium on plant nutrient needs, supply, efficiency and policy issues : 2000:2005 A.D.

Raghava Rao, D.V., Srinivasulu, B. and Sujatha, A. 2007. Five decades of coconut research at Agricultural Rsearch Station, Ambajipeta. *Indian Coconut J* .**37(9)**:15-19.

Sheeba, R.I., Nair, M.A. and Jayachandran, B.K.1999. Organic recyling in coconut. *Indian Coconut J.* **29(10)**:10-12.

Talashilkar, D.D., Nagwekar, S.R., Akhave, S.R. and Dabke, D.J. 2008. Effect of integrated use of manures and fertilizers on coconut yield and changes in available nutrient content and N, P and K fractions of inceptisols of Konkan region. *J. Plantn Crops.* **36(2)**:112-116.

Thampan, P.K.1995. "Organic Agriculture" Peekay Tree crops Development Foundation, Cochin-20.

Thomas George, V., Prabhu, S.R., Subramanian, P. and Rohini Iyer. 2001.Organic farming technologies in coconut. (English) – CPCRI Publication, 6p.

21

Potential for Biomass Recycling in Oil Palm

S. Sunitha, M.V.M. Shareef and M. Kochu Babu

INTRODUCTION

The importance of organic farming has increased due to its environmentally friendly approaches and growing consumer awareness of food safety. Producing palm oil by organic cultivation is highly possible since many of the agronomic practices for its cultivation have minimum environmental inputs and are moving towards even more eco-friendly methods.

Oil palm, one of the richest sources of vegetable oil on per hectare basis, is a nutrient-demanding crop as well. A plantation yielding 25 tonnes of FFB/ha is reported to remove 93.5kg N, 11.0 kg P, 92.7 kg K, 19.3 kg Mg and 20.3 kg Ca from soil every year (Turner and Gillbanks, 1988).

The bulk of wastes derived from the palm oil industry are basically cellulose and organic biomass with high nutrient content. With proper handling and management, the biomass can be converted into value-added by-products. At the same time, it is a beneficial source of organic matter and plant nutrients, especially for recycling in the field.

Annual availability of biomass was assessed from different components viz., pruned leaves (pinnae, rachis and petiole), male inflorescences, empty fruit bunches, mesocarp wastes and shell collected from a 16 year old plantation at NRCOP, Regional Station, Palode. The total biomass production of 27 year old plantation was also assessed by destructive sampling of representative palms by collecting various components viz., trunk, fronds, spears, inflorescences, leaf bases, bole mass and roots separately. The representative samples of all the components available annually as well as at felling were analysed for major and secondary nutrients and based on dry matter production, nutrient potential was worked out annually and also at felling stage.

Efforts were also made to recycle the nutrients available in the bio-wastes annually by way of composting. For this, compost was prepared from fronds and EFB, which is the major portion of palm wastes, by the technique of aerobic digestion (Sunitha and Varghese, 1999). Compost thus produced was used along with inorganic fertilizers in an integrated manner in a mature oil palm plantation.

Fronds contributed about 70% of the total dry matter through regular harvests and annual pruning and the remaining 30% was contributed by rest of the palm parts. Of the bio-wastes available to the tune of 15 tonnes every year from a mature plantation under rainfed condition, 132kg N, 11.5kg P, 155kg K, 92kg Ca and 38kg Mg could be expected (Table 21.1).

Table 21.1. Annual availability of biomass and nutrient contribution from a mature oil palm plantation

Palm parts	*Biomass production (kg/ha/year)*	*Nutrient contribution (kg/ha/year)*				
		N	*P*	*K*	*Ca*	*Mg*
Pinnae	3318	82.8	4.9	27.5	20.6	10.1
Petiole & Rachis	7222	19.2	2.1	79.7	66.2	17.8
Male Inflorescence	272	7.6	1.6	4.9	1.8	3.2
Empty Fruit Bunches	1301	12.5	1.7	33.5	2.7	4.0
Mesocarp wastes	2506	6.5	0.9	6.5	0.6	1.7
Shell	800.8	3.3	0.3	3.4	0.5	1.2
Total	15473	131.9	11.5	155.5	92.4	38.0

Considering a planting density of 143 palms in a hectare, 163 tonnes of total dry matter can be expected at the end of the crop. The trunk contributed more than 50% of the total dry matter followed by the underground bole mass. Varying figures were reported by different workers for the trunk weight at different ages. At the density of 148 palms per hectare and assuming that 85% of the palms are standing at the time of felling, the total weight of palm trunk amounts to about 75.5 tonnes ha^{-1} (Chan *et al.*, 1981).

At felling stage, nutrients to the extent of 1503 kg N, 129 kg P, 2345 kg K, 513 kg Ca and 438 kg Mg could be expected from the available biomass of 163 tonnes (Table 21.2) .

Chan *et al.* (1981) reported that about 90 t/ha of dry matter could be obtained at felling a mature plantation, which contained 518 kg N, 49 kg P, 721 kg K, 112 kg Mg and 182 kg Ca. Mohd. Hashim *et al.* (1993) quoted slightly lower values as 339 kg N, 32 kg P, 424 kg K, 76 kg Mg based on the measurements of palm residues of trunks and fronds only. Khan-Kiang Kee (2004) reported that with a palm density of 126/ha at replanting, total trunk biomass was estimated at nearly 67 t/ha. The potential nutrient pool that could be recycled from trunk biomass is enormous especially N and K at 187 kg and 770 kg, respectively. In the present study, all the values are high compared to the earlier reports, since in none of the earlier studies, the significant contributions from below ground biomass in terms of dry matter production or nutrient value is accounted.

Table 21.2. Availability of biomass and nutrient contribution at felling

Palm parts	*Biomass production (t/ha/year)*	*Nutrient contribution (kg/ha/year)* N	*P*	*K*	*Ca*	*Mg*
Trunk	85.80	729.30	60.06	1021.02	223.08	199.92
Leaflets	8.72	183.04	11.44	85.80	52.91	30.03
Rachis & Petiole	14.16	84.37	12.87	204.49	68.64	25.74
Spears	3.80	55.77	6.01	85.51	11.58	9.30
Inflorescence	0.12	2.86	0.43	2.86	1.00	0.43
Leaf bases	8.72	87.23	5.72	140.14	52.91	47.19
Bole mass	35.75	321.75	28.60	750.75	85.80	104.39
Roots	5.58	38.61	3.86	54.34	17.16	21.45
Total	162.65	1502.93	128.99	2344.91	513.08	438.45

Biomass Recycling

Decomposition processes play an important role in soil fertility in terms of nutrient cycling and formation of soil organic matter. On an average, most of the oil palm residues were found to decompose within 12–18 months while some of the hardier materials, particularly roots, took much longer than 18 months to decompose.

Nutrient release from oil palm residues showed different release pattern between residue types and nutrients. The rachis, trunks and roots all showed nutrient releases in the order of K> Mg = Ca>P>N (Khalid *et al.* 2000).

Proper residue management improves the spatial integration of nutrient release and uptake by the rooting systems of young palms. The supply of nutrient requirement that is partly provided by the recycling of biomass can reduce the use of inorganic fertilizers to optimize growth rates of the immature palms.

Limitation of Direct Mulching of Residues

The practical problems of using the biomass for direct mulching are: risk of fire, attracting of pests, favouring disease incidence, bulk in quantity, difficult in transporting and application, difficulty in harvesting and picking of loose fruits, temporary immobilization of nutrients etc.

Composting of Residues and Effect of Compost on Oil Palm Plantation

The compost prepared mainly from fronds and empty fruit bunches contained on an average 1.8-2% N, 0.2-0.3% P and 0.9-1.2% K. When the palms expressed the actual effect of nutrition after three years, all treatments except control had more or less equal number of bunches per palm, but for FFB yield, T_1, T_2 and T_3 were on par.

During next year, similar trend was observed for the number of bunches, whereas, FFB yields were on par for all the treatments except control. The cumulative average yield data for five years indicated no significant differences among T_1, T_2, T_3 and T_4 with respect to bunch number, but for bunch yield, T_1, T_2 and T_3 were on par and superior. The treatment, which received one-third dose as inorganic and two-third as organics, recorded the highest bunch yield (Table 21.3). Thus in an oil palm ecosystem, where there is much scope of nutrient recycling from the byproducts, an integrated approach would be beneficial for nutrient management.

Table 21.3. Number of bunches and fresh fruit bunch yield in different treatments (Average of five years)

Treatments	*No.of bunches*	*FFB yield(kg/palm)*
Full fertilizer dose as inorganic fertilizers	5.14	113.7
2/3rd dose as inorganic and 1/3rd dose as compost:	4.46	106.2
1/3rd dose as inorganic and 2/3rd dose as compost:	5.44	118.8
Full dose as compost.	4.57	93.35
Control	3.39	80.85
CD(P=0.05)	1.09	23.63

Conclusion

In the oil palm production chain, large quantities of biomass or by-products are produced which are hardly used for adding value to the production chain. Quiet often, byproducts are considered as a nuisance which may lead to environmental problems. As the biobased economy develops and markets demand organic products, these byproducts should be seen as resources. So oil palm is highly amenable for nutrient recycling which is environmentally safe, ecologically sustainable and economically feasible.

References

Chan, K.W., Watson, I. and Lim, K.C. 1981.Use of oil palm waste material for increased production. *The Planter* **57**:14-37.

Khalid, H. Zin, Z.Z. and Anderson, J.M. 2000. Decomposition processes and nutrient release patterns of oil palm residues. *Journal of Oil Palm Research* **12**:46-63.

Khan-Kiang Kee. 2004. Nutrient reserves and recycling from oil palm trunks at replanting. *Proc. of the Fourth International Crop Science Congress*, Sept. 2004, Australia.

Mohd. Hashim, T., Teoh, C.H., Kamaruzaman, A. and Mohd Ali. 1993. Zero burning–an environmentally friendly replanting technique. In: *Proc. of PORIM International palm oil Congress ' Update and Vision' Agriculture module.* (Eds. Jalani Sukaimi *et al*) Palm Oil Research Institute of Malaysia, pp.185-194.

Sunitha, S and Varghese, P.T. 1999. Composting of oil palm wastes for effiecient recycling of nutrients in palm plantations. *The Planter* **75**:677-681.

Turner P.D. and Gillbanks, R. A.1988. *Oil Palm cultivation and management.* Incorporated Society of Planters, Kuala Lumpur.

22

Studies on Organic and Biodynamic Agriculture in Tea

B. Radhakrishnan, R. Sasidhar and P. Sankaranarayanan

INTRODUCTION

Organic and biodynamic agriculture is as old as human civilization. In fact, primitive agriculture started with organic culture. It envisages agricultural production without the use of any synthetic agro-chemicals. However, the economic consideration calls for development of an appropriate technology for organic production through use of naturally occurring substances in a judicious blend, which will not only be environmentally safe and free from chemical residues but will also sustain the long term productivity at an economically viable level. Biodynamic agriculture first propounded by Rudolf Steiner (1924), is a method of organic farming that has its basis in a spiritual world-view, treats farms as unified and individual organisms, emphasizing balancing the holistic development and interrelationship of the soil, plants, animals as a closed, self-nourishing system. Regarded by some proponents as the first modern ecological farming system, biodynamic farming includes organic agriculture's emphasis on manures and composts and exclusion of the use of artificial chemicals on soil and plants. Methods unique to the biodynamic approach include the use of fermented herbal and mineral preparations as compost additives and field sprays and the use of an astronomical sowing and planting calendar.

The study was conducted in one of the tea estates in Munnar, Kerala planted with Assam seedling jat in 1924 at a spacing of 120 cm × 120 cm and pruned at a height of 60 cm from the ground level during April 2002. Three systems of tea cultivation viz., conventional intensive system of farming using chemical fertilizers, micronutrients, fungicides, insecticides and weedicides as per the recommendations of UPASI Tea Research Institute; Organic system of farming using organic fertilizers, fungicides and micro-nutrients as suggested by IMO; and Biodynamic system of farming using biodynamic formulations, calendar, and others as suggested by Biodynamic Association of India (BDAI) were adopted in large blocks each consisting of 500 tea bushes.

In the blocks under organic treatment, well-decomposed compost was applied @ 10 t/ha in the first year. In the second and third years, compost @ 5 t and neem cake @ 2.5 t/ha of was applied. Rock phosphate was applied @ 400 kg/ha in alternate years. Bio-fertilisers such as *Azospirillum* and phosphate solublilising formulations were applied @ 25 kg each/ha. To meet the potassium requirement, wood ash was applied @ 500 kg/ha.

For the control of mites, sulphur @ 1.0 kg/ha, neem formulations @ 1.0 l/ha were used and the fungal formulation *Paecilomycus fumosoroseus* were used. For the control of thrips, Verelac @ 1.5 kg/ha was used. Blister blight was not a serious problem in the experimental blocks. Copper oxychloride was applied @ 420 g/ha. With the permitted limit of 6 kg of copper oxychloride it was possible to control the disease. Weed control was only through manual weeding in both organic and biodynamic formulation treated blocks. Wherever large vacancies existed, cover crop such as *Mimosa invisa* was raised, which helped to suppress weed growth.

The biodynamic treatment blocks received biodynamic compost @ 10 t/ha containing preparations of the five fermented and potentised herb materials: Oak bark, Valerian, Dandelion, Nettle and Chamomile during the first week of August. Rock phosphate, biofertilisers and wood ash were used as in the case of organic treatment. Cow urine (1 L in 10 L of water) was applied once a month between January and March.

B D 500 (60 g in 40 L of water /ha) was applied thrice to soil every year. B D 501 (2.5 g in 40 L of water/ha) was applied twice every year over the canopy of bushes early in the morning. B D 500 (horn manure) was prepared by using cow horn and the lactating cow's dung. BD 501(horn silica) was prepared by using cow horn and powdered pure silica stones. Cow Pat Pit Manure (CPP) was applied once every month during April, May and September to November @ 5 kg/ha in 100 L of water/ha. It was prepared using cow dung, egg shell and rock dust. While making CPP manure, one set of Biodynamic preparations (BD-502 to BD-505) was used for every 30 kg of cow dung. The above formulations were potentised before application and applied during ascending/descending period. Similarly extracts of *Casuarina* and *Equisetum* were sprayed for the control of pests.

The results of the study carried out for one pruning cycle of three years (Table 22.1) indicated that the total yield obtained for 85 plucking rounds in organic and biodynamic system of farming was significantly lower than that of conventional system of farming. The yield decrease varied from 5–15 %.

Several authors had claimed that tea yield decreased drastically upon conversion to organic (Werf, 1990, Mohan and Bachi, 1999, Radhakrishnan *et al.*, 2006). During the transition from chemical based agriculture to natural system of farming, the yield levels definitely drop, sometimes precipitously (Ghosh Hajra, 2002). In the present study, the loss of crop in the different years after conversion varied between 5 and 15%. In south India, Werf (1990) reported that tea yield decreased by 21–33% in the first year of transition from conventional to organic cultivation. Similar trend was noticed in the present study also. Sudden withdrawal of inorganic fertilizers, severe zinc and magnesium deficiency symptoms and profuse weed growth might be the reasons for the depression in yield in organic and biodynamic blocks. This is in agreement with the findings of Mohan and Bachi (1999).

Table 22.1. Effect of different farming systems on tea yield

Treatments#	Yield (kg made tea/ha for 85 plucking Rounds) *
Conventional	2888.43b
Organic	2426.14a (19.05 %)
Biodynamic	2557.86a (12.92%)
CD (P=0.05)	168.41

Means followed by the same letter in a vertical column are not significantly different at 5% level. Figures in parenthesis indicate percentage reduction in crop compared to Conventional system of farming.

Observations made on the leaf samples from different treatment blocks revealed moderate to severe zinc and magnesium deficiency in the organic and biodynamic treatment blocks. Leaf analysis also showed that these nutrients were below critical level in the leaves (Table 22.2).

Fukuoka (1990) opined that once stability in ecosystem has been achieved, agro-practices adjusted/refined, the yield compares with chemical based farming system. Studies carried out in Makaibari showed that the yields begin to increase slowly from the fifth year (Ghosh Hajra, 2002).

Table 22.2. Deficiency symptoms recorded in the experimental site

Deficiency symptoms	*Severity*	*Critical values*
Nitrogen	Mild	6.00% (Normal)
Potassium	Moderate	1.50% (Deficient)
Magnesium	Severe	0.03% (Deficient)
Zinc	Severe	15 ppm (Deficient)

Mohan and Bachi (1999) reported when commercial tea estates switched over to organic farming, the usual tea fields lost 10% crop in the first year, 20, 30 and 40% in second, third and fourth year, respectively and stabilized to gain 20% in fifth year as a consequence of recycling organic matter and the resurgence of effective natural enemies of the crop pests. But Werf (1990) reported that in south India tea yield decreased by 21–33% in the first year of transition from conventional to organic cultivation. These studies are in agreement with the present observations.

In an experiment laid out at Tocklai on the comparisons of changeover of conventional plots to organic tea in a 25 year old clonal tea block, no drastic reduction in yield was observed. However, only marginal differences between the comparative plots manured with organics and inorganics were observed. This indicated that switch over to organic farming do not drastically reduce yield as is widely observed. However, the decline of yield towards the steady state can be thwarted if the initial yield had been significantly higher and if the required infrastructural base for organic farming is established well within the transition phase (Sivepalan, 1993). Singh *et al.* (1999) found that the yield losses could be avoided/minimised if new extension or uprooted area was adequately rehabilitated prior to planting. Further studies are needed to confirm this.

The organic matter, water holding capacity and cation exchange capacity of the soils were significantly higher in the biodynamic blocks (Table 22.3). The increased organic matter content of soil is conducive for the formation of stable soil aggregates with reduced bulk density. This seems to have increased the water holding capacity significantly as a result of increased total pore space.

Table 22.3. Impact of farming systems on certain soil properties

Treatments #	*O.M (%)*	*WHC (%)*	*CEC (%)*
Conventional	5.24 a	70.95 a	6.41 a
Organic	6.30 b	82.25 b	8.61 b
Biodynamic	6.34 b	88.25 b	9.71 b
CD(P = 0.05)	0.52	8.90	1.20

Means followed by the same letter in a vertical column are not significantly different at 5% level.

The significant increase in the cation exchange capacity of soils of biodynamic blocks is due to large number of adsorption sites for cations due to the increase in mean clay content of the soil and higher organic matter status. Thus the study showed that biodynamic system of cultivation is superior to other forms of cultivation in improving the physical properties and OM status, CEC and WHC of the soil.

In general bacterial populations were dominant followed by actinomycetes, yeast and fungi. Among the treatments Biodynamic blocks recorded higher bacterial population followed by organic and conventional. Similar trend was observed with actinomycetes and fungi (Table 22.4).

Table 22.4. Effect of different farming systems on soil properties and soil microbial population in tea

Treatment	*Conventional*		*Bio-dynamic*		*Organic*	
	Initial	*Final*	*Initial*	*Final*	*Initial*	*Final*
pH	4.1	4.4	4.0	4.5	4.0	4.5
EC	0.07	0.14	0.05	0.09	0.08	0.09
OM (%)	7.6	8.1	3.5	6.8	5.9	6.3
Total Bacteria	0.11×10^5	1.3×10^6	5.8×10^5	10.7×10^6	6.6×10^5	17.8×10^6
Total Fungi	0.9×10^4	23.5×10^4	1.3×10^4	10.7×10^4	1.9×10^4	15.4×10^4
Actinomycetes	1.27×10^5	2.5×10^4	3.0×10^5	4.5×10^4	3.0×10^5	4.5×10^4
Azospirillum	1.6×10^4	2.1×10^5	2.8×10^4	4.4×10^5	3.29×10^5	9.2×10^4
Phosphobacteria	4.0×10^4	6.24×10^4	6.5×10^4	12.2×10^4	5.0×10^4	14.02×10^4
Trichoderma	1.0×10^3	6.2×10^2	0.5×10^3	6.1×10^2	7.1×10^2	18.1×10^2

Studies conducted by Kamala Devi *et al.* (2001) in coffee and Selvaraj and Thiruvelavan (2001) in cabbage also showed higher population of all microbes in biodynamic and organic farming followed by conventional farming. The higher microbial population in biodynamic treatment may be due to the higher availability of essential nutrients.

The degree of infestation by various pests, diseased and weeds varied in the different treatment blocks (Table 22.5). It was noticed that the number of herbal/copper oxychloride sprays required for controlling pests and diseases was more in the blocks under organic and biodynamic treatments. It varied from 22–24 in these blocks. However, it was only 16 in blocks under conventional farming.

Table 22.5. Important pests, diseases and weeds recorded in the experimental site

Pests	*Diseases*	*Weeds*
Red spider and purple mites	Blister blight	*Calceloria mexicana*
Flushworm and Tea tortrix	Grey blight	*Bidens spilosa*
Aphids		*Ageratum conizoides*
Thrips		*Conyza ambigua*
		Crassocephalum crepidioides
		Basella alba

The number of rounds required for manual weed control in organic and biodynamic treatments was five in the first year and four in the subsequent years from pruning. The requirement for herbicidal sprays remained static at four rounds under the conventional treatments every year. In each round, the requirement of man days was triple under organic and biodynamic treatments compared to the conventional treatment.

The number of parasitoids and predators of Red spider mites, flushworm, aphids and tea tortrix were more in organic and biodynamic treatment blocks compared to conventional blocks (Table 22.6).

Table 22.6. Number of natural enemies of various pests in the different farming systems

Treatments	*Purple mites*	*Red spider*	*Tea Thrips*	*Flushworm*	*Tea tortrix*	*Aphids*	*Total*
Conventional	Nil	Nil	Nil	1	1	3	5
Organic	1	2	1	3	3	6	16
Biodynamic	1	2	1	4	3	7	18

The important natural enemies recorded for Red spider and purple mites, flushworm, tea tortrix, thrips and aphids are given in Table 22.7.

Table 22.7. Natural enemies of pests recorded during the present study

Pest species	*Predators*	*Parasitoids*
Purple mite,		Nil
Calacarus carinatus (Green)	*Amblyseius herbicolus* (Chant)	
Red spider mite,		Nil
Oligonychus coffeae(Nietner)	*Oligota pygmaea*	
(Acarina : Tetranychidae)	*Stethorus gilvifrons*	
Tea thrips		Nil
Scirtothrips bispinosus (Bagnall)	*Anthocoris* sp.	
(Thysanoptera : Thripidae)		
Flushworm	*Menochilus sexmaculatus* (Fab.)	*Apanteles aristaeus*
Cydia leucostoma Meyrick		Nixon
,		*Apanteles* sp. (*Glomeratus* group)
		Bracon nr *Hebetor* Say
	Tea tortrix	*Phytodietus spinipus* (Cameron)
Homona cofferia		*Palexorista solennis* (Walker)
Nietner	*Cantheconidea furcellata*	
(Lepidoptera : Tortricidae)	(Wolff)	
Tea aphid,		
Toxoptera aurantii(	*Aphidius colemani* Vierick	*Coccinella septempunctata* L.
Boyer de Fonscolombe*)*	*Trioxys indicus* Subba Rao and Sharma	*Episyrphus balteatus* (De Geer)
		Menochilus sexmaculatus (Fab.)
		Micromus timidus Hagen
		Paragus tibialis (Fallen)

The average cost/kg and cost of made tea/ha in organic treatment was Rs.73.15 and Rs. 1,46,297 respectively (Table 22.8). They were Rs. 72.25 and Rs. 1,58,955 respectively in the case of Biodynamic treatment. However, it was only Rs. 53.09 and Rs.1,32,732 in the conventional treatment.

Table 22.8. Cost of cultivation for organic, biodynamic and conventional farming system

Sl. No.	*Particulars*	*Cost (Rs/ha)*	*Cost (Rs/kg)*	*Cost (Rs/ha)*	*Cost (Rs/kg)*	*Cost (Rs/ha)*	*Cost (Rs/kg)*
		YPH 2000 kg & plucking average 26 kg (Organic tea)		*YPH 2200 kg & plucking average 27 kg (Biodynamic tea)*		*YPH 2500 kg &plucking average 29 kg (Conventional tea)*	
1.	Plucking	30786	15.39	31676	14.40	33773	13.51
2.	Pruning	705	0.35	705	0.32	705	0.28
3.	Disposal of prunings	1125	0.56	1125	0.51	1125	0.45
4.	Shade management	168	0.08	168	0.08	168	0.07
5.	Manuring	26280	13.14	29200	13.27	9034	3.61
6.	Dolomite	948	0.47	948	0.43	948	0.38
7.	Foliar application	972	0.49	972	0.44	3382	1.35
8.	Lane cutting	338	0.17	338	0.15	338	0.14
9.	Pest management	3410	1.71	3410	1.55	2058	0.82
10.	Disease management	3791	1.90	3791	1.72	2426	0.97
11.	Manual weeding	3263	1.63	3263	1.48	2326	0.93
12.	Cow pat and BD - 500 application	0	0.00	3572	1.62	0	0.00
	Total	71784	35.89	79166	35.98	56282	22.51
	Cost of manufacturing	26000	13.00	28600	13.00	32500	13.00
	Workers fringe benefits (60% of wages)	33512	16.76	36189	16.45	28950	11.58
	Overhead charges	15000	7.50	15000	6.82	15000	6.00
	Total cost/ha. and per kg mt	146297	73.15	158955	72.25	132732	53.09

YPH = Yield per hactare

References

Anonymous, 1995. Financial performance of tea companies 1973-74 to 1986-87. *Occasional paper,* 1, Indian Tea Association. Calcutta.

Crooke, W.M, 1964. The measurement of the cation exchange capacity of plant roots. *Plant and Soil.* XXI (1):43-49

Fukuoka, M. 1990. The one straw revolution. Friends Rural Centre Rasulia Hoshangabad, MO, India. :181p.

Ghosh Hajra, N. 2002. Organic Tea growing in Darjeeling hills: A case study. PLACROSYM **XV**: 437-47.

Kamala Bai, S., Paneerselvam, Muralidhara, H.R., and Raghuramulu, Y. 2001. Effect of Biodynamic manures on growth of coffee seedlings. In *Proceedings of the II International Conference on Biodynamic Agriculture.* 16-18. March, 2001,Bangalore, pp 43-45.

Mackay, J.M. 1990. Organic production systems. A means of reducing chemical inputs. *J.Sci.Food Agric.*, **53**:112-17.

Mohan, Brij and Bachi, Rainer. 1999. Organic tea. In: *Global Advancement in Tea Science* (ed. N. Jain). Aravali Books International, New Delhi, India, 813-16.

Muraleedharan, N. 2001. Organic tea in India: Problems and perspective. *Planters' Chronicle.* **97**:439-42.

Radhakrishnan, B. 2004. Package of practices for organic tea. *Planters' Chronicle.* **100**(1):19-39.

Radhakrishnan, B., Ranjit Kumar, Q., Ganapathy, M.N.K., and Hudson, J.B. 2006. Effect of conventional, organic and biodynamic farming systems in tea. *Journal of Plantation crops.* **34**(3):330-33.

Selvaraj, N., and Thiruvelavan. P. 2001. Effect of biodynamic treatments on yield and yield attributes of cabbage (*Brassica oleracea* var. capitato) cv. Questo. *Proceedings of the II international conference on Biodynamic Agriculture.* 16-18. March, 2001,Bangalore, pp 41-42.

Singh, I.D., Bhattacharya, A.K. and Bera, B.1999. Organic tea cultivation an ecofarming approach. *Assam Rev. & Tea News* **88**(9): 5-15.

Sivepalan, P. 1993. Organic farming of tea as an aid to maintaining environmental quality. *Proc. Intl. Symp. On Tea Science and Human Health. TRA, Calcutta, January 11-14,* India, pp 300-02.

Steiner, R., 1924. A series of lectures given in 1924. Biodynamic Farming and Gardening Association Inc., Kimberton, Pa. USA. (1993).

Takeo, T. and Oosawa, K. 1976. Photometric analysis and statistical evaluation of black tea infusion. *Bulletin of National Tea Research Institute for Tea (Japan)* **12**:125-81.

Waksman, S.A. 1952. Principles of soil microbiology. Baltimore: Williams and Wilkins, USA.

Werf, Erik J. van der. 1990. Organic tea cultivation at Singampatti Group of B.B.T.C., India. ETC – Foundation, Leuden, The Netherland.

23

Standardisation of Organic Farming Practices for Banana *Var*. Nendran

R. Pushpakumari, K.R. Sheela, C. Nandakumar and Annamma George

INTRODUCTION

Banana has emerged as the most important fruit crop of Kerala occupying an important place both in lowlands and uplands. Resource management in terms of nutrient application and pest and disease management for banana is very important for expressing the full genetic yield potential of the crop. There is growing concern on the ill effects of chemical fertilizers and pesticides. Unscientific and unbalanced use of chemical fertilizers and chemical pesticides in banana lead to the incidence of toxic residues in banana fruits causing health problems. Cost effective, nature friendly technologies are to be focused in response to changing environment for producing farm produce devoid of chemical contaminants Assessing the feasibility of organic farming practices in banana excluding the use synthetic pesticides, fertilizers and food additives will help to achieve sustainability in production and also to produce safe food for human consumption. Nutrient management for organic cultivation of Nendran variety of banana was standardiz at the College of Agriculture, Vellayani, Kerala.

The nutrient sources used were: NPK @ 190:115:300 kg plant^{-1}, FYM@ 15 kg + 4kg ash plant^{-1}, Poultry manure @ 5 kg + 4.5 kg ash plant^{-1} Vermi-compost @ 5 kg + 3.5 kg ash plant^{-1}, and Coir pith compost @ 6.5 kg + 3.5 kg ash plant^{-1}. The time of application included two (at planting and 4 MAP); three (at planting, 2½ and 5 MAP) and four splits (planting, 2, 4 and 6 MAP). The organic sources were quantified to satisfy the nutrient requirement of the chemical recommendation. Ash was used as the organic source to supply potassium.

The following organic practices were adopted uniformly for all treatments with a view to control pest and diseases and to ensure nutrient availability.

Selection of healthy, disease and insect free suckers for planting, hot water treatment for the control of nematodes, dipping of suckers in cow dung-ash slurry with 5% neem oil before planting for managing rhizome weevil, application of neem cake @ 1 kg plant^{-1} at planting, uniform spray of neem seed oil-garlic emulsion (5%) along with sticker at three and four months for managing pseudostem weevil, *in situ* manuring with cowpea around the banana sucker. 10–15 g seeds were sown around each sucker at planting, application of farm yard manure/compost @10 kg plan^{-1} tonnes along with 10g azoto bacter, keeping the field weed free and removing the dried leaf sheath at periodic interval.

Planting was done during March 2007 and harvest was completed by December 2007. The results indicated that different sources of nutrients did not impart any significant influence on yield attributes like number of hands bunch^{-1}, number of fingers hand^{-1}, length and girth of fingers, peel-pulp ratio and bunch weight (Table 23.1). Gomes *et al.* (1995) also observed no significant difference in yield between the different sources and quantity of organic manure in banana. However, coir pith compost registered the highest yield of 7.46 kg plant^{-1} followed by chemical source and poultry manure. The efficacy of coir pith compost and poultry manure in enhancing crop yield was reported earlier by Baiju (1998).

The different organic sources tried could effectively supply the nutrient required for banana as the quantity worked out was sufficient to substitute the chemical nutrient level. Use of vermi-compost as organic source along with incorporation of cowpea as green manure in enhancing the crop yield was also reported by Geetha and Nair (2000).Though the time of application did not influence the banana yield, application of nutrients in three splits registered the highest bunch weight of 7.50 kg/plant. The continued nutrient supply over a longer period might have favoured better availability and absorption of nutrients resulting in higher yield.

Economic analysis (Table 23.2) of different treatment combinations revealed that the highest net income was registered by the treatment receiving nutrients through chemicals in two splits. Among the different sources of nutrients, chemical fertilizers recorded the highest average net return of Rs.1.90 lakh per year. Compared to organic manures, the chemical fertilizers have relatively lower price resulting in higher net income. Among the different organic sources tested, poultry manure registered the highest net income of Rs. 1.78 lakh. The net return from application of vermi-compost and coir pith compost were low and Rs. 0.99 and 1.11 lakh/year, respectively. This could be due to the higher unit price of these purchased inputs. The net income from these sources can be enhanced by preparing these composts in the farm itself.

Table 23.1. Influence of organic farming practices on the yield attributes and yield of Nenrdan variety of banana

Treatment combinations (Quantity & time of application)	*Yield attributes and yield*					
	No. of hands/ bunch	*No. of fingers/hand*	*Fruit length (cm)*	*Fruit girth(cm)*	*Peel/ pulp ratio*	*Bunch wt/ plant (kg)*
190:115:300 kg NPK plant^{-1}/ planting and 4 MAP	4.67	7.67	25.50	13.33	2.80	7.60
190:115:300 kg NPK plant^{-1}/ planting, 2 ½ and 5 MAP	4.67	8.40	25.67	13.33	2.80	6.40
190:115:300 kg NPK plant^{-1}/ planting,2, 4 and 6 MAP	4.83	8.23	23.67	12.67	2.89	7.50
FYM 15 kg + 4kg ash plant^{-1}/ planting and 4 MAP	4.83	9.20	22.67	12.50	1.98	6.40
FTM 15 kg + 4kg ash plant^{-1}/ planting, 2 ½ and 5 MAP	4.67	8.97	21.17	13.00	2.32	6.90
FYM 15 kg + 4kg ash plant^{-1}/ planting, 2, 4 and 6 MAP	4.33	8.70	22.67	12.33	1.89	7.50
PM 5 kg + 4.5 kg ash plant^{-1}/ planting and 4 MAP	4.00	9.40	24.83	12.17	2.45	7.10
PM 5 kg + 4.5 kg ash plant^{-1}/ planting, 2 ½ and 5 MAP	3.83	3.27	25.33	12.83	3.61	6.90
PM 5 kg + 4.5 kg ash plant^{-1}/ planting, 2, 4 and 6 MAP	4.66	9.00	24.50	12.33	1.51	7.50
VC 5 kg + 3.5 kg ash plant^{-1}/ planting and 4 MAP	4.17	8.57	25.00	12.83	2.35	6.70
VC 5 kg + 3.5 kg ash plant^{-1}/ planting, 2 ½ and 5 MAP	4.33	8.17	25.17	12.67	2.07	6.40
VC 5 kg + 3.5 kg ash plant^{-1}/ planting, 2, 4 and 6 MAP	4.33	8.47	25.67	13.00	1.71	7.40
CPC 6.5 kg + 3.5 kg ash plant^{-1}/ planting and 4 MAP	4.33	8.83	24.67	13.50	2.45	7.40
CPC 6.5 kg + 3.5 kg ash plant^{-1}// planting, 2 ½ and 5 MAP	4.83	8.47	22.67	12.50	1.75	7.50
CPC 6.5 kg + 3.5 kg ash plant^{-1}/ planting, 2, 4 and 6 MAP	4.67	8.67	26.00	13.67	2.50	7.50
CD (P = 0.05%) Manure x Time	NS	NS		NS	NS	NS
Mean NPK @ 190:115:300 kg plant^{-1}	4.72	8.10	24.94	13.28	1.90	7.20
FYM@ 15 kg + 4kg ash plant^{-1}	4.61	8.96	22.17	12.61	2.06	6.95
Poultry manure @ 5 kg + 4.5 kg ash plant^{-1}	4.17	8.89	25.89	12.44	2.52	7.17
Vermi-compost @ 5 kg + 3.5 kg ash plant^{-1}	4.28	8.40	25.28	12.83	1.71	6.81
Coir pith compost @ 6.5 kg+3.5 kg ash plant^{-1}.	4.61	8.65	25.44	13.22	2.23	7.46
CD (P = 0.05%)	NS	NS	1.352	NS	NS	NS
Mean At planting and 4 MAP	4.40	8.73	24.53	12.97	2.41	7.04
At planting, 2 ½ and 5 MAP	4.47	8.45	24.00	12.87	1.95	6.82
At planting, 2, 4 and 6 MAP	4.57	8.61	24.50	12.80	1.90	7.50
CD (P = 0.05%)	NS	NS	NS	NS	NS	NS

NS = Not significant

FYM = Farm Yard Manure, PM = Poultry manure, VC = Vermi-compost and CPC = Coir pith compost

MAP = Months After Planting

Moreover, organic products fetches premium price in market which in turn could enhance the net return. Thus it could be inferred that different organic sources could be effectively used as a substitute of chemical fertilizers without any reduction in bunch yield of Nendran banana. Depending on the availability, any of the sources can be used by the farmer and going for two splits was more remunerative than three or four splits.

The sucker treatment with cow dung -ash + 5% neem oil was effective in containing the infestation of rhizome weevil. The organic pest management treatments were also effective in controlling the attack of pseudostem weevil. Among the total number of 180 plants treated as per schedule, only four plants exhibited symptoms of pseudostem weevil at harvest stage. This indicated that surveillance followed by weeding the base and interspaces of banana, removal of dried leaf sheaths and application of organic pest control methods could effectively manage the pseudostem weevil.

Table 23.2. Economics of organic farming in Nendran variety of banana*

Treatment combinations(Quantity & time of application)	*Cost of cultivation (Rs.)*	*Yield (t/ha)*	*Total @ Rs.18/kg income (Rs)*	*Net income (Rs)*
190:115:300 kg NPK plant^{-1}/ planting and 4 MAP	1,25,000	19.00	3,42,000	2,17,000
190:115:300 kg NPK plant^{-1}/ planting, 2 ½ and 5 MAP	1,32,000	16.00	2,88,000	1,55,000
190:115:300 kg NPK plant^{-1}/ planting,2, 4 and 6 MAP	1,40,000	18.75	3,37,000	1,97,500
FYM 15 kg + 4kg ash plant^{-1}/ planting and 4 MAP	1,32,000	16.00	2,88,000	1,55,500
FTM 15 kg + 4kg ash plant^{-1}/ planting, 2 ½ and 5 MAP	1,40,000	17.25	3,10,500	1,70,500
FYM 15 kg + 4kg ash plant^{-1}/ planting, 2, 4 and 6 MAP	1,47,000	18.75	3,37,500	1,85,000
PM 5 kg + 4.5 kg ash plant^{-1}/ planting and 4 MAP	1,37,000	17.75	3,19,500	1,82,500
PM 5 kg + 4.5 kg ash plant^{-1}/ planting, 2 ½ and 5 MAP	1,44,500	17.25	3,10,500	1,66,000
PM 5 kg + 4.5 kg ash plant^{-1}/ planting, 2, 4 and 6 MAP	1,52,000	18.75	3,37,500	1,85,500
VC 5 kg + 3.5 kg ash plant^{-1}/ planting and 4 MAP	2,00,000	16.75	3,01,500	1,01,500
VC 5 kg + 3.5 kg ash plant^{-1}/ planting, 2 ½ and 5 MAP	2,07,500	16.00	2,88,000	80,500
VC 5 kg + 3.5 kg ash plant^{-1}/ planting, 2, 4 and 6 MAP	2,15,000	18.50	3,33,000	1,18,000
CPC 6.5 kg + 3.5 kg ash plant^{-1}/ planting and 4 MAP	2,17,500	18.50	3,33,000	1,15,500
CPC 6.5 kg + 3.5 kg ash plant^{-1}/ planting, 2 ½ and 5 MAP	2,25,000	18.75	3,37,000	1,12,000
CPC 6.5 kg + 3.5 kg ash plant^{-1}/ planting, 2, 4 and 6 MAP	2,32,500	18.75	3,37,500	1,05,000

*data not statistically analysed

References

Baiju, B.R.1998. Yield maximization of banana (Musa AAB group Nendran) through organic and inorganic manuring. M. Sc (Ag) thesis, Kerala Agricultural University, Thrissur, pp.144.

Geetha, K. and Nair, R.R. 2001. Integrated plant nutrition systems (IPNS) for banana. *Ann. Agric. Res.*21 **(4)**:499-503.

Gomes, J.A., Nobrega, A.C. and Rocha, A.C.1995. Effect of organic matter sources and amounts in the parent plant, first and second followers of the banana cv. Para in Entries Espirato Sano state, high brinjal. *Acta Hort.* **370**:103-110.

24

Comparative Evaluation of Soil Health and Fertility Under Organic, Inorganic and Integrated Nutrient Management in Coconut Based Mixed Farming System

C. Palaniswami, George V. Thomas, P. Subramanian, R. Dhanapal and V. Selvamani

INTRODUCTION

Soil is the fundamental resource for the agriculture, being the medium for the crop growth. Soil functions to soil productivity, maintain environmental quality and provide plant human and animal health. Soil health is defined as the continued capacity of soil to function as a vital living system, by recognizing that it contains biological elements that are key to ecosystem function within land-use boundaries. These functions are able to sustain biological productivity of soil, maintain the quality of surrounding air and water environments, as well as promote plant, animal, and human health (Doran and Zeiss, 2000; Karlen *et al.,* 2001).

The intensive use of chemical fertilizers during the green revolution era led to the degradation of the soil health under agricultural land use system resulted with increasing input requirement with low factor productivity and environmental problems. The alternative farming system to improve and sustain the soil fertility, productivity and soil ecology for the long-term is the need of the hour, and one of the ways is organic farming.

Organic farming is agricultural production without the use of synthetic fertilizers, chemicals and growth regulators (USDA, 1980). Organic agriculture is a holistic way of farming, besides producing high quality produce it aims the conservation of the natural resources fertile soil, clean water and rich biodiversity. It is more helpful to view the soil fertility as an ecosystem concept integrating the diverse soil functions, including nutrient supply, which promote plant production (Swift and Palm, 2000). Hence, the organic farming which recognizes the complex relationships that exist between different system components and that the sustainability of the system is dependent upon the functioning of a whole integrated and inter-related system (Atkinson and Watson, 2000) is very appropriate for the soil fertility management. The important aspect of organic farming is that it promises the sustainability of soil health and environmental safety.

Organic farming includes different concepts to ensure the soil health like crop rotation, green manure, cover cropping, application of compost, mulching etc. Being perennial in nature the most promising organic farming concept can that be followed in plantation crops is inter-cropping/mixed farming. In plantation crops where the crop rotation is restricted, the mixed farming system comprising the dairy, poultry, fish components along with main and intercrops enables the system to self-sustain in nutrient supply under organic farming through following practices like addition of organic manures from animal component, recycling of organic wastes through composting, green manures and biological inputs like vermi-composts and bio-fertilizers, etc. Apart from supplying the nutrients in a balanced way the organics plays major role in soil health through improving the physical properties by increasing soil aggregation, water holding capacity, etc., and biological properties like increasing the beneficial micro and macro fauna by serving as a medium.

The effects of different farming and fertility management practices like monocropping of coconut and inter-cropping coconut with grass, application of nutrients through organic manures, through inorganic fertilizers as well as their combination on soil health and crop productivity were studied at the Central Plantation Crops Research Institute, Kasaragod, Kerala. The cropping system and nutrient management practices include:

T1– monocropping of coconut with recommended fertilizer as inorganics,

T2 – Coconut based mixed farming system comprising Coconut, Hybrid Bajra napier grass, pepper, dairy, poultry and pisiculture components (pepper intercropped in the basins and Hybrid Bajra Napier Grass (Co 3) in the inter spaces of coconut palms) along with application of 50% of the recommended nutrients (on available Potassium basis) as vermi-compost produced from the wastes from crop and animal components of the coconut based mixed farming system,

T3 – Coconut based mixed farming system with application of 100% of the recommended nutrient as vermi-compost obtained from waste recycling of the mixed farming system, T4 – Coconut based mixed farming system with application of 100% of the recommended nutrient as inorganic fertilizer.

Soil samples were analysed for organic carbon, total N, available P, K, Ca, Mg and micro-nutrient contents. Analysis of plant samples were done for macro and micro-nutrient contents. Physical (bulk density, hydraulic conductivity and water holding capacity) and biological properties (population level of bacteria, fungi and actinomycetes) of soil were also determined from the production system.

The treatments which received organic manures showed reduction in bulk density than the other treatments where the entire requirement of nutrients was met only through inorganic fertilizers (Table 24.1). This reduction in the bulk density is due to the application of vermi-compost which increased the soil aggregation and pore space, whereas the treatments where only the inorganic fertilizers are applied the bulk density was very high due to natural compaction of the red sandy loam soil. This particular limitation of the red soil is very well managed when the organic farming practices are followed. Similarly due to the reduced bulk density with high micro pores in the favorable soil structure developed by the application of vermi-compost, the water holding capacity of the soil is also increased (Logsdon and Linden, 1992). It was the highest in the treatment where 100 % organic nutrient sources are applied followed by the 50% organic and inorganic sources are integrated and it was the lowest in the 100% inorganic application. The hydraulic conductivity of soil was the highest where entire nutrients were supplied through inorganic source, but it was less in the treatment of 100% organic nutrient source application. Hydraulic conductivity was very low in monocropping of coconut than in the treatments were the grass was intercropped.

Table 24.1. Effect of organic cultivation on soil physical properties

Treatments	*Water holding capacity (%)*	*Bulk Density (g cm^{-3})*	*Hydraulic conductivity (mm second^{-1})*
Monocrop + Recommended fertilizer	19	1.56	0.0037
Mixed farming + 50% Organic + 50% inorganic	19	1.23	0.0064
Mixed farming + 100% organic	21	1.41	0.0044
Mixed farming +100% Inorganic	18	1.56	0.0105

The plant macro and secondary nutrient status of the main crop coconut and intercrop grass showed positive influence of organic nutrient application. Leaf N content was high (2.17%) when 100% nutrient recommendation was supplied through the organic source under mixed farming practices followed by full fertilizer application or integrated application when compared to monocropping. The high nitrogen content in coconut leaves grown under organic farming is due to the reduced leaching loss attained by the slow release of nitrogen from the vermi-compost than the inorganic nitrogen fertilizer. Similar kind of results were obtained by Lampkin (1990), who reported that there is some evidence to suggest that nitrate leaching may be less under organic than under conventional systems and nitrate leached below the root zone is effectively lost from the system.

The leaf P and K contents were high in the treatment where 50% inorganic and 50% organic sources of nutrients were applied than the rest of the treatments (Table 24.2) due to the readily available inorganic source and the availability of these inorganic sources were enhanced by the organic sources through its role on reducing the fixation loss of applied inorganic P and K. In the same way the Ca content of the was leaf also influenced by the application of 100% organic source resulting in the highest value among the treatments but Mg was found to be high in the treatment where 100% nutrient recommendation was applied through organics, which might be due to the slow release of nutrients in organic farming which reduced the very high leaching loss of the Mg when applied as inorganic fertilizer. The Mg content was high in the organic matter applied treatments and found very low in the inorganic fertilizer alone applied treatments due to the leaching loss of inorganic Mg source. The coconut leaf nutrient content was found to be low in the monocropping compared to the mixed cropping.

Table 24.2. Effect of organic cultivation on plant macro and secondary nutrient status

Treatments	*N (%)*	*P_2O_5 (%)*	*K_2O(%)*	*Ca (%)*	*Mg(%)*
Monocrop + Recommended fertilizer	1.83	0.12	1.01	0.394	0.187
Mixed farming + 50% Organic + 50% inorganic	1.97	0.14	1.14	0.519	0.199
Mixed farming + 100% organic	2.17	0.12	1.04	0.488	0.214
Mixed farming +100% inorganic	1.92	0.13	1.01	0.498	0.189
Grass	1.36	0.21	1.45	0.361	0.336

The content of Zn in coconut leaves was very high in the organic farming treatment (13.65ppm) than the other treatments (Table 24.3). Apart from Zn, the other micro-nutrients like Fe, Cu and Mn were found to be numerically low in the organic farming treatment than in the nutrient management using inorganic fertilizers. Except Zn, all the other micronutrients were high in 100 % inorganic source applied treatments (T1 and T4) followed by T2.

Table 24.3. Effect of organic cultivation on plant micro nutrient status

Treatments	*Fe(ppm)*	*Cu(ppm)*	*Zn(ppm)*	*Mn(ppm)*
Monocrop + Recommended fertilizer	172.85	7.00	7.30	357.50
Mixed farming + 50% Organic + 50% inorganic	151.20	8.00	9.45	235.50
Mixed farming + 100% organic	138.80	6.35	13.65	156.75
Mixed farming +100% inorganic	170.85	7.10	11.50	292.50
Grass	147.70	17.40	24.00	334.00

The microbial distribution in the coconut basin and the interspaces in the coconut garden showed that population of bacteria, fungi and actinomycetes were found to be very high in the 100 % organic farming treatment where vermi-compost was applied followed by the 50 % organic substituted treatment and the population was low in inorganic fertilizer alone applied plots (Table 24.4). The organic manure vermi-compost served as a good medium for the microbial growth due to growth promoting substances and enzymes present in it. Microbial population was comparatively high in the interspaces where grass was grown than the coconut basin due to high root biomass available in the grass cropped area, ultimately the root turn over in the rhizosphere and root exudates contributing highly conducive environment for the microbial population to multiply (Gunadi *et al.*, 1999 and Masciandaro *et al.*, 2000).

Table 24.4. Microbial distribution in coconut based cropping system

Treatments	Bacteria (10^5 cfu/g soil)		Fungi (10^3 cfu/g soil)		Actinomycetes (10^5 cfu/g soil)	
	Coconut basin	Inter - space	Coconut	Inter - space	Coconut	Inter - space
Monocrop + Recommended fertilizer	13.45	10	6.86	4.16	9.62	6.16
Mixed farming + 50% Organic + 50% inorganic	18.22	21.64	6.6	7.23	7.83	8.35
Mixed farming + 100% organic	23.17	29.22	18.16	19.31	11.33	14.26
Mixed farming +100% inorganic	11.76	14.66	7.12	8.35	7	12.94

When the inorganic monocropping and the organic farming under mixed farming system were compared on the available nutrient content of soil (Table 24.5), the inorganic monocropping showed high nutrient content than the mixed under organic nutrient management due to the high uptake of nutrients in the mixed farming system. Especially the K content was very low in the mixed farming because the rigorous potassium uptake nature of both coconut and grass components.

Table 24.5. Impact of organic farming on available nutrient content in soil

System	Depth (cm)	Available nutrients (kg/ha)			Org. C (%)
		N	P_2O_5	K_2O	
Inorganic Monocrop	0-25	271.49	223.14	422.35	0.43
	25-50	189.88	89.28	293.82	0.28
	50-100	159.78	26.26	426.96	0.25
Organically managed Coconut basin (mixed farming)	0-25	229.88	189.64	197.40	0.41
	25-50	181.80	108.98	192.96	0.27
	50-100	178.16	52.08	211.18	0.27

The present results showed that the organic farming practices under mixed farming system has a positive impact on soil health by improving the physical, chemical and biological properties when compared to the inorganic nutrient management practices under mixed farming system. The coconut based mixed farming system is highly sustainable in maintaining soil health than the monocropping of coconut combined with inorganic nutrient management apart from high economic return from mixed farming system.

The findings also suggested that organic managed soils could be considered as more conservative systems which is in line with the comparative study of organic, conventional and integrated apple production systems in Washington State, which found that organic and integrated systems had higher soil quality and potentially lower negative environmental impact than the conventional system. The data indicated that the organic system ranked first in environmental and economic sustainability, the integrated system second and the conventional system last (Reganold *et al.,* 2001). The organic farming practices in the coconut based mixed farming system promises to sustain the soil health through *in situ* nutrient management in an eco friendly manner.

References

Atkinson, D. and Watson, C.A. 2000. The research needs of organic agriculture-distinct or just part of agricultural research? The BCPC Conference- *Pests and Diseases* 151-58.

Doran, J.W. and Zeiss, M.R. 2000. Soil health and sustainability: managing the biotic component of soil quality. *Applied Soil Ecology* 15, 3-11.

Gunadi, B, Blount, C. and Edwards, C.A. 1999. The growth and fecundity of *Eisenia fetida* (Savigny) in cattle solids pre-composted for different periods. *Pedobiologia.*, **46**(1):15-33.

Jackson, M.L. 1973. *Soil Chemical Analysis.* Prentice Hall of India Pvt. Ltd., New Delhi.

Karlen, D. L., Andrews, S. S. and Doran, J. W. 2001. Soil quality: Current concepts and applications. *Advances in Agronomy* 74:1-40.

Lampkin, N. 1990. *Organic Farming.* Farming Press Books, Ipswich UK.

Logsdon, S.D. and Linden, D.R. 1992. Interactions of earthworms with soil physical conditions influencing plant growth. *Soil Sci.,* 154:330-337.

Masciandaro, G., Ceccanti, B. Ronchi, V. and Bauer, C. 2000. Kinetci parameters of dehydrogenase and inorganic fertilizers. *Biol. Fer. Soils,* **32**(6):579-87.

Reganold, J.P., Glover, J.D., Andrews, P.K. and Hinman, H.R. 2001. Sustainability of three apple production systems. *Nature* **410**:926-30.

Swift, M.J. and Palm, C.A. 2000. Soil fertility as an ecosystem concept: A paradigm lost or regained? In: *Accomplishment and changing paradigm towards the 21 st century.*

USDA 1980. In: *Report and Recommendations on Organic farming.* United States Government Printing Office, Washington DC.

SECTION – V
SOIL HEALTH MANAGEMENT

25

Coconut Leaf Vermiwash for Soil Health Improvement and Quality Yield of Vegetable Crops —Validation in Farmers' Fields

Murali Gopal, Alka Gupta, B.T. Rayudu and George V. Thomas

INTRODUCTION

Vermiwash is a by-product obtained from vermi-compost production technology that can be used as a liquid organic fertilizer for improving crop growth and yield. Vermiwash production can be done basically by two methods. One method involves soaking soil + cow dung + earthworms substrate in excess water in plastic tub and siphoning the wash periodically from the bottom of the tub (Ismail, 1997) while another one involves releasing the earthworms in lukewarm water and agitating them gently so as to shock them to secrete higher amount of body fluids and mucus (Kale, 1988). This liquid organic fertilizer is rich in nutrients and plant growth hormones (Ismail, 1997) and its application has been reported to stimulate anthurium (Karuna *et al.,* 1999), increase soil nutrient status and yield of paddy (Thangavelu *et al.*, 2003), biological productivity of marigold (Sivasubramanian and Ganeshkumar, 2004), *Aloe vera* (Saha *et al.*, 2005), fruit quality of tomatoes (Zaller, 2006), yield of spinach, onion and potato (Ansari, 2008) and cowpea (Zambare *et al.*, 2008).

The Central Plantation Crops Research Institute had developed a simple farmer friendly technology for production of vermi-compost from the high lignin containing coconut leaves using an indigenous isolate of earthworm *Eudrilus* sp. (Prabhu *et al.*, 1998). The technology became popular and rapidly spread to many of the coconut producing states viz. Karnataka, Tamil Nadu, Andhra Pradesh, Maharashtra, Bihar, Orissa and West Bengal in a span of 10 years. Experiments were initiated to standardize vermiwash production as a value addition to coconut leaf vermi-compost production technology. Vermiwash produced from actively vermi-composting substrates of coconut leaf + cow dung by *Eudrilus* sp. is clear brown liquid (Fig.25.1) possessing alkaline pH, containing major and minor nutrients, growth hormones, humic acid and plant beneficial bacteria particularly the fluorescent pseudomonads. Application of appropriately diluted coconut leaf vermiwash (CLV) increased germination and seedling vigour index of cowpea and paddy seeds in laboratory bioassays. Field trials conducted on cowpea, maize and okara in Institute farm during 2004-05 showed its capacity to enhance the biomass and yield of the crops accompanied by increase in soil microbial populations, enzyme activities and organic carbon content of soil (Gopal *et al*, 2005).

Fig. 25.1. Lush growth of amaranthus (A) with profuse production of fine root hairs (B) in the vermiwash treated plot at Edneer

To validate the results obtained in the in-house studies, field trials were initiated in collaboration with the Krishi Vigyan Kendra, Kasaragod at farmers' plots at Majal and Edneer area in Kasaragod District during 2006-2007. Two leading farmers, one from Majal and another from Edneer panchayath of Kasaragod District were selected for the field trials. Required quantities of coconut leaf vermiwash produced at CPCRI was supplied to them for application in the test crops by diluting with water at 1:5 ratio once in a week followed by regular irrigation with water. Poultry manure and neem cake was applied as basal organic inputs during field preparation followed by inorganic fertilizer (17-17-17 complex) during the middle stage of crop growth, which is the regular practice followed by farmers. This was compared with CLV treatment. Irrigation to the crops was done depending on the situation. Five rows of amaranthus, cow pea and chillies and ten pits for bitter gourd were set apart for evaluating the effect of CLV while the rest of the plots received the inorganic fertilizer application.

Soil samples were collected before the application of treatments and at the end of the experiment from all the crops at both the fields and analysed for general microbial communities of heterotrophic bacteria, filamentous actinomycetes and fungi, and plant beneficial microbial communities of free-living nitrogen fixers, phosphate solubilizers and fluorescent pseudomonads.

The results of the yield obtained from respective vegetable crops in both the experimental locations, as recorded by the farmers, are given in Table 25.1. CLV application yielded 75kg and 200kg cow pea and bitter gourd, respectively, at farm in Majal and 55kg, 155kg and 10kg of amaranthus, cowpea and dried chillies, respectively, at Edneer farm.

Table 25.1. Yield response of vegetable crops to application of CLV (Total yield from 5 rows/10 pits)

Place	*Crop*	*Yield(kg)*	*Farmer's remarks on effect of vermiwash on growth and yield quality of vegetables*
Majal	Cowpea	75(81)*	Good quality of produce, better plant growth, good storage quality of the produce, softer beans
Majal	Bitter gourd	200 (218)*	Very good quality of produce, healthier look of the plants, early bearing, delayed ripening of fruits
Edneer	Amaranthus	55(59)*	Bigger leaf size, deep leaf colour, no pest and disease attack, more hairy roots, harvested plants showed wilting after longer duration
Edneer	Cowpea	155 (156)*	Better plant growth, produce was of good quality and more tastier
Edneer	Chilli(dried)	10(12)*	Rich green colour of the leaves, better plant growth, good pungency

*Yield from plots receiving Farmer's inputs of poultry manure + neem cake as basal application followed by inorganic fertilizer mix

It is clearly seen that the application of CLV produced yields on par or slightly less compared to plot receiving farmer's regular inputs of organic and inorganic fertilizers. Similarly, Mäder *et al.* (2002) reported results from a 21 year study of agronomic and ecological performance of biodynamic, bioorganic, and conventional farming systems in Central Europe. They found crop yields to be 20% lower in the organic systems, although input of fertilizer and energy was reduced by 34 to 53% and pesticide input by 97 %. Though the yield of the vegetables obtained was lower or on par compared with the farmer's inputs, yet, both the farmers had opined that application of CLV helped in production of better quality vegetables (Table 25.1). They reported that application of vermiwash produced amaranthus possessing dense root hairs, broad and tender leaves with deep colour and minimum pest and disease incidence (Fig. 25.1a and b). The crop stand of bitter gourd (Fig. 25.2a), chilli (Fig. 25.2b) and cowpea (Fig. 25.2c) was also very lush. Saha *et al.* (2005) also reported that organic source of fertilizer in the form of vermi-compost and liquid vermiwash was found to be effective and comparable with inorganic source of fertilizer in increasing content of gel moisture, gel ash and aloin in *Aloe vera*. The organic *Aloe vera* thus produced was expected to be a better marketable product. Zaller (2006) reported that foliar spraying of aqueous extracts of vermi-compost on three tomato varieties had a positive effect on the fruit qualities. It increased or decreased the peel firmness of the tomato fruit,

Fig. 25.2. Luxuriant growth and good bearing of bitter gourd at farmer's field in Majal (A) chillies (B) and cowpea (C) at Edneer

increased the fruit firmness in all the three varieties. He also reported that vermi-compost extract application had markedly reduced the incidence of *Phytophthora infestans* disease. He suggested that vermi-compost extract could be used as foliar spray for fertilization and biological disease prevention of vegetable crops. Application of vermi-compost + vermiwash diluted at different ratio with water had improved the yield and quality of spinach, onion and potato (Ansari, 2008). Production of high quality yields can also be attributed to reduced nematode damage as it had been described that CLV was able to have a deleterious effect on the juveniles of burrowing nematode, *Radopholus similis*, root-knot nematode, *Meloidogyne incognita*, root lesion nematode, *Pratylenchus coffeae* and spiral nematode, *Helicotylenchus multicinctus* which are some of the major nematode pests of crops in coconut based high-density multi species cropping system (Banu and Iyer, 2006).

Studies on soil microflora and use of enzyme activity measurements as indicators of soil functionality and thus indicators of soil quality have been extensively discussed (Nannipieri *et al.*, 2002). Application of CLV had stimulated the population of all the soil microbial communities, particularly the plant beneficial groups of free living nitrogen fixers, phosphate solubilizers and fluorescent pseudomonads, which form an important fraction of the plant growth promoting rhizobacteria (Fig. 25.3).

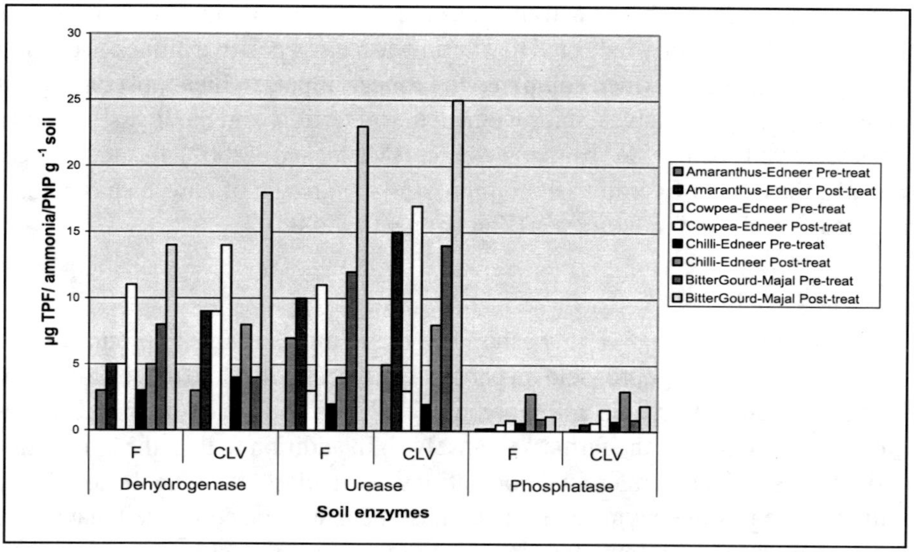

Fig. 25.3. Effect of coconut leaf vermiwash on rhizospheric microbial communities in different vegetable crops

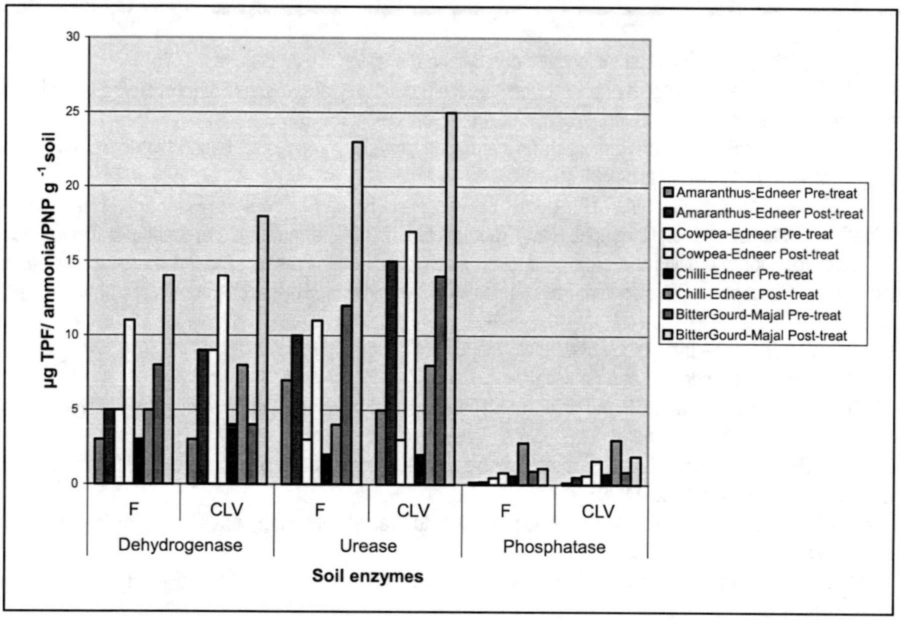

Fig. 25.4. Effect of coconut leaf vermiwash on soil enzyme activities in rhizosphere of vegetable crops

Similarly, the activities of three important soil enzymes viz. dehydrogenases, phosphatases and ureases increased sharply in the vermiwash applied plots compared to the poultry manure + neem cake and inorganic fertilizer applied plots (Fig. 25.4). Thus, we observed in vermiwash treated plots, a clear correlation between increase in microbial population with increased enzyme activities, as much of the soil enzymes originate from microbial biomass. This clearly indicated that vermiwash had a positive influence in building up the soil quality and health when compared to farmer's inputs, which must have resulted in production of quality vegetables. Improvement in soil fertility and biodiversity in organic farming has been well documented by many workers (Mäder *et al.*, 2002). In another recently reported study, the effect of vermiwash in improving the growth of cowpea has made it as a potential eco-friendly soil conditioner (Zambare *et al.*, 2008).

CONCLUSIONS

Application of coconut leaf vermiwash as soil drenching was able to produce cowpea, amaranthus, chilli and bitter gourd yield on par or slightly less compared to the plot receiving farmer's regular inputs of organic and inorganic fertilizers. However, the produce was of superior quality as opined by the farmers themselves. In addition to this, the application of vermiwash increased the soil microbial populations, particularly the plant beneficial ones, along with increase in soil enzyme activities. Thus, it can also be concluded that CLV can be used as a component of organic farming for sustainable soil health.

References

Allen, O.N. 1959. Experiments in Soil Bacteriology, 3rd edn. Minneapolis, MN, USA, Burgess, 117p.

Ansari, A.A. 2008. Effect of vermi-compost and vermiwash on the productivity of spinach (*Spinacia oleracea*), onion (*Allium cepa*) and potato (*Solanum tuberosum*). *World J. of Agril. Sciences* **4**:554-57.

Banu, J.G. and Iyer, R. 2006. Effect of vermiwash on nematode prevalent in coconut based high-density multispecies cropping system. *Indian J. Nematology* **36**:195-99.

Becking, J.H. 1959. Nitrogen fixing bacteria of the genus *Beijerinckia* in South African soils. *Plant Soil* **11**:193-206.

Gopal, M., Gupta, A., Palaniswami, C., Dhanapal, R and Thomas, G.V. 2005. Vermiwash from coconut leaf vermi-compost: An eco-technology for improving the crop production efficiency of soil. Presented in *International Conference on Human Centred Sustainable Development Paradigm* at M S Swaminathan Research Foundation, Chennai. August 7-10th, 2005.

Ismail, S.A. 1997. Vermicology – the biology of earthworms. Orient Longman, India, 192p.

Kale, K.E. 1998. Earthworm: Cinderella of organic farming. Bangalore: Prism Books Pvt. Ltd. 88p.

Karuna, K., Patil, C.R., Narayanaswamy, P and Kale, R.D. 1999. Stimulatory effect of earthworm body fluid (vermiwash) on crinkle red variety of *Anthurium andreanum* L. *Crop Research* **17**:253-57.

King, E.O., Ward, M.N., Raney, D.E .1954. Two simple media for the demonstration of pyocyanin and fluorescein. *Journal of Laboratory Clinic and Medicine* **44**:301-307.

Klein, D.A., Loh, T.C. and Goulding, R.L. 1971. A. rapid procedure to evaluate dehydrogenase activity of soils low in organic matter. *Soil Biol. Biochem.* **3**:385-87.

Mäder, P., Fliebach, A., Dubois, D., Gunst, L., Fried, P and Niggli, U. 2002. Soil fertility and biodiversity in organic farming. *Science* **296**:169-1697.

Martin, J.P. 1950. Use of acid, rose Bengal and streptomycin in the plate method for estimating soil fungi. *Soil Sci.* **69**: 215-32.

Nannipieri, P., Kandeler, R.P. 2002. Enzyme activities and microbiological and biochemical processes in soil. *In*: Burns, R.G and Dick, R.P. (Eds). *Enzyme in the Environment. Activity, Ecology and Application.* Dekker. New York, pp 1-33.

Pikovskaya, R.I.1948. Mobilization of phosphorus in soil in connection with vital activity of some soil microbial species. *Mikrobiologiya* **17**:362-70.

Prabhu, S.R. and Subramanian, P., Biddappa, C.C. and Bopaiah, B.M. 1998. Prospects of improving coconut productivity through vermiculture technologies. *Indian Coconut Journal.* **29**(4):79-84.

Saha, R., Palit, S., Ghosh, B.C. and Mittra, B.N. 2005. Performance of *Aloe vera* as influenced by organic and inorganic sources of fertilizer supplied through fertigation. *Acta Horticulturae.* **676**:171-75.

Sivasubramanian, K and Ganeshkumar, M. 2004. Influence of vermiwash on the biological productivity of marigold. *Madras Agric. J.* **91**:221-25.

Tabatabai, M.A. 1994. Soil enzymes. *In*: Weaver, R.W. *et al* (Eds). Methods of soil analysis. Part 2. Soil Sci. Soc. America. Madison, Wisconsin, USA, pp 775-833.

Tabatabai, M.A. and Bremner, J.M. 1969. Use of p-nitrophenol phosphate for assays of soil phosphatase activity. *Soil Biol. Biochem.* **1**:301-07.

Thangavel, P., Balagurunathan, R., Divakaran, J. and Prabakaran, J. 2003. Effect of vermiwash and vermicast extract on soil nutrients, growth and yield of paddy. *Advances in Plant Sci.* **16**:187-90.

Zaller, J.G. 2006. Foliar spray of vermi-compost extracts: Effects on fruit quality and indications of late-blight suppression on field-grown tomatoes. *Biological Agriculture and Horticulture* **24**:165-80.

Zambare, V.P., Padul, M.V., Yadav, A.A and Shete, T.B. 2008. Vermiwash: Biochemical and microbiological approach as eco-friendly soil conditioner. *ARPN J. of Agrl. Biol.Sci.* **3**:1-5.

26

Plant Growth Promoting Rhizobacteria from Selected Clones of *Hevea brasiliensis*

Kochuthresiamma Joseph, T.V. Ajitha Mohan, Shaji Philip, Amith Abraham, Roshini Susan Elias, Geetha Jose, V.K. Arunkumar and C. Kuruvilla Jacob

INTRODUCTION

Many of the rhizobacteria associated with agricultural crops are efficient in promoting the growth of plants by increasing the nutrient availability and controlling the phytopathogens (Bloemberg and Lugtenberg, 2001). Some of this plant growth promoting rhizobacteria (PGPR) can also enter root interior and establish as endophytic populations (Kloepper, *et al.*, 1999; Hallmann, 2001; Gray and Smith, 2005). They improve the plant availability of nutrients by fixing nitrogen (Saubidet *et al.*, 2002), solubilising phosphate (Mohan *et al.*, 2008) and producing plant growth promoting hormones (Arshad and Frankenberger, 1998). They also control plant pathogens either by pre emptive colonisation, direct antagonism by metaboliters or by induced systematic resistance (Compant *et al.*, 2005). Natural rubber (*Hevea brasiliensis*) is an important perennial cash crop of India. Continuous use of chemical fertilisers and pesticides for better growth and yield of this crop may create adverse effect on environment and human health. Use of naturally occurring rhizobacteria from rubber plantations that promote and protect plant growth by colonising and multiplying in the root could be an alternative for reducing the use of these chemicals.

Enumeration and Isolation of Rhizobacteria

Root samples of *H. brasiliensis* for isolation of rhizosphere bacteria and endophytes were collected from the mature plants of the five newly developed clones viz. RRII 414, RRII 422, RRII 430, RRII 417, RRII 429 and the popular clone RRII 105 planted at Central Experiment Station, Chethackal, Pathanamthitta District, Kerala. Sixty bacterial isolates from rhizosphere and 55 endophytes were screened for phosphate solubilisation using apatite agar medium by spot inoculation. After incubating for 5-6 days for the development of clear zone of phosphate solublisation around the colony was observed. Ten isolates showing more solubilisation of phosphate in apatite medium were studied for their ability to solubilise aluminium, ferric and tricalcium phosphate using Pikovskya's medium incorporated with these phosphates (Subha Rao,1977). After incubation for 10 days at 28+/- 2 °C the phosphate released was estimated using calorimetric methods and compared with that in the uninoculated control medium.

Screening of Rhizobacteria for Antagonism

Dual culture assay was performed for the identification of bacterial antagonists against the soil borne plant pathogens of *H. brasiliensis* viz., *Phytophthora meadii* causing abnormal leaf fall, shoot rot and patch canker and *Phellinus noxius* causing brown root disease. The bacterial isolates and the pathogen were coinoculated in potato dextrose medium. The plates were incubated for seven days at 28 +/-2 °C. The growth and the zone of growth inhibition of the pathogen in the coinoculated plates were observed and compared with that in plates inoculated with pathogen alone. Production of antifungal metabolites like siderophores (Yeole *et al.*, 2001),volatile compounds (Bhatia *et al.*, 2003) and HCN (Anandaraj and Sarma, 2003) were also tested *in vitro*.

Assay for Plant Growth Improvement

The ability of the 115 isolates to stimulate plant growth was evaluated using seedlings of *H. brasiliensis* in small cups in three batches keeping separate control for each batch. The germinated seeds from germination beds were transplanted and 3 ml of each bacterial culture (10^8 cfu/ml) were applied to the planting pit at the time of transplanting. The uninoculated seedlings with 3ml of culture free broth were maintained as controls. Three replicates were maintained for each isolate. The plants were harvested after 45 days growth and various growth parameters were recorded.

Molecular identification of isolates showing antagonism and growth promotion was carried out by 16S rDNA amplification, sequencing and BLAST analysis against the Ribosomal Database Project II.

The rhizosphere bacterial population in the clones varied from 46to 85 × 10^6 cfu/g soil (Fig. 26.1). The popular clone RRII 105 harboured the highest population of rhizosphere bacteria followed by RRII 430, RRII 414, RRII 417, RRII 429 and RRII 422. The endophytic bacterial population in these clones varied from 5 to 19 × 10^3 cfu/g of root tissue. The highest endophytic bacterial population among the clones was recorded for RRII 417 followed by RRII 430, RRII 105, RRII 422, RRII 429 and RRII 414. Sixty rhizosphere and 55 endophytic isolates were selected based on morphology for further studies.

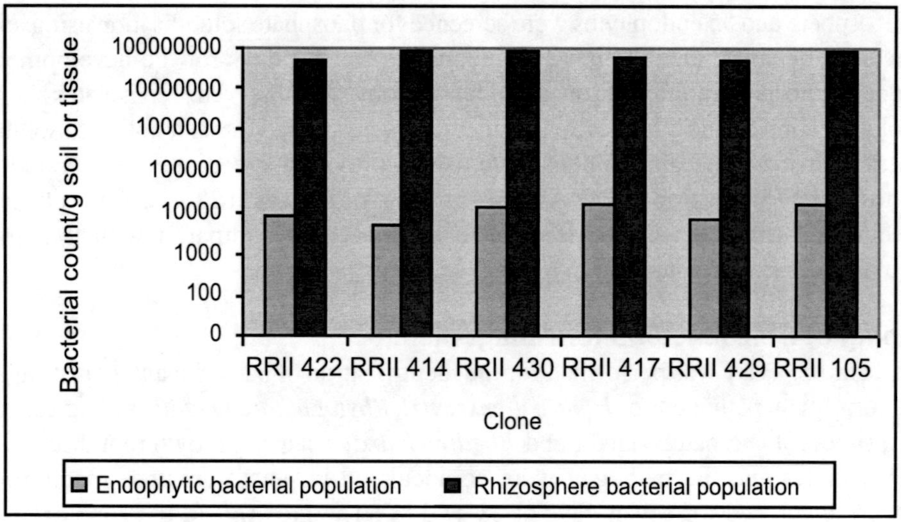

Fig.26.1. Population of rhizobacteria in different clones

Out of the 115 isolates screened for phosphate solubilisation in apatite agar medium 44 showed clearance zone around the colony (Fig. 26.2). Of these, isolates RB 6 and RH 15 showed the maximum clearance zone of 2.5cm.

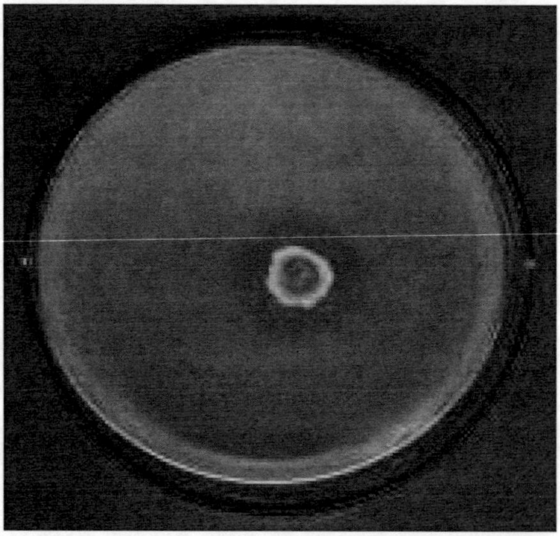

Fig. 26.2. Solubilization of phosphates by rhizobacteria

The solubilisation of different forms of insoluble phosphate by the 10 selected isolates in Pikovskaya's medium is given in Table. 26.1. The phosphate solubilsing capacity of the isolates varied with the source. All the isolates solubilised aluminium phosphate, high solubilisation (46.5 µg/100 ml) was recorded for isolate RH 26. The pH of the media was reduced to 2.3 while the pH of the control remained at 5. There was a positive correlation between the reduction in pH and the quantity of phosphorous released.

Table 26.1. Solubilization of different phosphates by selected rhizobacteria

Isolate no.	*Aluminium phosohate*		*Ferric phosphate*		*Tricalicium phosphate*	
	Phosphate released (µg/100ml broth)	*pH of the medium*	*Phosphate released (µg/100ml broth)*	*pH of the medium*	*Phosphate released (µg/100ml broth)*	*pH of the medium*
RB 6	39.5	3.2	83.00	5.30	730	5.71
RB 8	21.5	3.5	202.5	5.40	708	5.70
RH 15	13.5	4.0	102.5	4.10	625	5.55
RH 26	46.5	2.3	75.00	5.45	600	5.00
RH 31	31.5	3.0	204.5	5.27	300	5.42
RB 5	18.5	4.1	150.0	4.39	720	6.82
RB 2	27.0	3.5	202.5	4.65	730	6.38
RH 32	3.50	4.2	50.00	5.80	735	6.31
RH 36	27.0	3.6	332.5	4.27	620	5.33
RB 13	10.5	4.8	345.0	5.19	700	5.30
Control	–	5.0	–	8.00	–	8.00

All the then isolates showed solubilisation of ferric phosphate. Isolates RB 13 showed high solubilisation of ferric phosphate (345 microgram/100 ml). Reduction in media pH up to 4.10 was observed while the control had a pH of 8. The isolates also showed higher solubilisation of tricalcium phosphate and the phosphate released ranged from 300–735 µg/100ml. Maximum solubilisation was recorded for RH 32 and the pH was reduced to 6.31 from 8.

Among the 115 isolates tested for their antagonism against *P.meadii* by dual culture, 32 showed growth inhibition of the pathogen. Five isolates showed more than 3 cm zone of inhibition and the widest zone of 3.8 cm was recorded by the isolate RE 8 (Fig 26.3a). Twenty five isolates showed antagonistic activity against *P.noxius* in dual culture. Among them 6 isolates showed zone of inhibition of more than 3cm. The highest zone of 3.5 cm was recorded for the isolate RH 37 (Fig 26.3b).

1. *Siderophore:* Among the antagonistic bacteria 21 showed siderophore production, by $FeCl_3$ test. Higher production was shown by isolates RH 45 and RH 46.
2. *Volatile compound production*: The percentage reduction in the growth of *P.meadii* and *P.noxius* by the production of volatile organic compounds by the different antagonistic bacteria are given in the Table 26.2. All the ten isolates selected were

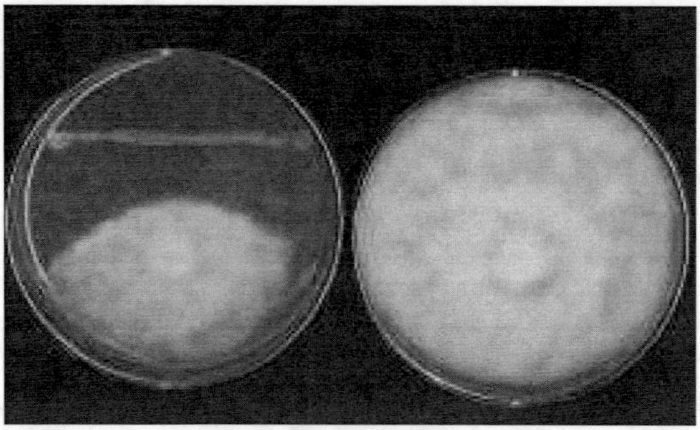

Fig.26.3a. Growth inhibition of *P. meadii*

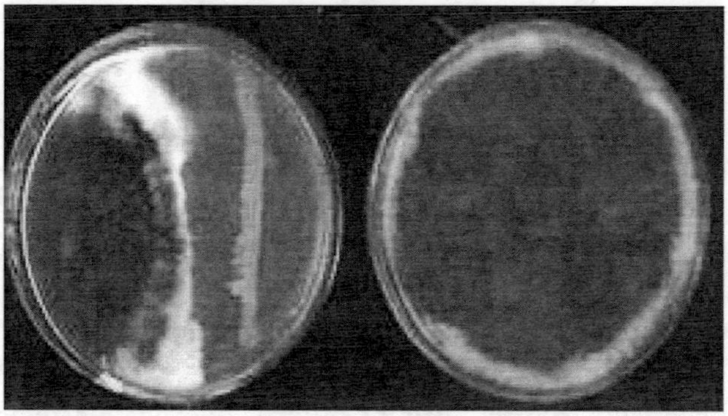

Fig. 26.3b. Growth inhibition of *P. noxius*

capable of producing volatile organic compounds that reduced the growth of the pathogens. Their effect resulted in 10 to 60% growth reduction compared to the growth in control. RB 62 showed maximum growth reduction (60 %) of *P.meadii*. Maximum growth reduction of *P.noxius* by rhizobacteria was 33.3%.

Table 26.2. Effect of volatile compounds on growth of two pathogens

Isolate No.	*Growth reduction of P. meadi (%)*	*Isolate No.*	*Growth reduction of P. noxius(%)*
RH 52	24	RH 18	33.3
RB 62	60	RH 34	22.2
RB 60	10	RH 22	22.2
RB 8	54	RB3	15.6
RH 63	52	RB 37	11.1

3. *HCN production*: The isolates producing HCN were grouped into three categories: low, medium and high based on the intensity of the colour developed on the filter paper treated with picric acid reagent placed over the culture. High level of HCN is

indicated by brown colour, medium production by orange yellow and lower production by yellow colour on the filter paper. Among the isolates higher HCN production was shown by RH 52, RH 18, RH 22, medium by RB 3, RH 63 and RH 37 and low by RB 60, RB 8 and RH 34. Plant growth enhancement by rhizobacteria.

Fig. 26.4. (a) Effect of inoculation on seedlings

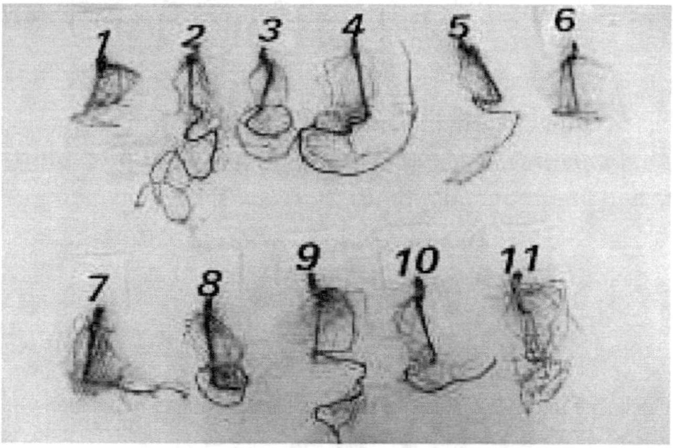

Fig. 26.4. (b) Effect of inoculation on root growth 1-control; 2-11 inoculated with rhizobacteria

Out of the 115 isolates studied for their effect on growth enhancement of *H. brasiliensis* seedlings, 65 showed significant improvement in any of the growth parameters viz., shoot length, shoot girth, shoot weight, etc., studied. Eight isolates from three batches showing significant enhancement in more than four parameters are presented in Table 26.3. Four cultures were from rhizosphere (RH) and four were root endophytes (RB). It was observed that during the initial period of growth (45 days) the PGPR isolates have more influence on various growth parameters like root volume, root branching and root weight than shoot parameters.

Table 26.3. Effect of selected rhizobacterial inoculation on growth of rubber seedlings

Isolate	*Shoot*			*Root*				*No. of leaves*
	Length	*Girth*	*Dry weight*	*Length*	*Volume*	*Tips*	*Dry weight*	
RH 21	42.4	1.4	1.3	35	2.9	50.5	0.2	5
RH 77	40.3	1.5	2.0	21.7	2.7	127.3	0.2	4.3
Control	28	1.2	0.3	17	1.5	35	0.1	2
CD(P = 0.05)	NS	NS	0.45	10.23	0.85	22.78	0.22	1.26
RB 85	37.25	1.5	1.0	33.0	2.5	62	0.5	4
RB 86	43.4	1.4	1.1	27.5	2.5	62	0.5	5
RB 88	38.3	1.3	0.8	24.1	2.3	59	1.1	4.7
RB 93	40.0	1.5	1.3	22.8	2.3	59	0.3	4
Control	35	1.2	0.2	21	1.2	30	0.2	2
CD(P = 0.05)	NS	NS	0.37	NS	0.85	21.68	NS	1.45
RH109	26.5	1.8	0.75	12.25	3	61.5	1.0	4
RH104	47.25	1.6	1.45	32	2.25	106.5	0.45	3.5
Control	27.5	1	0.9	15	1.2	31	0.2	3
CD(P = 0.05)	16.26	0.61	NS	11.09	0.80	22.41	0.17	NS

Molecular Identification

Three of the selected isolates showing antagonism and growth promotion were identified as *Pseudomonas pseudoalcaligenes* (root endophyte) and *Bacillus subtilis* (2 isolates, one rhizosphere bacteria and one root endophyte) by 16S rDNA sequencing (Fig. 26.5).

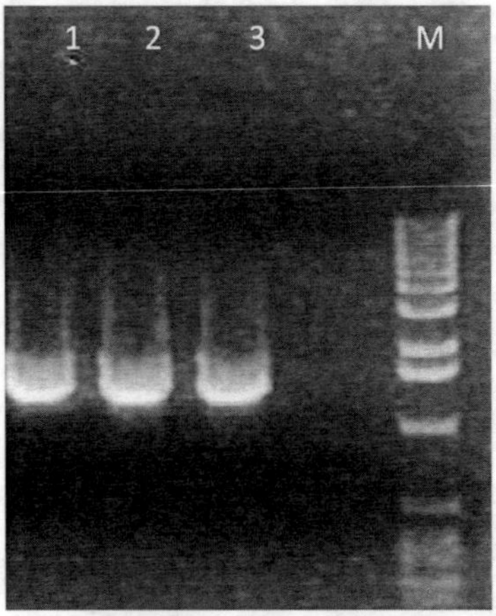

Fig. 26.5. Amplified 16Sr DNA

The region around the root is rich in nutrients which supports large and active microbial population. Many of these rhizobacteria are capable of exerting beneficial effects on plant growth by direct and indirect methods which include increasing the availability of nutrients, producing growth hormones and reducing the deleterious effects of plant pathogens.

The rubber growing soils of Kerala are inherently deficient in available P due to high fixation of applied P by hydroxides of Fe and Al. Inorganic P fractions in the traditional rubber growing soils comprising less than 25% of the total P is dominated by Fe-P followed by Al-P (Karthikakuttyamma *et al.*, 1991). In the present study 44 isolates collected from the root and rhizosphere of *H. brasiliensis* were found to be solubilising insoluble phosphates supplied in the growth medium. All the ten isolates selected were also found to solubilise aluminium and ferric phosphate. The pH of the medium amended with any of the three phosphates was found to be low corresponding to their solubilisation. Phosphate solubilising microorganisms are known to produce different organic acids (Subha Rao, 1977). Pareek and Gaur (1973) showed the release of phosphate from tricalcium phosphate and rock phosphate by organic acids and reported that aliphatic acids were more effective in solubilisation of phosphate than phenolic acids.

The role of PGPRin the biological control of root diseases are reported by many authors (Weller and Thomashow, 1993; Kim *et al.*, 1997). Many of the rhizobacteria isolated from *H. brasiliensis* were found to be antagonists against the soil borne pathogens *viz.*, *P. meadii* and *P. noxius*.

Antagonistic bacteria mediate biocontrol by one or more of the several mechanism of disease suppression. A primary mechanism of pathogen inhibition is by the production of secondary metabolites and other factors such as siderophore and microbial cyanides and lytic enzymes (Velusamy and Gnanamanickam, 2008). The antagonists from *Hevea brasiliensis* were also found to produce antifungal metabolites like siderophore, cyanides and other volatile organic compounds (VOC).

PGPR produce siderophores having greater iron binding potential. Although various bacterial siderophores differ in the ability to sequester iron, in general, they deprive pathogenic fungi of this essential element since the fungal siderophores have lower affinity (Compant *et al.*, 2005) Bacterial VOC have been shown to affect both mycelial growth and enzyme activity. When exposed to VOC produced by a series of bacterial isolates laccase activity in *Phanaerochaete magnoliae* ceased completely (O'Sullivan and O'Gara, 1992). Fernando *et al.* (2005) also reported that the bacteria isolated from canola and soybean plants produced VOC that inhibited sclerotia and ascospore germination and mycelial growth of *Sclerotinia sclerotiorum.*

Several studies indicate that PGPR may act as natural elicitors for improving the growth and yield of crops. It also enhances the rate of seedling emergence, tiller formation and plant dry weight. These plant growth responses are often variable depending upon the bacterial strain, crop variety and experimental site. In the present study 115 isolates obtained were inoculated to rubber seedlings and 64 were found to enhance plant growth, some of them were found to be phosphate solubilising and siderophore producing. They enhanced length, weight, volume and branching of roots and weight and girth of shoots. An increase in root surface area and the volume of soil foraged by the root, which leads to an enhanced nutrient uptake, is the most commonly proposed explanation for the beneficial effects of PGPR on plant growth. The improvement in mineral nutrition would explain the promotion of shoot growth. PGPR may enhance mineral uptake, not only as a consequence of the increase in root surface but also the stimulating the ion uptake system (Mohan *et al.*, 2005). *Pseudomonas* is one of the most extensively studied PGPR that has been reported to improve plant growth and yield. In this study the selected isolates belonged to *Pseudomonas* and *Bacillus* groups.

From the present study it can be concluded that application of PGPR as seed treatment to *H. brasiliensis* is beneficial and could be used as a potential agro-management practice for rubber plantations. They could serve as eco-friendly and sustainable alternative to hazardous chemicals used for growth promotion and protection from diseases.

References

Anandaraj, M. and Sarma,Y.R. 2003. The potential of PGPRs in disease management of spice crops. In: *Sixth International PGPR workshop* (eds. M.S. Reddy, M. Anandraj, S.J. Eapen.,Y.R. Sarma and J.W. Klopper). Indian Institute of Spice Research, Calicut Kerala, India pp 27-39.

Arshad, M. and Frankenberger, W.T. 1998. Plant growth promoting substances in the Rhizosphere: Microbial production and function. *Advances in Agronomy*, **62**:45-151.

Bhatia, S., Bhastia, S., Dubey, R.C. and Maheswari, D.K. 2003. Antagonistic effect of fluorescent pseudomonads against *Macrophomina phaseolina* that causes charcoal rot of ground nut. *Indian Journal of Experimental Biology*, **41**:1442-46.

Bloemberg, G.V. and Lugtenberg, B.J.J. 2001. Molecular basis of plant growth promotion and biocontrol by rhizobacteria. *Journal of Biological control*, **4**:343-50.

Compant, S., Duffy, B., Nowak, J., Clement, C. and Barka, E.A. 2005. Use of plant growth promoting bacteria for biocontrol of plant diseases: principles, mechanisms of action and further prospects. *Applied Environmental Microbiology*,**71(9)**: 4951-51.

Fernando, W.G.D., Ramarathnam, R., Krishnamooothy, A.S., Savehuk,S.C. 2005. Identification and use of potential bacterial organic antifungal volatiles in biocontrol. *Soil Biology and Biochemisty* **37**: 955-964.

Gray, E.J. and Smith, D.L. 2005. Intracellular and extra cellular PGPR: Commonalities and distinctions in the plant –bacterium signaling processes. *Soil Biology and Biochemistry*, **37**:395-412.

Hallmann, J., Quadt- Hallmann, A., Mahaffee,W.F. and Kloepper, J.W. 1997. Bacterial endophytes in agricultural crops. *Canadian Journal of Microbiology*, **43**:895-914.

Hallmann, J. 2001. Plant interactions with endophytic bacteria. In: *Biotic Interactions in Plant- pathogen Association* (eds. M.J. Jeger and N.J. Spence) CAB International, pp 87-119.

Karthikakuttyamma. M., Nair, A.N.S., Krishnakumar, A.K., Potty, S.N. and Mathew, M. 1991. Important organic P factors in rubber growing soils. *Indian Journal of Natural Rubber Research* **4(1)**:72-76.

Kim, D.S., Cook, R.J. and Weller, D.M. 1997. *Bacillus* sp Z 324-92 for biological control of three root diseases of wheat grown with reduced tillage. *Phytopathology*, **57**:551-58.

Klopper, J.W., Rodriguez-Ubana, R., Zehnder, G.W., Murphy, J.F., Sikora, E. and Fernandez, C. 1999. Plant root –bacterial interactions in biological control of soil borne diseases and potential extension to systemic and foliar diseases. *Australian Plant Pathology*, **28**:21-26.

Mohan,V., Narolia, V.K., Chhangani, S., Kushwaha, U., Sharma, M. and Rao, V.M. 2008. Plant growth promoting rhizobacteria (PGPR): A Review. In:*Biofertilizers* (ed. P.C..Trivedi). Pointer publishers, Jaipur, India, pp 180-192.

O'Sullivan, D.J. and O. Gara, F. 1992. Traits of fluorecent *Pseudomonas* spp involved in suppression of plant root pathogens. *Microbiology Review*, **56**:662-77.

Pareek, R.P. and Gaur, A.C. 1973. Release of phosphate from tricalcium phosphate by organic acids. *Current Science*, **42**: 278-79.

Saubidet, M.I., Fatta, N. and Barneix, A.J. 2002. The effect of innoculation with *Azospirillum brasilense* on growth and nitrogen utilization by wheat plants.*Plant and Soil*, **245**:215-22.

Subha Rao, N.S. 1977. S*oil Microorganisms and Plant Growth*, Oxford and IBH Publishing Co. Pvt. Ltd, New Delhi, p. 289.

Velusamy, P. and Gnanamanickam, S.S. 2008. The effect of bacterial secondary metabolites on bacterial and fungal pathogens of rice. In: *Secondary metabolites in Soil Ecology* (eds. P.Karlovsky). Springer- verlay, Berlin, Heidelberg,Germany.

Weller, D.M. and Thomashow, L.S. 1993. Use of rhizobacteria for biocontrol. *Plant Pathology*, **78**:1094-1100.

Yeole, R.D., Dave, B.P. and Dube, H.C. 2001. Siderophores production by fluorescent Pseudomonads colonizing roots of certain crop plants. *Indian Journal of Experimental Biology*, **39**:468-69.

27

Biomass and Nutrient Addition in a Cycle of Rubber Plantation

K.K. Ambily, A. Ulaganathan, Annie Philip, Sankar Meti
A. N. Sasidharan Nair, M.D. Jessy and N. Usha Nair

INTRODUCTION

Rubber (*Hevea brasiliensis*), the important commercial source of natural rubber, is a quick growing sturdy tree with tall trunk and thick canopy. Rubber plants take four to five years for canopy closure and attain full growth in 15 to 20 years. Natural rubber, having a perennial tree crop ecosystem, covers about 5.5 lakhs ha area in Kerala and parts of Kanyakumari. Rubber belonging to the family Euphorbiacea is deciduous in nature, having annual leaf litter fall during December–January. Cultivation practices in rubber include the establishment of cover crop in the immature phase (1–4 years) and systematic manuring by inorganic fertilizers during immature and mature phases. There is a regular biomass addition through annual litter fall and also at the time of felling of trees for replanting. Quantification of biomass recycling and nutrient addition through natural sources and inorganic fertilizers in one cycle of cultivation is important to assess the sustainability of the rubber ecosystem.

The earlier studies conducted at RRII and elsewhere have been used to quantify the biomass and nutrient addition through cover crop (*Mucuna bractiata*), litter fall and rain fall (Philip *et al.*, 2005); Karthikakuttyamma *et al,* (2000); Shorroks (1965) and Meti *et al.* (2008). Regarding nitrogen fixation by the cover crop, a conservative estimate of 300 kg N per hectare (Watson 1964) was adopted for the present study.

The biomass and nutrient addition at the time of tree felling was recorded by uprooting four trees of RRII clones of *Hevea* at an age of 30 years from the experimental field at

Central Experiment Station, Chethackal, Ranni of Rubber Research Institute of India. The uprooted trees were partitioned into different morphological units viz. trunk, branches, leaf and roots, and total weight of the plant parts were recorded. During felling the top portion of the tree including leaves, twigs and associated very small branches are left in the field and it undergoes decomposition. Therefore, these portions were also included in the leaf biomass. Estimation of leaf and root biomass and nutrient addition were carried out by recording total biomass and fresh weight and dry weight of the sub sample. Sub-samples were analysed for nutrient content and the nutrient addition to soil from the decomposition of leaf and root residue at the end of the planting cycle was calculated. It is reported that nutrients from the leaf and root residues will be completely decomposed within 2 years (Kheong 2001).

In rubber, legume cover is established during the immature phase as a soil cover and to reduce soil erosion in addition to nitrogen fixation. Rubber tree sheds its leaves during December - January resulting in recycling of nutrients. Another source of nutrient addition in rubber plantation is through rainfall. The estimated biomass addition in one cycle of rubber plantation is given in Table 27.1.

Table 27.1. Biomass addition in one cycle of rubber plantation (t/ha)

Source of biomass	*Quantity added to soil (t/ha)*
Cover Crop	4-6 *
Leaf Litter	6 **
Leaf, twig and very small branches tree felling- Root	1740

* Philip *et al.* (2005) **Karthikakuttyamma *et al.* (2000) and Shorrok(1965)

Apart from these, there is biomass addition during replanting to the extent of 17 t/ha through leaves and twigs as well as 40 t/ha through root residues. Usually during every replanting time, the top portion of the tree including canopy with petiole, twigs and very small branches are left in the field for decomposition. Therefore, leaf biomass at the time of felling includes these portions also. This is different from the annual leaf fall wherein the leaf fall occurs during December – January.

During immature phase there is addition of 236 kg N, 15kg P and 85kg K /ha through the cover crop residues (Philip *et al.,* 2005). About 300 kg nitrogen is added by the cover crop through nitrogen fixation process (Watson, 1964). Nutrient addition through the annual leaf fall is to the extent of 88kg N, 2.5kg P and 45 kg K/ha (Philip *et al.,* 2003). Through leaf fall there is recycling of nutrients annually. Hence, for calculating nutrient addition through leaf fall, it was accounted only once and not every year. From the present study, the nutrient release through decomposition of leaves, twigs and root residues after felling trees was estimated as 2,16 kg N, 250 kg P and 1,354 kg K per ha.

Comparison of annual nutrient addition through natural sources and by the inorganic fertilizer indicated that about 133.2 kg N, 2.5 kg P and 58.2 kg K per ha were added through leaf fall and rain fall (Philip *et al.*, 2003; Meti *et al.*,2008)). The corresponding nutrient addition through inorganic fertilizer is 30 kg each of N, P and K. This clearly shows that the nitrogen and potassium addition through leaf fall and rainfall are much higher than the inorganic fertilizer addition every year. However, the major portion of P addition was from fertilizers and it is higher than the P addition through leaf fall.

The percentage of annual addition of nutrients by natural sources and inorganic fertilizers shows that 82% N, 8% P and 66% K are added through leaf fall and rain fall. In the case of P, the application through external fertilizers is to the extent of 92% followed by 34% for K and 18% for N. From this it is clear that N and K addition through leaf fall and rain fall in rubber plantation is more than addition through inorganic form through external fertilizers.

The estimated total addition of nutrients through natural sources and inorganic form through external fertilizer in one cycle of rubber plantation is presented in Table 27.2. About 2,744 kg N, 268 kg P and 1,748 kg K per ha is added to the soil through organic form whereas, 840 kg N, 840 kg P and 754 kg K per ha only is added as inorganic form in one cycle of rubber cultivation.

Table 27.2. Total addition of nutrients in one cycle of rubber cultivation (30 years) kg/ha

Nutrients	*Natural sources **	*Inorganic fertilizers*
N	2744	840
P	268	840
K	1748	754

*(cover crop + Nitrogen fixation by cover crop + Rain fall + leaf fall + leaf and root at tree felling).

The analysis of nutrient addition through natural sources and inorganic fertilizers in one cycle of rubber cultivation indicates that organic form accounts for 77, 24 and 70 percentage of total N, P and K whereas, it is only 23, 76 and 30 percentage through inorganic form (Table 27.3).

Table 27.3. Percentage of total nutrient addition in one cycle of rubber cultivation

Nutrients	*Natural sources*	*Inorganic form(chemical fertilizers)*
N	77	23
P	24	76
K	70	30

This also clearly indicated that in one cycle of rubber cultivation the total nutrient addition through organic form is more than the inorganic form except for P.

From this study it is clear that substantial quantities of nutrients are added annually through leaf litter and rain fall in rubber ecosystem. Annual addition of N and K through leaf litter and rainfall are more than that through chemical fertilizers. In the case of P, addition is more through chemical fertilizers. The rubber ecosystem is more dependent on natural sources for nutrients compared to external sources like chemical fertilizers and also higher biomass addition and nutrient recycling is recorded in one cycle of rubber cultivation.

References

Karthikakuttyamma, M., Joseph, M., and Nair, A.N.S. 2000. Soils and Nutrition. In: *Natural Rubber: Agromanagement and Crop Processing* (Eds. P.J.George and C.Kuruvilla Jacob). Rubber Research Institute of India, Kottayam, India, pp. 170-98.

Karthikakuttyamma, M., Sathisha, G.C., Suresh, P.R., and Aiyer, R.S. 2004. Biomass production and nutrient budjeting of Hevea beasiliensis in South India. *Natural Rubber Research,* **17**(2):108-14.

Philip, A., Philip, V., George, E.S., Punnose, K.I., and Mathew, M. 2003. Leaf litter decomposition and nutrient release in a fifteen year old rubber plantation. *Indian Journal of Natural Rubber Research,* **16**(1&2):81-84.

Philip, A ., George, E.S. and Punnose, K.I. 2005. Comparative evaluation of dry matter production and nutrient accumulation in the shoots of Pueraria phaseoloides Benth and Mucuna bracteata D.C. grown as cover crops in an immature rubber (*Hevea brasiliensis*) plantation. *Natural Rubber Research,* **18**(1):87-92.

Meti, S. and George, S. 2008. Rainfall distribution, chemistry and nutrient flux in Mature rubber plantation- Submitted for publication in *Journal of Plantation Crops.*

Shorroks, V.M.1965. Mineral nutrition,growth and nutrient cycle of Hevea brasiliensis: 2. Nutrient cycle and fertilizer requirements. *Journal of the Rubber Research Institute of Malaya,* 19(1):48-61.

Watson, G.A. 1964. Maintenance of soil fertility in the permanent cultivation of *Hevea brasiliensis* in Malaya. *Outlook on Agriculture,* **4**:103-12.

Kheong, Y.F. 2001. Impact of Zero Burning on Biomass and Nutrient Turnover in Rubber Replanting. www.e-msss.com/mjss/vol01.htm.

28

Phosphorus Mobilization by Leguminous Cover Crops Grown in Rubber Plantations

P. Prasannakumari, M.D. Jessy, Annie Philip, A.N.S. Nair and N. Usha Nair

INTRODUCTION

Phosphorus deficiency is a major factor limiting crop production in tropical and sub-tropical regions. The peculiar mineralogy of tropical soils with high sesquioxide content favour strong retention of phosphate ions limiting phosphorus availability to crop plants (Kamh *et al.*, 1999; Horst *et al.,* 2001). Widespread deficiency of available phosphorus in soils in the traditional rubber growing areas was reported by George (1961). A survey of the rubber growing tracts of Kerala and Kanyakumari district of Tamil Nadu indicated that 75 % of the area have low soil available P status (NBSS and LUP,1999).

Nitrogen fixing legumes which have a net positive excess of cations over anions entering their roots will acidify the rhizosphere by releasing H+ ions to maintain the cation - anion balance. This nitrogen fixation related rhizosphere acidification is reported to enhance the availability of P from rock phosphate (Hinsinger, 1998). Leguminous cover crops are also reported to possess a remarkable capacity to mobilize soil P through root exudates which can directly shift the equilibrium between soil P fractions towards plant available fractions, which could be beneficial to the less P efficient crops grown in association (Jungk *et al.*, 1993; Kamh *et al.*, 1999; Hinsinger,2001; Horst *et al.*,2001). If P efficiency of the crop is due to mobilization of P from the usually less available soil P fractions, then depletion of readily available soil P might be reduced (Hinsinger,2001). Adams and Pate (1992) suggested that root exudation should also increase the solubility of organic P compounds such as phytins and thereby the pool of plant available P.

Leguminous cover crops are established and maintained in rubber plantations for the purpose of conserving soil, improving or maintaining the soil structure, fertility, water holding capacity and other soil physical properties (Soong and Yap,1976). The most widely grown cover crops in rubber plantations are *Pueraria phaseoloides* and *Mucuna bracteata*. Besides contributing much to the nitrogen requirement through symbiotic nitrogen fixation, cover crops helps in better growth during immature phase and higher yield during mature phase (Watson *et al.*, 1964; Yogaratnam *et al.*, 1984).

The P mobilization efficiency of *Pueraria phaseoloides* and *Mucuna bracteata*, the two commonly cultivated legumes in rubber plantations was studied at the Rubber Research Institute of India, Kottayam, Kerala.

Profuse weed growth was observed in the beds with natural cover and the predominant weeds were *Borreria*, *Chromolaena*, *Corchorus, Clerodendron* etc.

Organic carbon showed an increase from the initial value of 1.8% in all the treatments, but there was no significant difference among the three ground covers (Table 28.1). Available P and available K status was significantly higher in the plots where legume cover crops were established. Between cover crops, soil under *Pueraria* recorded higher available P content.

Soil pH was significantly lower in the rhizosphere of *Pueraria* and *Mucuna* compared to the natural cover (Table 28.2), indicating rhizosphere acidification by these cover crops. Acid phosphatase activity in the rhizosphere of legumes (63.4 and 61.3 g nitrophenol/g soil for *Pueraria* and *Mucuna*, respectively) was significantly higher compared to natural cover (46.6 g nitrophenol/g soil).

The distribution of soil P among the different inorganic P fractions indicated that Al-P was significantly lower in the rhizosphere of *Mucuna* compared to *Pueraria* and natural cover, and also compared to pre-treatment value. Fe-P was also significantly lower in the rhizosphere of *Mucuna* compared to *Pueraria.* Ca-P content was significantly lower in the rhizosphere of *Pueraria.*

N accumulation in biomass was significantly higher for legumes (140.8 and 140.5 kg/ha for *Pueraria* and *Mucuna,* respectively) compared to natural cover (61.0kg/ha) (Table 28.4). Phosphorus accumulation did not show significant difference among ground covers, whereas K accumulation was significantly higher for *Pueraria*.

The study indicates the ability of *Pueraria* and *Mucuna* to mobilize sparingly plant available soil P fractions. This capability is related to root exudation of H^+, OH^-/HCO_3^-, low molecular weight organic acid anions and phosphatase enzymes which strongly influence the bioavailability of soil P in the rhizosphere (Hinsinger, 1998; Kamh, 1999; Neumann and Romheld, 1999 and Horst, 2001). The influence of root exudations in changing the pH of the rhizosphere was reported earlier (Hinsinger, 1998). In the present study also, significant reduction in pH was observed in the rhizosphere of legumes, compared to natural cover, indicating rhizosphere acidification. Nitrogen fixing legumes which have a net positive

Table 28.1. Organic carbon, available P and available K in soil(0-30cm)

P levels(kg/ha) *Pre-treatment*	*OC (%)* *1.80*				*Available P (ppm)* *21.0*				*Av.K(ppm)* *82.0*			
	Pueraria	*Mucuna*	*Natural cover*	*Mean*	*Pueraria*	*Mucuna*	*Natural cover*	*Mean*	*Pueraria*	*Mucuna*	*Natural cove*	*Mean*
0	2.38	2.36	1.88	2.21	35.5	28.3	21.1	28.3	88.5	94.0	74.7	85.7
15	2.44	2.15	2.54	2.38	33.0	31.7	25.2	29.9	103.1	83.7	50.0	78.9
30	2.53	2.20	2.32	2.35	32.2	28.4	25.5	28.7	88.0	95.0	63.6	82.2
45	2.06	1.87	1.55	1.83	33.4	29.5	26.3	29.4	115.5	101.0	69.9	95.5
Mean	2.35	2.15	2.07		33.5	29.5	24.5		98.8	93.4	64.6	
CD (P = 0.05) (crop)		NS				1.33				9.17		
CD (P = 0.05) (P level)		0.32				NS				NS		
CD (P = 0.05) (Crop x P level)		NS				NS				18.3		

NS = Not Significant

Table 28.2. pH and acid phosphatase activity of rhizosphere soil

Pre-treatment	*pH* 4.68				*Acid phosphatase activity (g nitrophenol/g soil)* 35			
P levels(kg/ha)	*Pueraria*	*Mucuna*	*Natural cover*	*Mean*	*Pueraria*	*Mucuna*	*Natural cover*	*Mean*
0	4.38	4.57	4.72	4.56	55.0	63.0	44..5	54.2
15	4.53	4.58	4.62	4.58	70.5	64.0	49.0	61.2
30	4.61	4.64	4.72	4.66	59.0	58.0	53.0	56.7
45	4.64	4.53	4.70	4.62	69.0	60.0	40.0	56.3
Mean	4.54	4.58	4.69		63.4	61.3	46.6	
CD (P = 0.05) (Crop)	0.092	4.66						
CD (P = 0.05) (P level)	NS	NS						
CD (P = 0.05) (Crop x P level)	NS	NS						

NS = Not Significant

Table 28.3. P fractions and total inorganic P in rhizosphere soil

Pre-treatment	Al – P (ppm) 42.25				Fe – P (ppm) 33.36				Ca – P 24.18				Saloid P 2.73				Total inorg. P (ppm) 102.52			
P levels(kg/ha)	Puer-aria	Muc-una	Nat. cover	Mean	Puer-aria	Mu-cuna	Nat. cover	Mean	Puer-aria	Mu-cuna	Nat. cover	Mean	Puer-aria	Mu-cuna	Nat. cover	Mean	Puer-aria	Mu-cuna	Nat. cover	Mean
0	43.4	35.2	43.7	40.8	36.4	21.0	32.3	29.88	13.1	16.9	17.9	16.0	2.7	4.0	3.3	3.3	95.5	77.2	97.2	90.0
15	64.1	38.8	55.6	52.8	28.1	39.6	25.8	31.2	18.7	26.0	21.4	22.0	3.7	4.0	4.2	3.9	114.4	108.4	106.9	109.9
30	61.3	36.1	64.1	53.8	31.1	34.64	34.7	33.5	20.7	24.5	19.8	21.7	3.8	3.3	3.6	3.5	116.7	98.5	122.1	112.5
45	46.4	36.1	63.9	48.8	68.3	20.5	33.2	40.7	15.3	15.5	17.8	16.2	3.6	3.9	3.6	3.7	118.7	76.1	118.6	104.5
Mean	53.8	36.6	56.8		40.9	28.9	31.5		16.9	20.7	19.2		3.4	3.8	3.6		111.4	90.7	111.2	
CD (P = 0.05) (Crop)			3.19				9.48				2.86				0.23				8.66	
CD (P = 0.05) (P level)			3.69				NS				3.30				0.27				9.99	
CD (Crop X P Level)			6.39				18.95				NS				0.47				17.32	

NS = Not Significant

Table 28.4. Biomass and nutrient accumulation in biomass

P levels(kg/ha)	*Biomass (kg/ha)*				*Nutrient accumulation in biomass (kg/ha)*											
					N				*P*				*K*			
	Puer-aria	*Muc-una*	*Nat. Cover*	*Mean*	*Puer-aria*	*Muc-una*	*Nat. Cover*	*Mean*	*Puer-aria*	*Muc-una*	*Nat. Cover*	*Mean*	*Puer -una*	*Muc-una*	*Nat. Cover*	*Mean*
0	3910.0	4126.7	3366.0	3800.9	112.0	140.8	45.0	99.3	7.7	9.0	5.3	7.3	58.7	36.1	58.7	51.2
15	4631.3	3988.7	4352.0	4323.7	133.3	107.8	63.9	101.7	10.3	6.8	8.8	8.6	69.4	44.9	52.9	55.7
30	4420.0	4201.0	4778.7	4466.6	136.2	134.7	73.6	114.8	9.6	8.0	8.5	8.7	95.7	56.5	47.7	66.6
45	4984.0	4548.0	3900.0	4477.3	181.2	178.7	61.6	140.5	12.3	10.9	7.7	10.3	83.2	70.8	55.7	69.9
Mean	4486.3	4215.9	4099.2		140.8	140.5	61.0		10.0	8.7	7.6		76.8	52.1	53.7	
CD (P = 0.05) (Crop)	NS	35.13	NS	16.86												
CD (P = 0.05) (P Level)	NS	NS	NS	NS												
CD (P = 0.05) (Crop X P Level)	NS	NS	NS	NS												

NS = Not Significant

excess of cations over anions entering their roots (alkaline uptake pattern) will acidify the rhizosphere by releasing H^+ to maintain the cation-anion balance (Augilar and Diest, 1981). This nitrogen fixation related rhizosphere acidification is reported to enhance the availability of P from rock phosphate (Hinsinger,1998).

Enhanced acid phosphatase activity in the rhizosphere of legumes is an indicator of the efficiency of legumes to mobilize P from less available soil P fractions. Acid phosphatase is known to catalyse the hydrolysis of organic P into inorganic P (Hinsinger, 1998; Radersma and Grierson, 2004). Very high acid phosphatase activity observed in the rhizosphere of legumes, allow the legumes to access poorly available organic and inorganic P fractions, and thus increase the pool of soil available P, which could be available to the crops grown in association with legumes (Jungk *et al.*, 1993; Horst *et al.*, 2001).

Root exudates play a significant role in the acquisition of inorganic soil P (Hinsinger, 2001). Ae *et al.* (1990) showed that the peculiar ability of pigeon pea to use poorly soluble Fe phosphate was related to the exudation of piscidic acid, and some of its derivatives, that it enhanced P solubilisation via the chelation of Fe. Hinsinger (2001) also suggested that chelation of Fe and Al by root exuded organic acids result in solubilization of Fe and Al bound phosphates and hence in an increased bioavailability of inorganic P in the rhizosphere. In the present study, depletion was observed for Al-P and Fe-P in the rhizosphere of *Mucuna* and Ca-P in the rhizosphere of *Pueraria,* indicating the efficiency of these legumes to acquire P from these stable residual P fractions.

As the P efficiency of the legumes is due to mobilization from unavailable soil P fractions, such as organic P and the inorganic P fractions, the depletion of readily available P in the soil might be reduced (Horst *et al.*, 2001). A significant increase in available P was observed in the legume soil compared to natural cover, indicating mobilization of P by the legumes.

The better P acquisition capacity of legumes can be beneficial to rubber when grown in association with legumes, as the mobilized P could be available to the main crop, especially in the strongly P fixing soils of the traditional rubber growing tract. Cover crop adds to the P nutrition of rubber also through the recycling of the mobilized P *via* plant residues and enhanced colonization with P mobilizing rhizosphere micro-organisms.

Although cover crop has long been a component of *Hevea* cultivation, its importance has diminished in recent years due to a shift towards intercropping in immature phase. However, for farming systems to remain productive and sustainable in the long term, it will be necessary to replenish the resources of organic matter and nutrients which are removed or lost from the soil through renewable resources.

References

Adams, M.A. and Pate, J.S. 1992. Availability of organic ond inorganic forms of phosphorus to lupins (Lupinus Spp.) *Plant and Soil*, **145**:107-13.

Ae, N., Okada, K., Yoshihara, T. and Johansen, C. 1990. Phosphorus uptake by pigeon pea and its role in cropping systems of the Indian subcontinent, *Science* **248**:477-80.

Augilar, A.S. and Diest, V. 1981. Rock phosphate mobilization induced by the alkaline uptake pattern of legumes utilizing symbiotically fixed nitrogen. *Plant and Soil*, **61**:27-42.

Bagayoka, M., Alvey, S., Neumann,G and Buerkert, A. 2000. Root induced increase in soil pH and nutrient availability to field grown cerealsand legumes on acid sandy soils of Sudano- Sahelian West Africa, *Plant and Soil*, **225**:117-27.

Bray, R.H and Kurtz,L.T. 1945. Determination of total, organic and available forms of phosphorus in soils. Soil Sci. **59**:39-45.

Gardner, W.K and Boundy, K.A. 1983. The acquisition of phosphorus by *lupinus albus* l. iv. the effect of interplanting wheat and white lupin on the growth and mineral composition of the two species. *Plant and Soil* **70**:391-402.

George, C.M. 1961. Mature Rubber Manuring : Effect of fertilizers on yield. *Rubb. Bd. Bull.* **7**:37-40.

Hinsinger, P .2001. Bioavailability of soil inorganic P in the rhizosphere as affected by root induced chemical changes-A Review, *Plant and Soil*, **237**:173-95.

Hinsinger, P. 1998. How do plant roots aquire mineral nutrients? chemical processes involved in the rhizosphere. *Advances in Agronomy* (Ed. Sparkes, D.L) Academic Press, London, pp. 225-65.

Horst, W.J, Kamh, M., Jibrin, J.M, and Chude,V.O. 2001. Agronomic measures for increasing P availability to crops, *Plant and Soil*, **237**:211-23.

Jungk,A., Seeling, B, and Gerke, J. 1993. Mobilisation of different phosphate fractions in the rhizosphere, *Plant and Soil*, **155/156**:91-94.

Kamh, M., Horst, W.J., Amer. F., Mostafa, H. and Maier, P. 1999. Mobilisation of soil and fertilizer phosphate by cover crops, *Plant and Soil*, **211**:19-27.

NBSS and LUP. 1999. Resource soil survey and mapping of rubber growing soils of Kerala and Tamil Nadu. *National Bureau of Soil Survey and Land Use Planning*, Indian Council of Agricultural Research, New Delhi, 111p.

Neumann, G and Romheld,V. 1999. Root exudation of carboxylic acids and protons in phosphorus deficient plants. *Plant and Soil*, **211**:121-30.

Peterson, G.W and Corey, R.R. 1966. A modified chang and jackson procedure for routine fractionation of inorganic phosphates, *Soil Sci. Soc. Am. Proc.*, **30**:563-65.

Piper, C.S. 1966. *Soil and Plant Analysis*, Hans Publishers, Bombay, 368p.

Radersma, S and Grierson, P. 2004. Phosphorus mobilization in agroforestry:organic anions, phosphatase activity and phosphorus fractions in the rhizosphere. *Plant and Soil*, **259**:209-19.

Soong, N.K, and Yap, W.C. 1976. Effect of cover management on physical properties of rubber growing soils. *J. Rubber Research Institute of Malaya*, ***24***(3):145-49.

Tabatabai, M and Bremner, J. 1969. Use of P- nitrophenyl phosphatefor assay of soil phosphatase activity. *Soil Biol. Biochem.* **1**: 301-07.

Watson, G.A., Weng, D.P and Narayana, R. 1964. Effect of cover plants on soil nutrient status and growth of hevea: IV. Leguminous creepers compared with grasses micania cordata and mixed indigenous covers on four soil types. *J. Rubber Research Institute of Malaya*, **18**(3):123-45.

Yogaratnam, N., Perera, A.M.A and De Mel, G.J. 1984. Management of ground cover for optimum production. *Proceedings of The International Rubber Conference*, 1984 Colombo, Sri Lanka, **1**(2):521-27.

29

Evaluation of Soil Quality in Vegetable Based Cropping System Under Organic Practices

P. Panneerselvam, A.N. Ganeshamurthy, B.S. Prabhakar, S.S. Hebbar and P.R. Ramesh

INTRODUCTION

Soil quality provides an overall picture of soil functionality, which cannot be measured directly, but measuring certain specific soil properties viz, physical, chemical, biological and biochemical properties can infer it. There are number of scientific evidences well documented that organic farming practices is one of the possible alternatives to sustain the soil quality for a long run. In vegetable production, intensive use of inorganic fertilizers is the key factor in the high yields associated with "modern" agriculture. At present, the indiscriminate and imbalanced usage of inorganic fertilizers very much deteriorates the soil quality. It is high time to study and optimize agro ecosystems to be less dependent on high inputs, to maintain productivity and soil fertility, and minimize negative effects on the environment. To overcome these problems, the only possible alternative is 'Organic farming' which is gaining much importance in the international trade due to global awareness on the ill effects of continuous imbalanced usage of inorganic fertilizers which are considered to be ecologically hazardous.

The variation in soil quality in terms of microbial, biochemical, chemical properties and yield between organic and conventional practices in vegetable based cropping system was studied at the Indian Institute of Horticultural Research, Bangalore during 2005-06. At the time of initiation of experiment, soil had initial pH of 6.39, 0.46% OC, 113 kg N/ha, 19.7 kg P /ha and 115 kg K/ha. In this experiment, two set of practices were adopted ie., 50 t FYM ha^{-1} (organic) and 25t FYM plus recommended NPK ha^{-1} (conventional). Both organic and conventional practices comprised six treatments each, wherein each treatment was considered as one cropping sequence and details are given in Table 29.1. The six cropping sequences, i.e., T_1–T_6 were imposed in organic practices; similarly the same cropping sequences, i.e., T_7 –T_{12} were imposed in conventional practices. Vegetable crops such as Cabbage (Krishna), Tomato (Arka Ananya) and Onion (Arka Niketan) were cultivated based on the cropping sequence. The range of nutrient content of FYM used in this experiment was as follows 0.76-0.82% N, 0.182-0.198% P and 0.85-0.92% K.

Table 29.1. Cropping sequences adopted in this experiment

Sl. No.	*Monsoon*	*Winter*	*Summer*
1.	Cabbage	Tomato	Onion
2.	Cabbage	Onion	Tomato
3.	Onion	Cabbage	Tomato
4.	Onion	Tomato	Cabbage
5.	Tomato	Cabbage	Onion
6.	Tomato	Onion	Cabbage

Population of bacteria, fungi and actinomycetes was analysed from soil samples by using serial dilution plating technique. Soil enzyme activities viz., urease (Tabatabai and Bremner, 1972), dehydrogenase (Thalmann, 1968) ß-glucosidase (Tabatabai, 1982) and phosphatase activities (Tabatabai and Bremner, 1969) were estimated by adopting standard methods. Soil pH, organic C, available N, P and exchangeable K, soil respiration and mineralizable N (Elkarim and Usta, 2001) were estimated by using the standard methods. The yield of cabbage, tomato and onion were recorded after completion of each season.

Microbial population of bacteria (132.7–152.2 × 10^8 cfu g^{-1}), fungi (4.6–6.2 × 10^4 cfu g^{-1}), actinomycetes (3.8–7.1 × 10^5 cfu g^{-1}) and total diazotrophs (9.8–12.1 × 10^4 cfu g^{-1}), *Azospirillum* (11.0–16.0 × 10^4 cfu g^{-1}), *Azotobacter* (5.3–9.2 × 10^4 cfu g^{-1}), phosphobacteria (12.9–16.2 × 10^4) and *P. fluorescens* (9.4–13.4 x 10^4 cfu g^{-1}) were found to be higher in 50t FYM ha^{-1} received treatments compared to conventional practices, ie., 25t FYM plus recommended NPK application (Table 29.2). The soil respiration and mineralizable nitrogen were recorded in the range of 6.79–7.38 C mg/kg of soil/h and 36.75–87.50 mg N/kg of soil, respectively in all the treatments in organic practices, which was significantly higher as compared to conventional practices. This finding clearly indicated that, the increase in microbial population in organic amended treatments might have increased the soil respiration and mineralizable nitrogen content. In organic practices, the enhanced organic C content in soil as a result of organic manure application might have supplied energy and nutrient sources to microorganisms, which would have increased the microbial population when compared to conventional practices. Several studies showed that increasing amount of organic manures application stimulated aerobic bacterial growth (Vallini *et al.*, 1993). The microbial population *viz.*, bacteria, fungi and actinomycetes conspicuously increased with the application of different organic N source compared to the control (Chang *et al.*, 2007).

The soil enzyme activities *viz.*, dehydrogenase (143.5 -189.8 μg of TPF released g^{-1} of soil h^{-1}, ß-glucosidase (134.0–158.4 μg g^{-1} of soil h^{-1}), acid phosphatase (112.0–121.6 μg of p-nitrophenol released g^{-1} of soil h-1), alkaline phosphatase (136.0-142.8 μg of p-nitrophenol released g^{-1} of soil h-1) and urease (28.7–41.3 μg of NH_4 formed g^{-1} of soil h^{-1}) were found to be higher in treatments which received 50 t FYM ha^{-1} than 25 t FYM ha^{-1} plus recommended dose of inorganic fertilizer treatments(Table 29.3). However, the significant variations in following enzyme activities *viz.*, dehydrogenase, ß-glucosidase and urease were observed with in organic as well as conventional practices.

The enzymes in soils originate from animal, plant and microbial sources. Hence, various cropping system and altered microbial population would have caused the variation in soil enzyme activities with in organic and conventional practices. There are evidences to support that organic manure amended treatments influences microbial population, hence higher microbial population in organic experiments might have increased soil enzyme activities. The higher level of enzyme activities were reported by many researchers in soil treated with organic manure compared with soil treated with mineral fertilizer (Marinari *et al.*, 2000; Krishnakumar *et al.,* 2005; Chang *et al.*, 2007).

The results of the soil chemical properties (Table 29.3) showed that among the treatments there was no much variation in pH of soil with in organic as well as conventional practices, the pH was recorded in the range of 6.9–7.3. The organic carbon content in soil was registered significantly higher in treatments T_1–T_8 (2.0 2.3%) as compared to T_9- T_{12} (1.3–1.8%). In general, the available soil nitrogen and phosphorous were higher in 50 t FYM ha^{-1} applied treatments (T_1–T_8), ie., in the range of 741–849.5 and 55.6–98.0 kg ha^{-1} respectively compared to 25 t FYM ha^{-1} received treatments (T_7-T_{12}), however, the variations in nutrient status of soil were observed with in T_1-T_6 as well as T_7-T_{12} treatments. This result was corroborated with earlier findings of Bolan, et al., (1991) and Kemmitt *et al. (*2006). They reported that the application of chemical N fertilizer resulted in a significant lowering of soil pH. Similar type of results were observed by Chang *et al.* (2007), who reported that high application rates of compost resulted in an increase in soil total NO_3^--N and extractable K, Ca and Mg.

The yield response of vegetables under organic and conventional practices revealed that the yield reduction of vegetable, ie., 27% and 20% was recorded in organic plots during winter and monsoon season respectively as compared conventional plots, where as the yield reduction was very narrow during summer season. In general, the overall yield reduction of vegetable was 16.30% in organic plots compared to conventional practices, where as it was 27.0% after one year of experimentation. The yield reduction of vegetables in organic plots may be due pest and disease problems. This study amply inferred that an organic practice in vegetable based cropping system improves the soil quality as compared to conventional practices.

Table 29.2. Soil microbial population, respiration and mineralizable nitrogen in organic vegetable based cropping system

Treatment	*Microbial population*								*Soil respiration*	*Mineralizable N*
(10^8 cfu g^{-1})	*Bacteria (10^4 cfu g^{-1})*	*Fungi (10^4 cfu g^{-1})*	*Actionomy cetes (10^5 cfu g^{-1})*	*Total Diazotrophs (10^4 cfu g^{-1})*	*Azospirillum (10^4 cfu g^{-1})*	*Azotobacter (10^4 cfu g^{-1})*	*PSB (10^4 cfu g^{-1})*	*P.fluorescens (10^4 cfu g^{-1})*	*C mg/kg of soil/hr*	*mg (NH_4+ NO_3 + NO_2) N /kg soi*
CTO - Organic	152.2	5.7	6.4	11.4	16.0	8.9	14.6	13.4	7.38	40.25
COT - Organic	141.4	4.6	5.9	10.6	13.8	7.6	13.8	9.4	6.84	50.25
OTC - Organic	146.3	6.2	7.1	9.8	14.3	9.2	12.9	11.8	7.06	77.0
OCT - Organic	148.6	5.8	4.3	12.1	14.2	5.3	16.2	9.8	6.79	36.75
TCO - Organic	139.7	4.8	6.4	12.0	13.0	9.0	13.1	13.0	7.25	38.50
TOC - Organic	132.7	5.3	3.8	10.9	11.0	8.1	16.0	11.0	7.28	87.50
CTO - Conventional	137.8	6.4	4.1	11.0	10.9	4.9	12.6	10.2	5.40	21.00
COT - Conventional	131.5	5.0	5.7	9.4	11.4	7.2	11.8	10.3	4.54	17.50
OTC - Conventional	116.7	4.6	5.4	5.3	7.6	6.4	10.1	7.6	5.52	26.25
OCT - Conventional	109.8	4.7	4.1	7.6	6.0	5.8	9.8	8.4	4.60	21.00
TCO - Conventional	123.2	4.2	4.9	7.0	7.1	5.3	7.9	6.1	5.92	21.00
TOC - Conventional	111.6	3.8	3.6	6.4	6.3	6.9	9.0	6.8	4.59	17.50
SE d	7.88	0.33	0.33	0.57	0.70	0.44	0.73	0.61	0.36	2.84
CD (P = 0.05)	16.45	0.68	0.69	1.20	1.45	0.92	1.52	1.28	0.76	5.92

Table 29.3. Soil enzyme activity and chemical properties in organic vegetable based cropping system

Treatment	*Soil enzyme activity*					*Chemical properties*				
	Dehydrogenase (μg of TPF released g^{-1} of soil h^{-1})	*ß Glucosidase (μg g^{-1} of soil h^{-1})*	*Acid Phosphatase (μg of p-nitroph enol released g^{-1} of soil h^{-1})*	*Alkaline Phosphatase (μg of p-nitroph-enol released g^{-1} of soil h^{-1})*	*Urease (μg of NH_4 formed g^{-1} of soil h^{-1})*	*pH*	*OC (%)*	*Av. N (kg/ha)*	*Av. P(kg/ ha)*	*Exc. K(kg/ha)*
CTO - Organic	188.2	156.8	121.6	142.8	41.3	7.0	2.2	813.2	98.0	658.0
COT - Organic	189.8	153.7	121.2	136.0	34.3	6.9	2.1	795.7	60.8	602.0
OTC - Organic	143.5	141.2	113.8	139.2	28.7	6.9	2.2	813.2	83.4	854.0
OCT - Organic	161.2	156.3	114.1	136.3	35.0	6.9	2.3	849.5	55.6	840.0
TCO - Organic	151.7	158.4	113.8	136.0	35.7	7.0	2.2	831.3	61.7	644.0
TOC - Organic	147.4	134.0	112.0	140.3	32.2	7.0	2.0	741.0	89.3	616.0
CTO - Conventional	133.7	157.4	105.8	135.2	30.8	7.1	2.0	759.1	33.6	686.0
COT - Conventional	137.6	129.2	90.2	132.2	36.4	7.2	2.1	776.9	41.2	602.0
OTC - Conventional	91.1	136.7	105.7	132.9	36.0	7.3	1.4	542.1	48.4	700.0
OCT - Conventional	91.9	105.7	105.7	132.1	47.6	7.2	1.6	596.2	26.3	574.0
TCO - Conventional	92.3	139.8	99.87	131.4	55.3	7.1	1.8	686.5	54.8	574.0
TOC - Conventional	96.8	129.8	101.89	132.8	56.0	7.1	1.3	488.0	34.0	924.0
SE d	8.39	8.20	6.39	8.03	2.29	0.42	0.11	44.03	3.79	43.83
CD (P = 0.05)	17.51	17.11	NS	NS	4.78	0.87	0.23	91.86	7.90	91.43

Table 29.4. Comparison of vegetable yield (t/ ha) between organic and conventional practices (second year)

Cropping sequences	*Season (Conventional practices)*			*Season (Organic practices)*		
	Kharif	*Winter*	*Summer*	*Kharif*	*Winter*	*Summer*
CTO	28.65*	50.86	32.80	18.14	35.66	27.80
COT	20.31	23.15	29.99	17.72	19.45	23.74
OTC	24.50	38.85	48.83	25.19	27.74	47.83
OCT	28.00	53.04	29.55	21.53	35.70	28.38
TCO	26.16	39.97	31.90	19.56	26.94	28.20
TOC	24.90	23.27	39.85	19.60	21.29	52.73

C – Cabbage, T – Tomato, O – Onion   * All the values are mean of two replications

References

Bolan, N.S., Hedley, M.J. and White, R.E. 1991. Processes of soil acidification during nitrogen cycling with emphasis on legume based pastures. *Plant and Soil* **134**:53-63.

Chang, E.H., Chung, R.S. and Tsai, Y.H. 2007. Effect of different application rates of organic fertilizer on soil enzyme activity and microbial population. *Soil Science and Plant Nutrition* **53**:132-40.

Elkarim, A.K.H.A. and Usta, S. 2001. Evaluation of some chemical extraction methods used as indices of soil nitrogen availability in Polath state farm soils in Ankara Province. *Turk. J. Agric.* **25**:337-45.

Kemmitt, S.J., Wright, D., Goulding, KWT., and Jones, D.L. 2006. pH regulation of carbon and nitrogen dynamics in two agricultural soils. *Soil Biol. Bichem.* **38**:888-91.

Krishnakumar, S., Saravanan, A., Natarajan, S.K. Veerabadran, V., and Mani, S. 2005. Microbial population and enzymatic activity as influenced by organic farming. *Research Journal of Agriculture and Biological Sciences* **1(1)**:85-88.

Marinari, S., Masciandaro, G., Ceccanti, B. and Grego. S. 2000. Influence of organic and mineral fertilizer on soil biological and physical properties. *Biores. Technol.* **71**:9-17.

Tabatabai, M.A. 1982. Soilenzymes. In methods of soil analysis part 2 chemical and microbiological properties. Ed. ALpage, Academic Press, New York, 903-947.

Tabatabai, M.A. and Bremner, J.M. 1969. Use of p-nitrophenylphosphate for assay of soil phosphatase activity. *Soil Biol.Biochem.* **1**:301-307.

Tabatabai, M.A. and Bremner, J.M. 1972. Assay of urease activity in soil. *Soil Biol.Biochem. 4*:479-84.

Thalmann, A. 1968. Zur methodik der Bestimmung der Dehydrogenase aktivit At im Boden mittels triphenytetrazolium chloride (TTC). *Landwirtsch. Forsch.* **21**:249-58.

Vallini, G., Pera, A., Avio, L., Valdrigni, M.M. and Giovonnettin, M. 1993. Influence of humid acids on laurel growth, associated rhizospheric microorganisms and mycorrhizal fungi. *Biology and Fertility of Soils,* **16**:1-4.

30

Windrow Composting of Sugar and Distillery Industrial Wastes: Evolution of Temperature, Chemical Transformation and Phytotoxicity

G.C. Satisha and L. Devarajan

INTRODUCTION

Energy is the key limiting factor for attaining higher crop yields per unit area per unit time. The cost of non-renewable sources has, raised sharply during the last decade, increasing the prices of fertilizers. The recycling of renewable organic wastes and industrial by-products as fertilizer for maintenance of soil health by hygienic methods has, therefore, been a major activity in many countries for increasing crop production and the welfare of mankind. India has huge biomass crop residues (100 to 115 million tonnes) like sugarcane trash, straw, bagasse, coir pith waste, cotton waste, farmland waste, agro-industrial wastes, aquatic weeds etc., but the potentiality of these organic resources was not fully tapped.

India is the second largest sugar producing country in the world. Sugar industries generate large quantities of liquid and solid wastes. The industries release annually about 7.5 and 5.0 million tonnes of molasses and press mud, respectively as wastes. In India, about 700 million litres of alcohol are produced annually from the molasses. In the distillery industry, for every litre of alcohol produced, about 15 litres of waste water known as *spent wash* or *distillery effluent* are discharged; in total about 10 to 11 billion litres of spent wash is produced annually in India. Yeast sludge, a by-product from distillery industry, is also referred as molasses solubles and /or molasses sludge, is a promising source of protein and water soluble carbohydrate besides other nutrients and also contains B-complex vitamins (Aruna, 1999). Bagasse is another waste material from sugar industry found to be the best substrate for the growth of cellulolytic as well as ligninolytic microorganisms (Poonam and Prabhu, 1986) and could serve as excellent carrier for bacterial inoculants (Subba Rao and Tilak, 1984).

The recycling and utilization of these wastes and by-products through development of an economically viable, socially acceptable and environmentally sensible technology is urgently needed for the reason that the large scale accumulation of these wastes will pose disposal and pollution problems. Besides, the substantial quantities of mineral nutrients as well as humus in these wastes are of great importance for maintaining soil organic matter especially in semi-arid tropics of southern India. Composting is the most widely applicable process of handling these diverse wastes as a means of converting raw waste organic matter into usable humus.

Collection of Raw Materials for Composting and Analytical Composition

Fresh pressmud (65–70% w/w) of sulphitation process and bagasse were collected from M/s Sakthi Sugars Ltd., Sakthinagar, Tamil Nadu, India. The distillery spentwash after biomethanation (primary treatment process) was collected from the distillery unit of M/s Sakthi Sugars Ltd., at different intervals in well cleaned 200 litres capacity plastic containers as per the procedure outlined by Ramteke and Moghe (1988) and dried yeast sludge was also obtained from distillery unit of M/s Sakthi Sugars Ltd., Erode, India. Coir pith and sugarcane trash were collected from one of the farmer's field and water hyacinth (*Eichhornia crassipes*) from a river-bed at Bhavani, India. The samples thus collected were analysed for initial physico-chemical properties by standard procedures and the results are presented in Tables 30.1 and 30.2.

Table 30.1. Physico-chemical properties of raw materials used for composting process

Parameters	*Methods*	*Press mud*	*Coir pith*	*Yeast sludge*	*Bagasse*	*Sugarcane trash*	*Water hyacinth*
pH (1:10 w/v)	Jackson (1973)	7.18	5.91	5.45	4.74	6.95	6.95
EC (dSm^{-1})	Jackson (1973)	2.96	1.43	9.30	0.33	0.63	6.15
Organic carbon (%)	Walkley & Black (1934)	32.60	37.26	9.52	37.98	36.00	27.67
Nitrogen (%)	Piper (1966)	1.20	0.42	1.67	0.45	0.44	0.81
C:N ratio		27.17	88.71	5.70	84.40	81.82	34.16
Phosphorus (%)	Piper (1966)	1.15	0.02	0.78	0.06	0.17	0.07
Potassium (%)	Pratt (1965)	0.62	0.54	4.48	0.14	0.46	1.43
Calcium (%)	Piper (1966)	4.14	0.46	7.66	0.70	0.24	1.76
Magnesium (%)	Piper (1966)	1.10	0.31	0.77	0.30	0.18	1.40
Lignin (%)	Sadasivam & Manickam (1996)	16.40	49.80	5.31	20.82	12.20	28.24
Cellulose (%)	Sadasivam & Manickam (1996)	13.52	34.90	7.00	21.00	20.20	24.35
Hemicellulose (%)	Sadasivam & Manickam (1996)	19.38	7.30	6.80	16.00	8.00	10.61
Total phenols(mg $100g^{-1}$)	Sadasivam & Manickam (1996)	23.30	116.70	22.20	62.00	89.60	83.20

Table 30.2. Characteristics and composition of treated distillery spent wash

Parameters	*Methods*	*Range*	*Mean†*
pH	Jackson (1973)	7.80-8.10	7.97 ± 0.06
Moisture (% w/w)	A.O.A.C.(1962)	88.33-93.80	91.81 ± 0.94
Density (Mg m^{-3})	A.O.A.C.(1962)	1.026-1.042	1.034 ± 0.003
Total solids (mg L^{-1})	Ramteke and Moghe (1988)	80800-81600	81350 ± 143
COD (mg L^{-1})	Standard Methods (1965)	41000-44000	43000 ± 548
BOD (5 d at 20ºC) (mg L^{-1})	Standard Methods (1965)	5000-5825	5475 ± 140
Organic carbon (g L^{-1})	Walkley and Black (1934)	23.76-25.26	24.57 ± 0.31
Total N (mg L^{-1})	Piper (1966)	1,720-1,770	1,743 ± 8.56
C:N ratio	–	13.81-14.43	14.10 ± 0.12
Total P (mg L^{-1})	Piper (1966)	428-465	448 ± 6.76
Total K (mg L^{-1})	Toth and Prince (1949)	10,110-11,350	10,654 ± 201
Total Ca (mg L^{-1})	Piper (1966)	1,250-1,450	1,342 ± 33.60
Total Mg (mg L^{-1})	Piper (1966)	1,550-1,775	1,665 ± 41.50

† Average of five samples ± standard error of mean

Composting Technology

The basic organic material used for composting was press mud. Rock phosphate and micro-nutrients viz., $ZnSO_4$ and $FeSO_4$ were used as inorganic additives. The organic additives used with press mud were yeast sludge, bagasse, coir pith, water hyacinth and sugarcane trash. There were four treatment combinations containing pressmud, organic and inorganic additives as shown in Table 30.3.

Table 30.3. Changes in organic carbon (g/kg^{-1}) and total N (%) at different periods of composting

Treatments	*Composting period (days)*			
	30	*60*	*90*	*120*
		Organic carbon (g/kg^{-1})		
C1 – Press mud (PM) only	312	294	261	230
C2 – PM + Organic Additives (OA)	329	298	256	220
C3 – PM + OA + Rock phosphate (RP) @5%	301	275	231	194
C4 – PM + OA + RP@5% + $ZnSO_4$ & $FeSO_4$ each@1.5%	300	276	221	192
LSD (Pd"0.05): composts 0.72; days 0.59; composts x days 1.44.				
		Total nitrogen (%)		
C1 – Press mud (PM) only	1.22	1.27	1.32	1.36
C2 – PM + Organic Additives (OA)	1.30	1.33	1.40	1.45
C3 – PM + OA + Rock phosphate (RP) @5%	1.31	1.35	1.46	1.52
C4 – PM + OA + RP@5% + $ZnSO_4$ & $FeSO_4$ each@1.5%	1.32	1.36	1.47	1.50
LSD (Pd"0.05): composts 0.04; days 0.03; composts x days NS.				

The entire treatments received treated distillery spent wash at 1.3 (w/v) press mud-effluent ratio; LSD- NS-Not significant

Press mud windrows of two tonnes capacity, which were 4.5 m in length, 1.5 m in width at the base and 1.5 m in height were constructed on the cement platform of Compost Yard, M/s Sakthi Sugars Ltd., Sakthinagar, Tamil Nadu, India as outlined by Kuhlman (1990).

The organic additives were added, each at 2% of the fresh weight of press mud and mixed thoroughly using aerotiller. Then, weighed quantities of inorganic additives viz., rock phosphate as Mussoorie rock phosphate at the rate of 5 per cent and micronutrients ($ZnSO_4$, $7H_2O$ and $FeSO_4$, $5H_2O$ each at 1.5% w/w basis) were spread over the windrows and again mixed well. After two days, composite microbial culture at 0.1 per cent level was added and turnings were given to facilitate for complete contact with the decomposing materials. From the fifth day onwards, primary treated effluent at 1:3 (w/v) pressmud-effluent ratios was uniformly added for all the treatments at weekly intervals up to 90 days. After each addition of effluent to the windrows, the decomposing materials were mixed thoroughly by mechanical means using aerotiller and aeration was achieved by turning over the compost mass. The number of turnings-over required was established from Kochtitzky *et al.* (1969) who showed that 6 to 8 separate turnings-over are enough to ensure an adequate oxygen supply for a composting period of 8-10 weeks (bio-oxidative phase). The optimum level of moisture between 40 and 60 per cent was maintained during the whole bio-oxidative phase (Spohn, 1970).

Collection and Analysis of Decomposing Materials

Samples were collected from the left, middle, and right side of the windrows at 30, 60, and 90 cm depths periodically during the composting process at an interval of 15 days up to 120 days. Samples from each side of the windrows at three depths were combined and mixed to generate a single composite. The resulting three composite samples were air-dried and ground in a hammer mill followed pulverization using a vibrating cutter, to pass through a 1-mm sieve and utilized for analysis of various parameters. Temperature was monitored at weekly intervals using a digital thermometer probe (± 0.1 °C) inserted at different locations and depths in windrows (Bhoyar *et al.,* 1979). Each reading was the average of three measurements taken at symmetrical locations at each level.

The pH, electrical conductivity, organic carbon, mineral N (NH_4^+-N and NO_3^--N) content, total N, P, K ,Ca, Mg, Zn, Cu, Fe, Mn, S and cation exchange capacity of the composite samples were determined.

The phytotoxicity test was performed by seed germination with radish (*Raphanus sativus*) seeds in petridish containing 20 ml of water and 20 g of composts from different treatments on filter paper medium using the method described by Tam and Tiquia (1994).

Changes in Temperature During the Composting Process

Temperature evolution is an indicator of microbial activity during the composting process and it may be considered as a good indicator of the end of the bio-oxidative phase. The temperature of the compost mixtures was 42–50 °C at 15th day of composting. As composting proceeded, the temperature of the decomposing wastes was increased rapidly, reaching the maximum temperature during 60th day of composting which was maintained at relatively high level till the day 90 (Thermophilic stage) (Fig. 30.1). After the thermophilic stage, the temperature of all the composts decreased gradually and became stable. As observed in Fig. 30.1, the temperature evolution rate in C_1 compost (press mud alone) was considerably higher than the other composts. The temperature of 47 °C measured on 15th day reached 67 °C on 60th day, which was maintained in the range of 58–67 °C till 90th day. Further, it was observed that the passage of the thermophilic phase to the mesophilic took place in all the compost mixtures between 90 and 105 days of composting. From this time onwards, the decrease in temperature was noted until stabilization from 105–120 days of composting. This indicates that when the metabolic rate decreases, because of exhaustion of available substrate, the temperature starts to fall and the thermophilic microflora is replaced by mesophilic one, which continues the partial degradation of some bio-resistant compounds, fundamentally cellulose and, to a lesser degree, lignin (De Bertoldi *et al.*, 1983).

Several authors concur that the thermophilic phase has to be maintained for at least 10 days and the optimum level should be between 60 and 70 °C to destroy thermosensible pathogens and to permit the hatching parasites eggs, whose larvae are destroyed by the thermal exposure (De Bertoldi *et al.,* 1983). Previous studies on composting have shown that temperature can be used as a parameter to indicate the rate and extent of composting

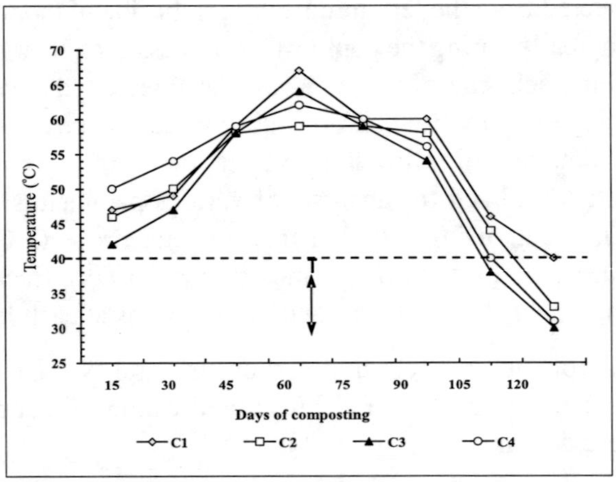

C1- Pressmud (PM) alone; C2 - PM+Organic additives (OA); C3-PM+OA+Rock phosphate (RP) @ 5%; C4 - PM+OA+RP @ 5%+ZnSO4 & FeSO4 each @ 1.5%. T: Thermophilic phase.

Fig. 30.1. Changes in temperature during the composting process and maturity of composts (Tiquia, 1996; Tiquia *et al.,* 1997). They pointed out that compost materials could be considered mature when the temperature in the heap had cooled to the ambient level.

Changes in Physico-Chemical Properties

The pH of the decomposing wastes increased gradually from 7.40–7.63 on 30th day to 7.68–8.02 on 90th day and there after decreased considerably as noticed in Fig. 30.2. The pH was almost neutral and stabilized at maturity phase, which may be due to the buffering nature of humic substances. The increase in pH of the compost mixtures during initial stage of composting (30th–90th day) was mainly because of the metabolic degradation of organic acid or loss by volatilization and/or intensive proteolysis of liberating ammonia compounds due to protein degradation. Later, the pH of the composts was significantly decreased in all the treatments on 120th day, as a consequence of new synthesis of organic acids and production of phenolic compounds. This coincided with a decreased in the NH_4^+-N and an increase in NO_3^--N content of the mixes on day 90. The pH of composting mixes which received inorganic additives recorded significantly lower pH values throughout the period of composting compared to other treatments; may be due to the degradation of easily hydrolyzed polysaccharides (Jiménez and Garcia, 1991). The findings in the present study concurred with the results reported by Verdonck (1988) and Tiquia *et al.,* 1997.

The C: N ratio affects the speed of the composting process and the volume of material finished. In other words, the rate at which organic matter decomposes during composting is principally dependent upon the C: N ratio of the materials (Harada *et al.*, 1981; Hirai *et al.*, 1983). The rate of mineralisation of organic matter during composting was

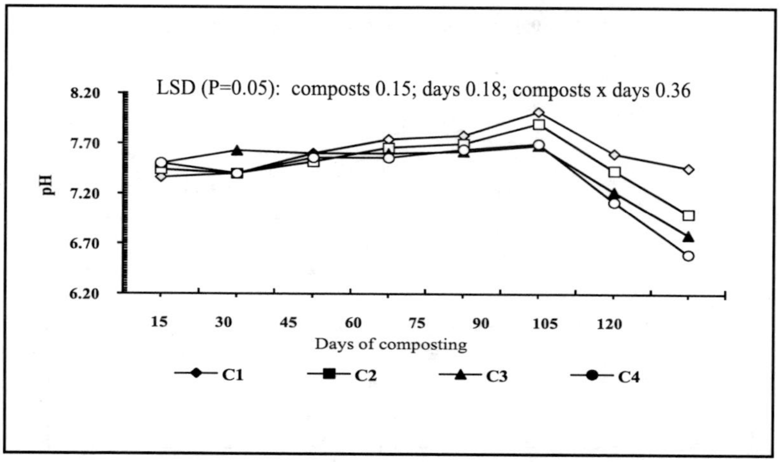

C1 - Pressmud (PM) alone; C2 - PM + Organic additives (OA); C3 - PM + OA + Rock phosphate (RP) @ 5%; C4 - PM + OA + RP @ 5% + ZnSO4 & FeSO4 each @ 1.5%.

Fig. 30.2. Changes in pH during the composting process

measured by determining the C: N ratio of the decomposing mixture. The C: N ratio of the mixtures was high at the initial stages, which decreased substantially up to 90th day and became comparatively stable later on (Fig. 30.3). The composts, which received rock phosphate (C3 and C4), showed higher potentialities in narrowing the C: N ratio than in the non-phosphate treatments. This coincides with the significant reduction in organic carbon (Table 30.3) indicating a rapid biodegradation rate. The reduction in organic carbon through bio-oxidation of carbon to CO_2 could be attributed to the positive effect of rock phosphate and organic additives in increasing the rate of decomposition during composting. This increased rate of biodegradation could also be due to increased availability of N (Table 30.3) and other nutrients from degradable organic molecules as energy source for micro-organisms involved in the process (Tiquia and Tam, 2000). Further, incorporation of yeast sludge (rich in N) and bagasse (substrate for microbes) as additives might have helped in increasing biological activity (Poonam and Prabhu, 1986). In addition, coir pith might have provided favourable atmosphere (vent) for proliferation of aerobic microbes, because, the magnitude of gaseous losses during composting is determined by the availability of energy for microbial proliferation (Srikanth, 1997).

The C: N ratio is considered to be one of the simple indices to evaluate any organics for its fitness for application and it is the index traditionally used to establish the maturity degree of compost (Bernel *et al.*, 1998). The relevance of the C: N ratio relies on the fact that a decrease in the ratio implies an increase in the degree of humification of organic matter (Allison, 1973). A C: N ratio below 20 is indicative of an acceptable maturity degree

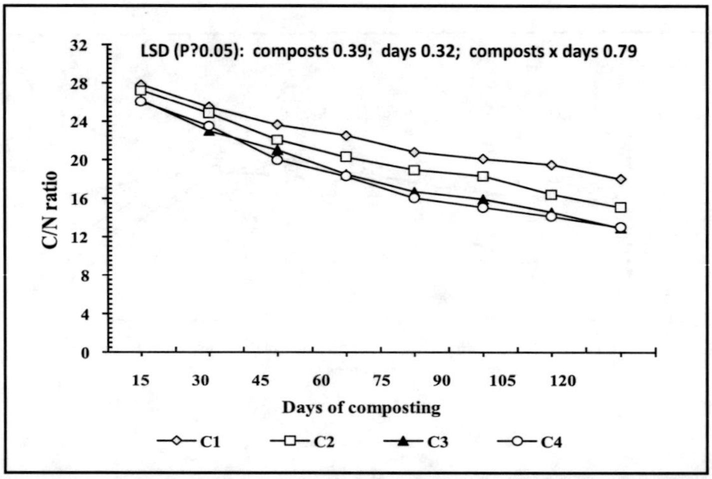

C1 - Pressmud (PM) alone; C2 - PM + Organic additives (OA); C3 - PM + OA + Rock phosphate (RP) @ 5%; C4 - PM + OA + RP @ 5% + ZnSO4 & FeSO4 each @ 1.5%.

Fig. 30.3. Changes in C: N ratio during the composting process

(Golueke, 1981), a ratio of 15 or less is preferable (Morel *et al.,* 1985). In the present study, the C: N ratios of all composts were well within 15 at the end of 120th day and this ensures that the composts attained complete maturity.

The NH_4^+-N content of the composts was increased significantly up to 90 days of decomposition and later decreased considerably at the end of composting (Fig. 30.4a). The higher content of NH_4^+-N during the initial stages of composting indicates continuation of higher rate of respiration in the composting mixtures. During this active phase of composting, the growing concentrations of NH_4^+-N appeared to coincide with greater microbial activity; ammonia concentrations increase as a result of the microbial metabolism of protein. The microbes first metabolize protein to its constituent amino acids (Eq. 1) and then deaminate the amino acides to produce ammonia (Eq. 3) (Jakobsen, 1992).

R
|
$HCNH_2 + O_2 + H_2O$ → → $RH + CO_2 + HCO_3^{ñ} + NH_3$ (RH = CH_4) *Eq. 1*
|
COOñ
|
$CH_4 + 2\,O_2$ → $CO_2 + 2\,H_2O$ *Eq. 2*

Amino acid + $3\,O_2$ → $2\,CO_2 + HCO_3^{ñ} + NH_3 + H_2O$ *Eq. 3*

The fall in NH_4^+-N contents after 90 days might have been due to the rapid conversion of NH_4^+-N to NO_3^--N. This coincides with the significant increase in NO_3^--N content of the composting mixes (Fig. 30.4b). Further, it is quite marked that NH_4^+-N contents were higher in all the treatments up to 90 days than NO_3^--N. This indicates that nitrification may be slow, because of the high density of ammonifiers, presumed to have inherited from pressmud, which helps in the ammonification to take place continuously. This confirms with the results of C_1 compost (pressmud alone) where the NH_4^+-N content was higher than NO_3^--N throughout the period of composting.

The results also show that the content of NH_4^+-N and NO_3^--N in composts which received additives either organic alone or both organic and inorganic, were higher than with the press mud alone (C_1). This ensures that the additives have significant effect on both ammonification and nitrification by maintaining airspaces within the composting matrix (Liao *et al.,* 1997). This is in agreement with the observations of Singh and Yadav (1986) during paddy straw decomposition. Further, it has been noted that the decrease in NH_4^+-N is an indication of both good composting and maturation process (Hirai *et al.,* 1983 and Tiquia, 1996).

Significant differences in nutrient contents were observed among the different composting mixes (Table 30.4). Significant increase in their contents was observed in all the composts (C_3 and C_4) which received rock phosphate. According to Tiwari *et al.* (1989), the application of RP at the rate of two per cent was beneficial in providing better quality

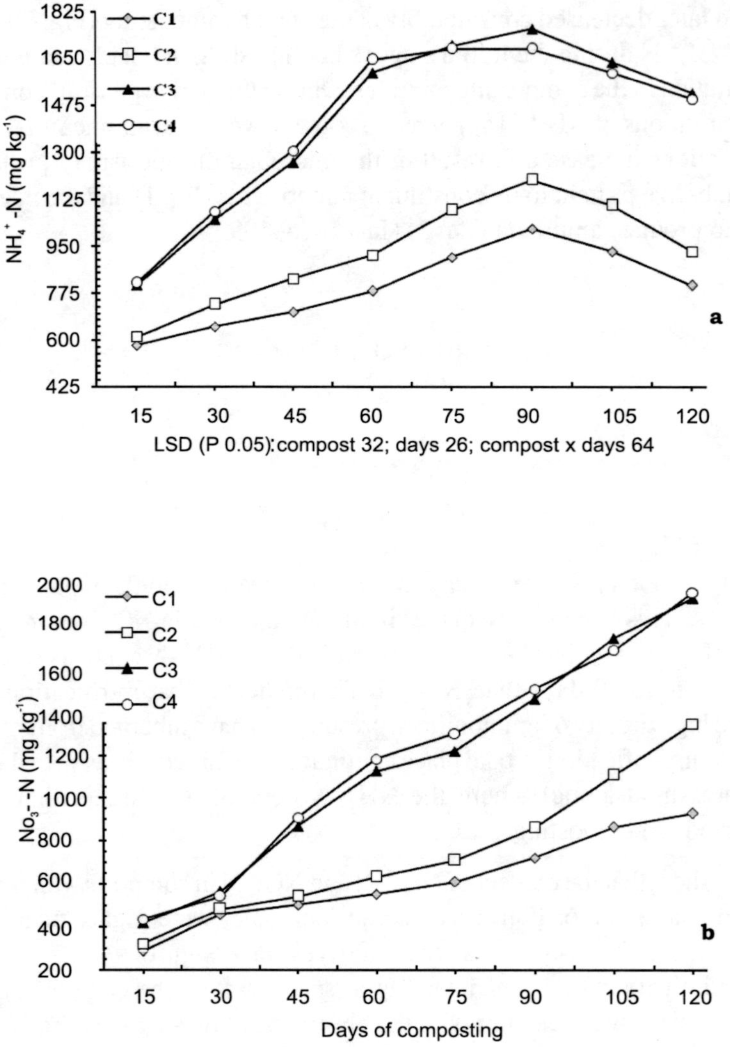

a) NH_4^+-N content. b) NO_3^--N content. C1 - Pressmud (PM) alone; C2 - PM + Organic additives (OA); C3 - PM + OA + Rock phosphate (RP) @ 5%; C4 - PM + OA + RP @ 5% + $ZnSO_4$ & $FeSO_4$ each @ 1.5%.

Fig. 30.4. Changes in mineral nitrogen content during the composting process

Table 30.4. Changes in mineral nutrients at different periods of composting

Treatments	*Total nutrients*																	
	P (%)		*K (%)*		*Ca (%)*		*Mg (%)*		*S (%)*		*Fe (ppm)*		*Mn (ppm)*		*Cu (ppm)*		*Zn (ppm)*	
	Composting period (days)																	
	30	120	30	120	30	120	30	120	30	120	30	120	30	120	30	120	30	120
C_1	1.23	1.41	1.00	3.00	4.27	5.12	1.22	1.73	1.28	1.63	3869	4735	143	192	79	138	198	238
C_2	1.22	1.40	1.12	3.13	4.38	5.39	1.24	1.78	1.28	1.54	4066	4859	158	208	94	142	210	252
C_{-3}	1.71	2.16	1.00	3.26	4.55	5.48	1.30	1.79	1.30	1.62	4002	4875	167	212	108	153	233	313
C_4	1.85	2.28	1.10	3.15	4.27	5.40	1.39	2.14	1.32	1.76	4530	5350	172	214	105	160	465	562

C_1- Press mud (PM) alone; C_2 - PM + Organic additives (OA); C_3-PM + OA + Rock phosphate (RP) @ 5%; C_4 - PM + OA + RP @ 5% + ZnSO4 & FeSO4 each @ 1.5%; all the treatments received treated distillery spent wash at 1.3 (w/v) press mud-effluent ratio.

compost. Higher amounts of K in these compost mixes may be due to the high content of K in effluent and the enhancement in the concentration of K, Ca, Mg and S with time was due to mineralisation of native carbon accompanied by a reduction in the total volume of the wastes (Metting, 1993). Calcium being a constituent of apatite structure, it might have resulted in higher content in rock phosphate enriched composts and incorporation of micronutrients ($ZnSO_4$ & $FeSO_4$) resulted in sulphur enrichment in C_4 compost. The composts which received $ZnSO_4$ and $FeSO_4$ (C_4 and C_6) obviously recorded higher content of Zn and Fe compared all other compost treatments (Table 30.6). Gupta *et al.* (1986) reported that the application of Zn along with press mud have resulted in increased availability of Zn due to the chelating effect of organic materials. This is further supported by the fact that enriching the compost with rock phosphate enhanced the concentration of micronutrients particularly of Fe and Mn, as reported by Mathur and Debnath (1983). Hence, incorporation of micro-nutrients along with RP to the composts can yield high value composts with higher levels of available micro-nutrients.

Table 30.5. Phytotoxicity of composts by germination test using radish (*Raphanus sativus*) as sensitive crop at different periods of composting

Treatments	Composting period (days)			
	60	90	105	120
		Per cent germination		
C_1 - Press mud (PM) only	63.70(46.70)	80.60(53.06)	89.83(57.74)	95.00(61.50)
C_2 - PM + Organic Additives (OA)	60.50(45.61)	84.40(54.79)	93.90(60.65)	98.35(66.70)
C_{-3} - PM + OA + Rock phosphate (RP) @5%	68.63(48.42)	85.00(55.08)	96.70(63.78)	98.60(67.06)
C_4 - PM + OA + RP@5% + $ZnSO_4$ & $FeSO_4$ each@1.5%	65.00(47.14)	86.36(55.76)	97.86(66.13)	100.00(71.57)
LSD (Pd"0.05): composts 2.06; days 1.69; composts x days 4.13.				

Figures in parenthesis are arcsine-transformed values.

The cation exchange capacity (CEC) of the composts has also been used to evaluate the degree of humification and/or maturity as proposed recently by Roig *et al.* (1988). The data on CEC of the composts show that the values increased steadily up to 90 days and then the increase was gradual (Fig. 30.5).

Harada *et al.* (1981) found an increase in the CEC values during the composting of city refuse, with a considerable decrease in the C: N ratio. They concluded that a CEC value more than 60 cmol (p+)kg^{-1} is the minimum needed to ensure an acceptable maturity degree. All the composts at 90th day registered more than the proposed minimum value of 60 cmol (p+)kg^{-1}, thus indicating the phase of maturation. At the end of 120 days of decomposition, the CEC of the composts ranged between 91.3–110.6 cmol (p+) kg^{-1}, the highest was recorded in C_4 compost. The increase in CEC during composting could be attributed to the progressive oxidation in these compost mixtures. As a result, the accumulation of materials bearing the negative charge, such as lignin-derived products (Lax *et al.*, 1986) and the increase of carboxyl and /or phenolic hydroxyl groups (Aleksandrova, 1969) might have contributed to higher values of CEC in the composts.

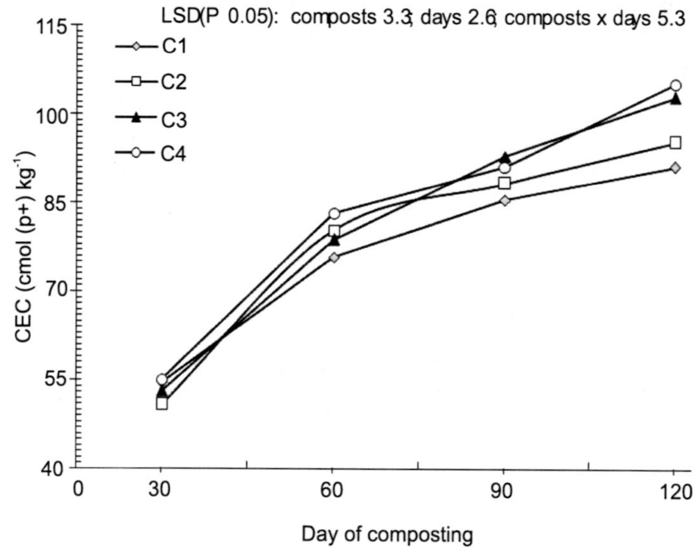

C_1 - Pressmud (PM) alone; C_2- PM + Organic additives (OA); C_3 - PM + OA + Rock phosphate (RP) @ 5%; C_4 - PM + OA + RP @ 5% + ZnSO4 & FeSO4 each @ 1.5%.

Fig. 30.5. Changes in cation exchange capacity during the composting process

The suitability of the compost for application to crop plant was tested *in vitro* condition with radish seeds. The germination index has been used to evaluate the toxicity of composts (Tam and Tiquia, 1994; Tiquia *et al.*, 1996). The compost samples collected at 60, 90, 105 and 120th day of composting were used as supporting medium for the germination test and the results of the test are presented in Table 30.5. The germination per cent of radish in compost samples collected on 60th day of composting was found to be low with values ranging between 63.70 and 70.15%. As composting proceeded, the per cent germination of radish increased considerably. It is apparent from the results that enriched composts showed comparatively higher values than unenriched composts. On 120th day, the values were found to range from 95–100%. According to Zucconi *et al.* (1981), a germination index value of 80 per cent has been used as an indicator of the disappearance of phytotoxicity in composts. Tiquia *et al.* (1996) used this value not only as an indication of the disappearance of the phytotoxicity but also as an indication of the maturity of spent litter compost. In the present study, the per cent germination of 80% was obtained in all the enriched composts on 90th day itself and subsequent increase in germination percentage of radish indicated that the composts were free from phytotoxic substances (Tiquia *et al.,* 1997).

CONCLUSIONS

The rise and fall of temperature may constitute characteristics features of the windrow composting of press mud with other biodegradables for establishing the biological stabilization and optimum degree of compost maturity. The rate of biodegradation process was higher in composts, which received rock phosphate than in non-phosphate treatments, result in the reduction of organic carbon with desirable C: N ratio in 90 days; the C: N ratio decreased from an initial value of 26–13 during composting. The changes in NH_4^+-N concentrations and CEC over time may be good indicators of the progress of the composting process. The compost mixes treated with rock phosphate and other additives recorded significantly higher amounts of mineral nutrients compared to other treatments.

References

A.O.A.C., 1962. Official and Tentative Methods of Analysis. Association of Official Agricultural Chemists, Washington, DC.

APHA-AWWA-AWPCF, 1965, Standard Methods for the Examination of Water and Waste Water, 1-93. Amer. Pub. Health Assoc., Amer. Water Works Assoc. Amer. Water Pollution Cont. Federation, New York.

Aleksandrova, L.V. 1969. Soil organic matter, In: Soil Science, (Ed. Kaurecher E.C.), Koloss Publishing House, Moscow, pp. 62-82.

Allison, F.E. 1973. Soil organic matter and its role in crop production. Elsevier Publishers, New York.

Aruna. 1999. Biorecycling of distillery yeast sludge for sustainable agriculture and aquaculture, *Ph.D. Thesis*, Tamil Nadu Agricultural University, Coimbatore, India.

Aruna. 1999. Biorecycling of distillery yeast sludge for sustainable agriculture and aquaculture, *Ph.D. Thesis*, Tamil Nadu Agricultural University, Coimbatore, India.

Bernal, M.P., C. Paredes, M.A. Sanchez and J. Cegarra. 1998. Maturity and stability parameters of composts prepared with a wide range of organic wastes. *Biores. Technol.* **63**:91-99.

Bhoyar, R.V., M.S. Olaniya and A.D. Bhide. 1979. Effect of temperature on mineralisation of nitrogen during aerobic composting. *Indian J. Environ. Hlth.* **21(1)**:23-34.

Bremner, J.M. and D.R. Keeney. 1966. Determination and isotope ratio analysis of different forms of nitrogen in soils. 3: Exchangeable ammonium, nitrate and nitrite by extraction distillation methods. *Soil Sci. Soc. Am. Proc.* **30**:577-582.

De Bertoldi, M., G. Vallini and A. Pera. 1983. The biology of composting: a review. *Waste Manage. Res.* **1**:157-176.

Golueke, C.G. 1981. Principles of biological resources recovery. *BioCycle*, **22**:36-40.

Gupta, A.P., R.S. Antil and V.K. Gupta. 1986. Effect of press mud and zinc on the yield and uptake of zinc and nitrogen by corn. *J. Indian Soc. Soil Sci.* **34**:810-814.

Harada, Y. and A. Inoko. 1980. The measurement of the cation exchange capacity of compost for the estimation of the degree of maturity. *Soil Sci. Plant Nutr.* **26**:127-134.

Harada, Y., A. Inoko, M. Tadaki and T. Izawa. 1981. Maturing process of city refuse compost during piling. *Soil Sci. Plant Nutr,* **27**:357-364.

Hirai, M., V. Chanyasak and H. Kubota. 1983. A standard measurement for compost maturity. *BioCycle.* **24**:54-56.

Jackson, M.L. 1973. Soil Chemical Analysis. Prentice Hall of India, New Delhi, India.

Jakobsen. 1992. Chemical reactions and air change during decomposition of organic matters. *Resour. Conserv. Recycl.* **6**:259-266.

Jiménez IE and Garcia VP (1991) Composting of domestic refuse and sewage sludge. I. Evolution of temperature, pH, C/N ratio and cation-exchange capacity. *Resour Conserev Recycl.* **6**:45-60.

Kochtitzky OW, Seaman WK and Wiley JS (1969) Municipal composting research at Johnson City, Tennessee. *Compost Sci.* **9**:5-16

Kuhlman, L.R. 1990. Windrow composting of agricultural and minicipal wastes. *Resour. Conserv. Recyl.* **4**:151-160.

Lax, A., A. Roig and F.Costa. 1986. A method for determining the cation-exchange capacity of organic materials. *Pl. Soil.* **94**: 349-355.

Liao, P.H., L. Jones, A.K. Lan, S. Walkemeyer, B. Egan and N. Holbek. 1997. Composting of fish wastes in a full-scale in vessel system. *Biores. Technol.* **59**:163-168.

Mathur, B.S. and D. Debnath. 1983. Studies on phosphorus in compost charged with rock phosphate. *J. Indian Soc. Soil Sci.* **31**:545-548.

Metting, F.B. 1993. Soil microbial ecology-Application of agricultural and environmental management. Marcel Dekker, New York.

Morel, J.L., F. Colin, J.C. Germon, P. Godin and C. Juste. 1985. Methods for the evaluation of the maturity of municipal refuse compost. ***In*:** *Composting of agricultural and other wastes* (Ed. Gasser. J.K.R.), Elsevier, London, pp. 56-72.

Piper, C.S. 1966. Soil and plant analysis. Inter Science Publications Inc., New York.

Poonam, N. and K.A. Prabhu. 1986. The effects of some added carbohydrates on cellulases and ligninase and decomposition of whole bagasse. *Agrl. Wastes,* **17**:293-299.

Pratt, P. 1965. Digestion with hydroflouric and perchloric acids for total K and Na, ***In*:** Methods of soil analysis, Vol. 2, Amer. Soc. Agron. Inc., Madison.

Ramteke, D.S. and C.A. Moghe. 1988. A course manual on water and waste water analysis. National Environmental Engineering Research Institute, Nagpur, India.

Roig, A., A. Lax, J. Cegarra, F.Costa and M.T. Hernandez. 1988. Cation exchange capacity as a parameter for measuring the humification degree of manures. *Soil Sci.* **146(5):**311-316.

Sadasivam, S. and A. Manickam. 1996. *Biochemical methods*, New Age International (P) Ltd., Publishers, New Delhi.

Singh, R.D. and D.V. Yadav. 1986. Evaluation of low grade rock phosphate composted with some agricultural waste for use in a crop rotation. *Agrl. Wastes.* **180**:73-79.

Spohn, E. 1978. Determination of compost maturity. *Compost Sci.* **19**:26-27

Srikanth, K. 1997. Enrichment of compost and its effect on soil properties and crop growth. *Ph.D. Thesis,* University of Agricultural Sciences, Bangalore, India.

Subba Rao, N.S. and K.V.B.R. Tilak. 1984. Importance of Rhizobial cultures in pulse production. *In*: *Pulse production constraints and opportunities*, Oxford & IBH Publ. Co., New Delhi, India.

Tam, N.F.Y. and S.M. Tiquia. 1994. Assessing toxicity of spent litter using germination technique. *Resour. Conserv. Recycl.* **11**:261-274.

Tiquia, S.M. 1996. Futher composting of pig-manure disposed from the pig-on-litter (POL) system in Hong Kong. *Ph.D. Thesis*, The University of Hong Kong, Pokfulam road, Hong Kong.

Tiquia, S.M. and Tam, N.F.Y. 2000. Microbiological and chemical parameters for compost maturity evaluation of spent pig litter disposed from the pig-on-litter system. *In*: Warman PR and Taylor BR (eds) *Proceedings of the International Composting Symposium* (ICS 1999). Volume II. Pp 648-669, Nova Scotia, Canada: CBA Press, Inc.

Tiquia, S.M., N.F.Y. Tam and I.J. Hodgkiss. 1996. Effect of composting on phytotoxicity of the spent pig manure saw dust litter. *Environ. Pollut.* **93**:249-256.

Tiquia, S.M., N.F.Y. Tam and I.J. Hodgkiss. 1997. Effect of turning frequency on composting of spent pig-manure sawdust litter. *Biores. Technol.* **55**:201-206.

Tiwari, V.N., A.N. Pathak and L.K. Lehri, 1989. Effect of cattle dung and rock phosphate on composting of wood waste. *Biol. Wastes.* **27**:237-241.

Toth, S.J. and A. Prince. 1949. Estimation of cation exchange capacity and exchangeable calcium, potassium and sodium contents of soils by flame photometer techniques. *Soil Sci,* **67**:439-445.

Verdonck, O. 1988. Composts from organic materials on substitutes for the usual horticultural substrate. *Biol. Wastes.* **26**:325-330.

Walkley, A. and C.A. Black. 1934. An examination of the Degtjareff method for determining soil organic matter and proposed modification of the chromic acid titration method. *Soil Sci.* **37**:29-39.

Zucconi, F., A. Pera, M. Forte and M. De Bertoldi. 1981. Evaluating toxicity of immature compost. *BioCycle.* **22**:44-46.

31

Production of Quality Biomanure from Non-Conventional Organisms – A Preliminary Study

Joycy Rodrigues and Parashurama Chandravanshi

INTRODUCTION

The history of composting dates back to the history of agriculture. The Israelites, Romans, and Greeks used organic waste directly or composted them (Martin and Gershuny, 1992). Early civilizations of South America, India, China Japan practiced intensive agriculture and used animal and human wastes are fertilizers (Howard, 1943). The concept of large scale composting in a methodical manner is often attributed to Sir Albert Howard and his Indore Process at the Institute of Plant Industry, Indore, Central India between 1934 and 1941 (Howard, 1935).

Animal wastes are considered as important resources that fertilize crop fields, supplement organic matter and improve soil condition but are a source of environment pollution too. Animal waste significantly contributes to increase in nitrogen and microbial load in addition to causing odour problems (Reinecke *et al.,* 1992). Currently the fertilizer value of animal waste is not fully utilized resulting in loss of potential nutrients. Composting is a widely employed method and composting by earthworms has driven much deserved attention and has in fact developed into technology called as 'vermitechnology' utilizing various rural, urban and other waste converting it into a product of economic value (Ndegwa and Jhomson, 2001). Attempts are also made to use other detrivorous organisms to utilize organic matter. Ashwini and Shridhar (2006) have reported compost preparation from pill millipede (*Arthrosphaera magna*) with various plantation crop wastes and have assessed compost quality. A systematic study of these plantation millipedes has been conducted by Ashwini (2003). The compost made using pill millipede was evaluated for its effect on growth and dry matter yield by Ashwini and Shridhar (2006) in black gram (*Phaseolus mungo*) and finger millet (*Eleusine coracana*).

The present work involves composting with millipedes of genus *Trigoniulus* sp (Fig.31.1 and 31.2), locally available red millipede in and around Mangalore area (Coastal Karnataka). The substrate used is goat pellets and hair. Goat dung can be easily collected, stored and transported and is excellent manure either used directly or composted. They enrich the soil with nutrients and also neutralize soil acidity. Their mechanism of neutralization of soil pH is described by Ano and Ubochi (2007). Agamuthu (1994) has shown that composting of goat manure with various additives improves fertilizer quality. The effect of goat manure on growth and yield of Okra (*Abelmoschus esculentus* (L) Moench has been reported by Awodun (2007). The millipede utilizes the goat waste and converts it into fine moist compost. The manure thus produced is tested for its effectiveness on green gram (*Vigna mungo*) and rice (*Oryza sativa*) plants.

Fig. 31.1. *Trigoniulus* sp., the red millipede

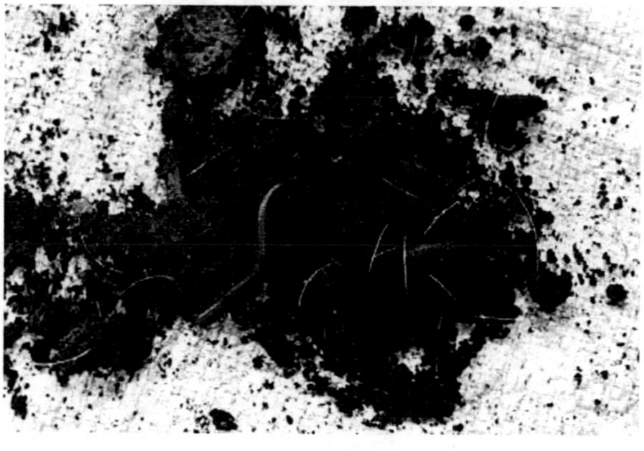

Fig. 31.2. Composting by millipedes using goat dung as substrate

Preparation of Compost from Millipedes

Goat manure was collected on daily or weekly basis and used to fill a box of dimension 45 × 45 × 45 cm (Fig. 31.3). Perforations are made in the box to prevent water logging, and to facilitate aeration. Moisture was maintained between 40 – 60 % throughout the composting period. Compost prepared (Fig.31.4) was collected and analysed for nutrients. The number of millipedes (adults and larvae) per 500 g of sample was counted done on a 15 day basis for about three months.

Fig. 31.3. Composting bin

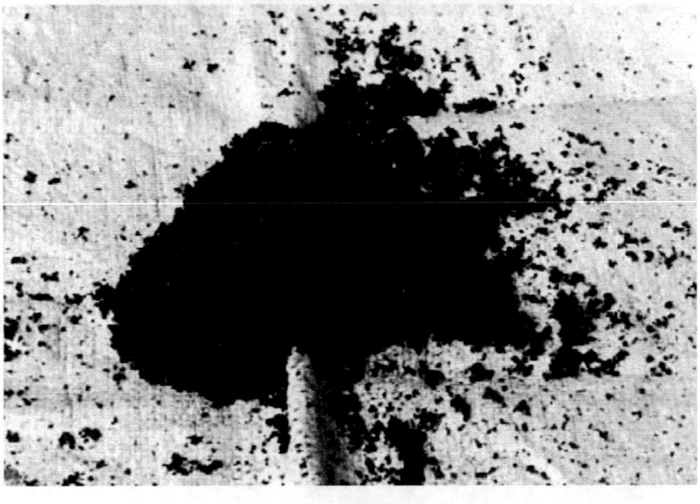

Fig. 31.4. Prepared compost

The compost was subjected to nutrient analysis to find out the quantity of micro and macro-nutrients along with that other parameters like Electrical Conductivity (EC) and pH. The nutrient content from the millipede compost were compared with other composts.

The average size of the compost particle was 0.97 ± 0.15 mm. It was observed that the millipede number was almost constant during the observation for three months for adults and larvae in 500 g of soil at 6 ± 1.84 and 22.16 ± 0.910, respectively.

Millipede compost was analyzed for nutrients in KVK, Navile, Shimoga. The N, P and K content was found to be 1.12, 0.65 and 0.42 percent per 100g of compost, respectively (Table 31.1). This manure composition was then compared with other manures as shown in Table 31.2.

Table 31.1. Aanalysis of millipede compost for nutrients and other parameters

Sl. No.	*Parameter with unit*	*Value*
1.	Moisture (%)	30
2.	pH	7.30
3.	Electrical Conductivity (EC) (desi seimens/meter	0.410
	MACRO-NUTIENTS	
1.	Nitrogen (%)	1.12
2.	Phosphorus (%)	0.65
3.	Potassium (%)	0.42
4.	Calcium (milli Eq./100g of compost)	0.29
5.	Magnesium (milli Eq./100g of compost)	0.22
	MICRO-NUTRIENTS	
1.	Zinc (ppm)	1000
2.	Copper (ppm)	50
3.	Iron (ppm)	3656
4.	Manganese (ppm)	495

Table 31.2. Comparison of millipede compost with other organic manures

Types of manure	*Expressed as percentage*		
	N	*P_2O_5*	*K_2O*
Milli-Compost	1.12	0.65	0.42
Farm yard manure	0.5	0.3	0.5
Goat dung/sheep dung	3.0	1.0	2.0
Cow dung	1.5	0.5	0.5
Cow dung slurry from biogas plant	1.8	1.0	1.0

Green gram and rice seeds are soaked for six to seven hours in distilled water and the seeds are sown in boxes of control and test. Germination and growth of seedlings was observed for about one week. The germination was 100% (Fig. 31.5) with the use of millipede compost (Table 31.5) and 90% of plants had leaves opened on 3rd day. Height of seedlings on 4th and 6th day after germination also showed higher values when millipede compost was used (Table 31.6).

Table 31.5. Germination in green gram

Parameters	*Expressed as percentage*	
	Soil only [control]	*Soil + manure (1:1) [test]*
Germination on second day	60	100
Plants in which leaves opened on the 3rd day	70	90

Table 31.6. Average height of green gram plants on a given day

Height of the plant	*Expressed in cm*	
	Soil only [Control]	*Soil + manure (1:1) [Test]*
Height of the plant on the 4th day	2.77±1.661	7.46±1.515
Height of the plant on the 6th day	12.52±1.6983	17.15±1.265

Results expressed as mean ± SD

Fig. 31.5. Germination and growth in green gram (*Vigna mungo*)

The germination in rice plants on 3rd day was 80 with millipede compost, whereas it was only 60 with soil alone (Fig.31.6). Height of seedlings on 3rd and 5th day after germination also showed higher values when millipede compost was used (Table 31.7).

Fig. 31.6. Germination and growth in rice plant (*Oryza sativa*)

Table 31. 7. Average height of rice plants on a given day

Parameter	*Height in cm*	
	Soil only (control)	*Soil + manure (test)*
Average height of the plant on the 3rd day	0.48 ± 0.072	0.84 ± 0.14
Average height of the plant on the 5th day	2.770 ± 0.24	4.3 ± 0.5

Results expressed as mean ± SD

CONCLUSION

Disposal of animal dung material is a serious problem. Currently the fertilizer values of animal dung are not fully utilized in India resulting in the loss of potential nutrients. Composting provides an alternative for recycling of dung material. The compost prepared from millipedes increases the nitrogen, phosphorus and potassium in the soil thus significantly reducing the addition of synthetic manures to the soil. This in turn suggests that millipedes can be considered as option for composting in addition to vermi-composting. It is observed over a period of a year that the millipedes are resistant to ant, bird, and rodent attacks. In spite of realizing the role of millipedes in soil fertility in tropical and temperate regions no serious attempts have been made to produce compost by millipedes on a large-scale. Hence, there is an urgent need to study the distribution, conservation strategies, lifecycle, and rearing and compost production by millipedes.

References

Agamuthu, P. 1994. Composting goat dung with various additives for improved fertilizer capacity. *World J. Microbiol Biotechnol* **10**: 194-198.

Ano, A.O., Ubochi, C.I. 2007: Neutralization of soil acidity by animal manures: mechanism of action. *Afr. J. Biotechnol.* **6**(4): 364-368.

Awodun, M.A. 2007: Effect of goat manure and urea fertilizer on soil, growth and yield of Okra (*Abelmoschus esculentus* (L) Moench). *Int. J. Agri. Res.* **2**(7): 632-636.

Ashwini, K.M. 2003. Ecological studies of Indian Pill millipede *Arthrosphaera magna.* Ph.D Thesis, Mangalore University, Karnataka, India.

Ashwini, K.M., Shridhar, K.R., 2006. Breakdown of plantation residues by pill millipedes (*Arthrosphraera magna*) and assessment of compost quality. *Current Science* **90**: 954-959.

Ashwini, K.M., Shridhar, K.R., 2006. Evaluation of Pill Millipede (*Arthrosphaera magna*) compost on plant growth and dry matter yield. *EJEAFche* **5**(2): 1323-1329.

Howard, A., 1935. The manufacture of humus by the Indore process. *J. Royal Society of Arts* **84**:26-29.

Howard, A., 1943: An agricultural Testament, oxford University Press, London.

Martin, D.L. and Gershuny, G., 1992: The Rodale Book of Composting. Rodale Press. Emmaus, P.A.

Ndegwa, P.M. and Jhomson, S.A. 2001. Integrating composting and vermicomposting the treatment and bioconversion of biosolids. *Biores. Technol.* **76**:107-112.

Reinecke, A.J., Viljoen, S.A., Saayman, R.J. 1992. The suitability of *Eudrilus eugeniae, Perionyx excavatus, Eisenia foetida* (Oligochaeta) for vermicomposting in Southern Africa in terms of their temperature requirement. *Soil Biol. Biochem.* **22**:1295-1307.

Van Vuren, J. P. G. 1949. Soil Fertility and Sewage. Faber and Faber Ltd, London.

32

Potential Phosphate Solubilising Bacteria from Various Soils of Kerala

Priya Nair, R. Poornima, R.S. Misra and K. Susan John

INTRODUCTION

Phosphorus ranks third after nitrogen and potassium as regards to the extent of its requirement by plants. A substantial portion of the P applied to soil through chemical fertilizers becomes unavailable to plants by forming calcium or magnesium phosphates in calcareous soils and iron or aluminium phosphates in acid environments as they are insoluble. The insoluble form of P is transformed into soluble form through the action of phosphate solubilizing microorganisms (PSM) and hence, becomes available (Fallah, 2006). The PSM present in the rhizosphere is more important than in non-rhizospheric soils (Kucey *et al.*, 1989) and the phosphate solubilizing bacteria is more important than that of fungi (Kucey, 1983). Moreover, a low recovery of applied phosphatic fertilizers (15–25%) due to fixation in the soil and the possibility that the manufactured phosphatic fertilizers will become expensive in the near future due to limited deposits of natural sources of sulphur and high grade rock phosphates, encourage the use of PSMs.

As a prelude to isolation and development of plant growth promoting rhizobacteria (PGPR), screening and isolation of bacteria and fungi from different soil types of Kerala especially Western Ghats were undertaken. Soil samples collected from different locations viz. Nelliampathy, Kanjirampally and other areas of Western Ghats were screened and a total of 340 bacteria and 80 fungi obtained were further screened for phosphate solubilization. Positive isolates were used for phosphorous quantification and from this the most efficient phosphate solubilizing bacteria and fungi were separated.

The bacterial isolates numbered from TCB1-TCB96 and TCB201-TCB444 and fungi numbered from TCF1-TCF80 were screened for the preliminary testing in agar medium of which 70 of the bacterial isolates (Fig. 32.1 and 32.2) and 25 of the fungal isolates (Fig. 32.3) showed positive results, producing clear zones around the colonies in the medium. This clear zone was due to the solubilization of calcium phosphate present in the medium. The diameters of the zones of all the isolates were measured as an indicator of the P-solubilization potential of the organism. The maximum diameter observed was 12mm. From the data generated, the percentage of P-solubilizing bacteria and fungi was found to be 20.58% and 31.25%, respectively.

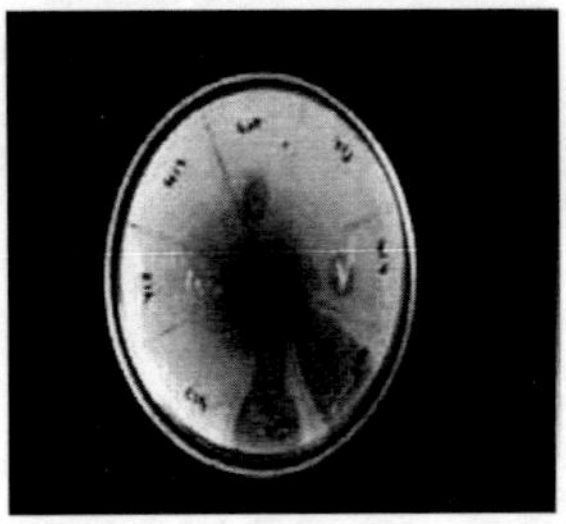

Fig. 32.1

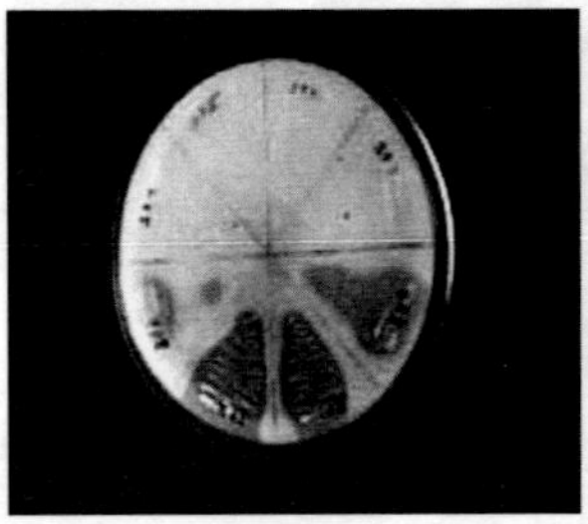

Fig. 32. 2

Phosphate solubilization by bacteria in agar medium

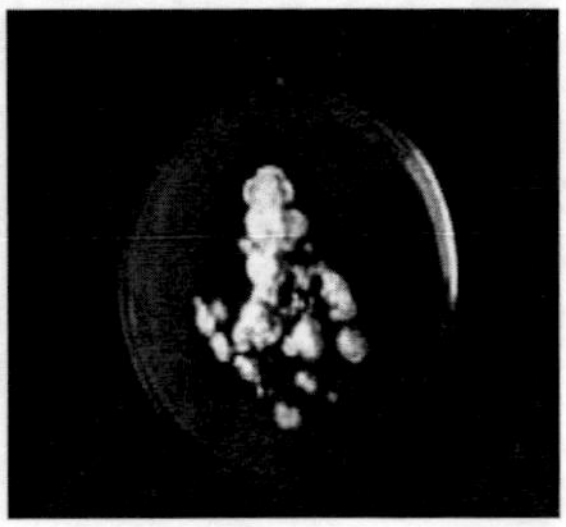

Fig. 32.3

Phosphate solubilization by fungi

The quantification of these strains towards P solubilization showed seven bacterial isolates that were more efficient in this process and gave concentrations higher than 70 ppm (Table 32.1).Tremendous variability in the phosphate solubilization capacity of the different strain was observed. Bacteria are efficient phosphate solubilizers than fungi. Among the different strains tested, TCB330 and TCB333 isolated from Kanjirampally zone recorded the highest concentration of 150 and 100 μg g^{-1} respectively.

Table 32.1. Phosphate solubilization potential of the bacterial isolates

Sl. No.	*Isolate No.*	*Optical density at 530nm*	*Concentration in ppm*
1	TCB24	0.029	71
2	TCB330	0.062	150
3	TCB333	0.041	100
4	TCB347	0.030	73
5	TCB400	0.029	71
6	TCB410	0.039	94
7	TCB421	0.039	95

These efficient strains after mass multiplication can be formulated into biofertilizer and by using this bio-fertilizer, the phosphorous fixation problem currently encountered in both acidic and alkaline soils can be overcome to a great extent.

References

Clesceri, L.S., A.E. Greenberg, and A.D. Eaton .1998. Standard methods for the examination of water and wastewater, 20th ed. APHA-AWWA-WEF. Washington, D.C.

Kucey, R.M.N., 1983. Phosphate-solubilizing bacteria and fungi in various cultivated and virgin Alberta soils. *Can. J. Soil Sci.*, **63**: 671-678.

Kucey, R.M.N., Jansen, H.H., and Leggett, M.E. 1989. Microbially mediated increases in plant-available phosphorous. *Adv. Agronomy* **42**: 199-228.

33

Studies on the Biology and Composting Efficiency of *Eisenia foetida* (Savigny) and *Eudrilus eugeniae* (Kinberg) in Tea Plantations

B. Radhakrishnan and P. Mahendran

INTRODUCTION

In India tea is grown in more than five lakh ha where the crop residues such as tea pruning, shade tree lopping, weeds and tea waste are available in plenty. Thus, the environmentally acceptable vermi-composting technology using earthworms can very well be adopted for converting these wastes into wealth. Recycling of wastes through vermitechnology reduces the problem of non-utilization of agro-wastes. Considerable work has been carried out on vermi-composting of various organic materials. It has been established that epigeic earthworms can hasten the composting process to a large extent, with production of a better quality of composts as compared to those prepared through traditional methods. Earthworms have been presented as key sources to combat the problems of organic waste disposal on a low input basis (Kale *et al.,* 1982; Butt, 1993; Elvira *et al.,* 1998; Dominguez, *et al.,* 2001; Garg and Kausik, 2005; Suthar, 2006). However, their utilization for waste recycling in tea plantations is not well-documented. In view of this, a study was carried out to assess the feasibility for utilizing plantation wastes through the commonly available non-native earthworm species viz., *Eisenia foetida* (the tiger or brandling worm or red worms) and *Eudrilus eugeniae* (The European night crawler).

Surveys were carried out in the tea growing regions of Kerala (Wynadu and Vandiperiar) and Tamil Nadu (Coonoor, Kotagiri, Ooty and Gudalur) to examine for earthworm population. During the surveys both the species of earthworm were collected from manure heaps and soil cow dung juncture of tea plantations with little soil moisture difference

(*Eisenia foetida*-56.51; *Eudrilus eugeniae*-58.45). *Eisenia foetida* was peregrine in South India almost found everywhere where the surveys were conducted. *Eudrilus eugeniae* is another species of commercial importance in vermi-compost industry, collected in all regions except in Vandiperiyar.

The abundance and distribution of earthworms depended on the general soil characteristics. Soil moisture, pH and organic matter played a major role in the distribution of earthworms. *Eisenia foetida* and *Eudrilus eugeniae* are epigeic species associated with high organic matter (25.92% and 31.01%, respectively) and live in a pH range of 7.0-8.65 (Table 33.1).

Table 33.1. Ecological category, distribution and habitat relationships of earthworms

Earthworm Species (Family)	*Colour*	*Ecological category and feeding niche*	*Habitat*						*Casting activity*	*Distribution*
			Collection on site	*Vertical distribution*	*OM (%)*	*pH*	*EC dSm^{-1}*	*SM (%)*		
Eisenia foetida (Lumbricidae)	Deeply pigmented anteriorly; red with yellow bands at tail region	Epigeic, living in organic dung matter heap	Cow	0-15	25.92	7.0-8.35	0.26	56.51	Non-burrower, surface casting, loose, granular	Nilgiri -Wyanaad, Wyanaad and Vandiperiyar
Eudrilus eugeniae (Eudrilidae)	Deeply pigmented (Deep red or violet)	Epigeic, living in organic matter	Soil cow dung juncture	0-15	31.01	8.65	0.39	58.45	Non burrower, surface casting,	Nilgiris, Nilgiri-Wyanaad and Wyanaad

In order to study the life cycle of earth worm species, plastic containers (19.5 cm length, 12.5 cm breadth and 4.5 cm depth) with suitable food (Culture medium) were used. *Eisenia foetida* (Savigny, 1826) and *Eudrilus eugeniae* (Kinberg) are surface living (epigeic) species inhabiting cow dung heaps. For rearing this species, air-dried, sieved (1mm mesh) and ground cow dung (200 g/container) free from any foreign cocoon was used as food. The cocoons after production were immediately isolated and were washed lightly in distilled water to remove debris adhering to the sticky hull and then the size and weight of the cocoons were measured and incubated. The old culture medium was replaced by same amount of fresh media at weekly intervals so that food was not a limiting factor. The moisture level in the culture media was maintained at 60% close to the field soil moisture level. Freshly laid cocoons were kept individually on moist filter paper with a little cow dung (80 % moisture) in a petri dish. They were housed in the same room as the experimental containers.

Mating was observed only in *Eisenia foetida* on the surface of the culture media during the day time. The mating pairs overlapped front ends ventrally by attaching through seminal grooves and each exchanged the sperm with the other. Cocoon production started one week after mating. Cocoons produced by *Eisenia foetida* were transparent and oval with a tube like projection at the anterior end and a hairy filament at the posterior end. Freshly laid cocoons were yellowish and turned reddish yellow during maturation or on emergence.

Cocoons of *Eudrilus eugeniae* were circular or oval with both ends pointed. Initially the cocoons were yellowish but turned dark yellowish brown during emergence. The size and morphological relationship of cocoons of both the species are presented in Table 33.2. The hatchlings of both the species emerged through the tube like pointed structure present at the anterior portion of the cocoons. Further, the hairy filament was used for climbing the leaf litter. Maximum cocoon production was observed in *Eisenia foetida*. The young ones started to emerge on the 2nd and 4th week in *E. foetida* and *E. eugeniae* respectively. The maximum number of hatchlings observed per cocoon was seven in *Eisenia foetida* and two in *Eudrilus eugeniae*. Stephenson (1930) found that 1–20 fertilized ova in each cocoon. Usually one or two young worms hatched from a cocoon, but they may be as many as six in *Eisenia foetida* (Talashilkar and Dosani, 2005). *Eisenia foetida* and *Eudrilus eugeniae* attained maturity after 59.73 ± 0.47 and 43.93 ± 0.33 days, respectively. Maximum growth rate and biomass was observed in *Eudrilus eugeniae*. Growth rate was rapid in both the species till the maturation of the worms thereafter a slight decrease in biomass gain was observed mainly after starting the cocoon production. Many workers found that the pattern of growth in many species of earthworms follow a sigmoid curve (Neuhauser *et al.*, 1988). They also reported similar trend in weight loss in *Eisenia foetida* on allowing these earthworms for longer duration in the treated sludge.

Table 33.2. Cocoon morphology of two earthworm species*

Sl. No.	Characters	Eisenia foetida	Eudrilus eugeniae
1.	Morphology	More or less oval shaped with soft surface layer	Circular or oval shaped with coarse surface layer
2.	Length (mm)	3.9 ± 0.14	5.92 ± 0.05
3.	Breadth (mm)	2.4 ± 0.05	2.97 ± 0.03
4.	Colour	Translucent yellow	Dark straw or yellow
5.	Ornamentation	Tube like projection at anterior end; a hairy filament at posterior end	Pointed ends at both the sides
6.	Fresh weight (mg)	12.19 ± 0.45	15.7 ± 0.12

* values given in the above table are mean ± SE of 15 replicates

Table 33.3. The growth rate, life cycle and other characteristics of *Eisenia foetida* and *Eudrilus eugeniae*

Characters	*Eisenia foetida**	Eudrilus eugeniae*
Maturation (days)	59.73 ± 0.47	43.93 ± 0.33
Growth rate (mg/worm/day)	7.94 ± 3.33	12.14 ± 5.67
Max Biomass (mg)	593.77 ± 8.36	1164.55 ± 28.95
Commencement of cocoon production after clitellum development (days)	10.27 ± 0.85	7.47 ± 0.19
Incubation period of cocoons (days)	23.47 ± 0.72	13.53 ± 0.29
Total number of cocoons produced/worm/cycle	33.47 ± 1.81	21 ± 0.48
Number of cocoons/worm/day	0.57 ± 0.03	0.60 ± 0.01
Avg. hatchlings/cocoon	3.64	1.29
Hatching rate (%)	93.33	85

* Mean ± standard error of 15 replicates

The growth and fecundity of *Eisenia foetida* and *Eudrilus eugeniae* in different bedding materials are given in Tables 33.4, 33.5 and 33.6.

Table 33.4. Growth of *Eisenia foetida* in different feeding materials*

Sl. No.	*Feeding Material*	*At Initial*		*Maximum Body Mass After 75 days*	
		Length (mm)	*Weight (mg)*	*Length (mm)*	*Weight (mg)*
1.	Cow dung (CD) alone	4.97 ± 0.11	1.66 ± 0.07	7.38 ± 0.18	597.27 ± 6.58
2.	CD:TP mixture (1:0.5)	4.88 ± 0.09	1.69 ± 0.07	7.84 ± 0.09	612.21 ± 12.28
3.	CD:SOL mixture (1:0.5)	4.82 ± 0.09	1.67 ± 0.18	7.74 ± 0.17	606.48 ± 8.53

*Mean ± standard error of 15 replicates

Table 33.5. Growth of *Eudrilus eugeniae* in different feeding materials*

Sl. No.	*Feeding Material*	*At Initial**		*Maximum Body Mass After 75 days**	
		Length (mm)	*Weight (mg)*	*Length (mm)*	*Weight (mg)*
1.	Cow dung (CD) alone	18.4 ± 0.16	35.53 ± 0.15	131.4 ± 1.49	958.06 ± 20.58
2.	CD: TP mixture (1:0.5)	18.53 ± 0.19	35.75 ± 0.28	133.67 ± 1.72	994.92 ± 27.03
3.	CD:SOL mixture (1:0.5)	18.87 ± 0.24	35.64 ± 0.29	138.73 ± 1.09	1092.41 ± 28.04

*Mean± standard error of 15 replicates

Table 33.6. Effect of different bedding materials on fecundity of two earthworm species

Bedding material	*Number of earthworms introduced*	*Number of cocoons laid after 75 days*	
		Eisenia foetida	*Eudrilus eugeniae*
Cow dung (CD) alone	50	127	82
CD: TP mixture (1:0.5)	50	353	198
CD: SOL mixture (1:0.5)	50	512	212

Maximum growth rate and biomass gain was observed in *Eisenia foetida* when fed with cow dung and tea pruning mixture. *Eudrilus eugeniae* achieved maximum growth and body mass when fed with cow dung and silver oak litter mixture at 1:0.5 ratio. It was further shown that there was a positive correlation between biomass and length of the worm during the study period. Maximum fecundity was observed in both the species when bedded with cow dung and silver oak litter mixture at 1:0.5 ratio. The changes in biomass and cocoon production differed depending on the substrates. In general the growth, biomass gain and fecundity of earthworms were found to be high when grown in cow dung and crop residue mixture than solely on cow dung. The availability of nutrients in the crop residues may be a reason for the better growth and cocoon production in CDTP mixture and CDSOL mixture.

Results from studies of Garg *et al.* (2006) on the growth and fecundity of *Eisenia foetida* in different animal wastes are available.

Circular plastic containers (40 cm diameter, 15 cm depth) filled with various media such as Cow dung (CD)-T1; CD + Tea pruning (TP) mixture at 1:0.5 ratio-T2; CD + silver oak litter (SOL) mixture at 1:0.5-T3; CD + Tea Waste (TW) at 1:0.5-T4 and CD+TP mixture at 1:3 –T5 were used to study the composting efficiency of different species. The moisture content was adjusted to 60- 80 % during the study period by spraying adequate quantity of tap water. The wastes were turned over manually every day for 15 days to eliminate volatile toxic gases. After this period, 50 earth-worms of each species were introduced into separate containers to assess the composting efficiency. For studying the effect of bedding materials on growth parameters of earthworms, the plastic containers (19.5 cm length, 12.5 cm breadth and 4.5 cm depth) filled with the bedding materials (as mentioned above) were used. Five days old juveniles were introduced individually into each container after taking the initial weight. All containers were kept at room temperature (25 ± 1 °C). Growth and biomass gain of the earth-worms were taken at weekly interval. The time taken for composting the different waste materials and the activity of earthworms were also noted. The composted materials were air-dried and analysed for pH, EC, moisture and nutrient content.

pH of *Eisenia foetida* processed samples ranged from mildly acidic to neutral/mildly alkaline while pH of *Eudrilus eugeniae* processed vermi-compost samples ranged more or less neutral. Electrical conductivity (EC-dSm^{-1}) was less than one in all the vermi-compost samples. The mildly acidic condition may be due to the bioconversion of organic material into organic acids and the nature of raw materials used.

During vermi-composting organic carbon content was reduced due to the loss of carbon as CO_2. Total Nitrogen content (%) was almost more than one in all the vermi-compost samples. The final TKN content in the vermi-compost is dependent on the initial nitrogen present in the feed material and the degree of decomposition. According to Viel *et al.* (1987) losses in organic carbon might be responsible for nitrogen addition. Addition of nitrogen in the form of mucus, nitrogenous excretory substances, growth stimulating hormones and enzymes from earthworms has been reported (Tripathi and Bharadwaj, 2004). Decrease in pH may be another important factor in nitrogen retention as this element is lost as volatile ammonia at higher pH values (Hartenstein and Hartenstein, 1981). Our studies also indicated the same trend. The C: N ratio is used as an index for maturity of organic wastes. C: N ratio was almost below 20 in all the vermi-compost samples except in the vermi-compost where the silver oak litter and cow dung mixture was used as raw material. According to Senesi (1989) a decline in C:N ratio to less than 20 indicated an advanced degree of organic matter stabilization and reflected a satisfactory degree of organic wastes. Lewi-Minzi *et al.* (1986) reported that the C: N ratio of farmyard manure decreased after storing for a period of three months. In our experiments, it was achieved within 20–32 days by the action of earthworms when processed with *E. eugeniae* and within 45–58 days when processed with *Eisenia foetida.*

The size of the worms and the nature of the raw materials contributed much to the variations in processing time. While using *E. eugeniae* the time was reduced and took 20 days for processing the cow dung, 25 days for CD:SOL mixture (1:0.5), 28 days for CD:TP mixture (1:0.5) and 32 days for CD:TP mixture when given at 1:3 ratio.

Eudrilus eugeniae processed samples possessed comparatively less P than *Eisenia foetida* processed samples. Increase in phosphorus during vermi-composting is probably through mineralization and mobilization of phosphorus by bacterial and faecal phosphatase activity of earth-worms (Edwards and lofty, 1972). On the other hand, potassium (TK) was slightly lower in all the vermi-compost samples. This might be caused by leaching of TK by excess water that drained through the feeds. Benitez *et al.* (1999) reported that the leachate collected during vermi-composting process had higher potassium concentrations.

Table 33.7. Effect of bedding materials on physico-chemical parameters of vermi-compost (*Eisenia foetida)*

Physico- chemical parameters	*CD*	*CD+TP (1:0.5)*	*CD+SOL (1:0.5)*	*CD+TW (1:0.5)*	*CD+TP (1:3)*
pH of 1% Suspension	7.40	7.95	8.15	6.35	6.95
EC of 1% Suspension dSm^{-1}	0.20	0.15	0.10	0.26	0.19
Organic Carbon (OC) % T	36.05	27.66	34.03	38.15	23.31
otal Nitrogen Content (N) %	1.96	1.51	1.34	2.35	1.43
Phosphorus as P_2O_5 %	1.07	1.96	2.67	0.78	6.11
Potassium as K_2O %	1.47	0.70	0.76	0.81	0.62
C:N ratio	18.39	18.32	25.40	16.23	16.30

Table 33.8. Effect of bedding materials on physico-chemical parameters of vermi-compost (*Eudrilus eugeniae)*

Physico- chemical Parameters	*CD*	*CD+TP(1:0.5)*	*CD+SOL (1:0.5)*	*CD+TP (1:3)*
pH of 1% Suspension	7.00	8.00	7.05	7.20
EC of 1% Suspension dSm^{-1}	0.12	0.172	0.12	0.08
Organic Carbon (OC) %	36.49	1.00	32.97	27.28
Total Nitrogen Content (N) %	1.82	1.29	1.34	1.22
Phosphorus as P2O5 %	1.32	1.24	0.37	1.40
Potassium as K2O %	0.382	1.46	0.45	0.31
C/N ratio	0.05	16.28	24.60	22.36

Effect of vermi-compost on the performance of tea plants was studied using different media such as red soil (Control), red soil: vermi-compost (3:0.5), red soil: vermi-compost (3:1) and red soil: Vermi-compost (3:2). Polythene sleeves of 150-gauge thickness, 30 cm length and 12.5 cm lay flat width were filled with the respective soil mixtures. Cuttings of clone UPASI-9 were harvested after requisite mother bush treatments and planted. Data collected on one year old plants revealed that general growth of cuttings grown in vermi-compost mixture was superior to those grown in normal nursery medium. Shoot length, number of leaves and branches per plant, fresh and dry weight of shoot and root were higher in the plants grown in red soil and vermi-compost at 3:1 ratio (Table 33.9). Mixing of vermi-compost with the growing medium improved the overall nutrient status. By using a richer nursery medium, it could be possible to raise the plants in a shorter period of eight to nine months compared to the normal period of 12 months. This also resulted in the production of healthier plants, cost reduction in nursery and better utilization of nursery space. The study has shown the possibility of recycling the organic waste in tea more profitably.

Table 33. 9. Effect of vermi-compost on growth of nursery tea

Treatments	*Length (cm)*		*No.of*	*No.of*	*Fresh weight (mg)*		*Dry weight (mg)*	
	Shoot	*Root*	*leaves*	*branches*	*Shoot*	*Root*	*Shoot*	*Root*
Red soil: Vermi-compost (3:2)	7.9	*	6.5	*	205	*	160	*
Red soil: Vermi-compost (3:1)	9.7	14.8	5.8	5.5	374	495	222	112.5
Red soil: Vermi-compost (3:0.5)	8.3	6.0	5.7	9.0	305	480	180	107.4
Red soil (Control)	5.1	13.0	5.0	8.0	182	435	146	106.7

*indicates no rooting in related treatments

CONCLUSION

Vermiculture can be effectively employed for the degradation of plant residues and waste from the tea processing units and the same can be utilized for raising the nursery tea plants in a shorter period. *Eisenia foetida* is the potential species for recycling plantation wastes. Although the European night crawler *Eudrilus eugeniae* processed the wastes faster than *Eisenia foetida,* it failed to thrive inside the tea waste. Prolonged exposure of this species to this waste resulted in the aggregation of earthworms above the culture bed and their death within 12 hours. In this sense, *Eisenia foetida* is a better species for recycling the tea waste. It survived well and processed the tea waste without showing any abnormal activities. Studies on the growth, biomass gain and fecundity of the two species also confirmed that addition of crop residues with cow dung enhanced all the parameters studied. It is concluded that agricultural by products are potential sources of plant nutrients which can be retrieved through the eco-friendly low cost vermitechnology.

References

Benitez, E., Nogales, R., Elvira, C., Masciandaro, G., and Ceccanti, B. 1999. Enzyme Activities as Indicators of the Stabilization of Sewage Sludge Composting with *Eisenia foetida, Biores. Technol.* **67**:297–303.

Butt, K.R. 1993. Utilization of solid paper mill sludge and spent brewery yeast as a food for soil-dwelling earthworms. *Bioresource Tech.,* **44**:105-107.

Dominguez, J., Edwards, C.A., Ashby, J. 2001. The biology and population dynamics of *Eudrilus eugeniae* (Kinberg) (Oligochaeta) in cattle waste solids. *Pedobiologia,* **45**:341-353.

Edwards, C.A. and Lofty, J.R. 1972. *Biology of Earthworms*, Chapman and Hall, London. In: Garg. V. K, Yadav. Y. K, Aleenjeet Sheoran, Subhash Chand and Priya Kaushik 2006. Livestock excreta management through vermi-composting using an epigeic earthworm *Eisenia foetida. Environmentalist*, **26**:269-276.

Elvira, C., Sampedro, L., Benitez, E., Nogales, R. 1998. Vermi-composting of sludges from paper mill and dairy industries with *Eisenia andrei*: A pilot scale study. *Bioresource Tech.,* **63**:205-211

Garg, V.K.; Kaushik, P. 2005. Vermistabilization of textile mill sludge spiked with poultry droppings by epigeic earthworm *Eisenia fetida. Bioresource Tech.,* **96**:1063-1071.

Garg. V.K., Chand. S., Chhillar. A and Yadav. A. 2005. Growth and reproduction of *Eisenia foetida* in various animal wastes during vermi-composting. *Applied ecology and environmental research,* **3** (2): 51-59.

Hartenstein, R., Neuhauser, E.F., and Kaplan, D.L. 1979. 'Reproductive Potential of the Earthworm *Eisenia foetida*', *Oecologia,* **43**:329–340

Kale, R.D., Bano, K., Krishnamoorthy, R.V. 1982. Potential of *Perionyx excavatus* for utilizing organic wastes. *Pedobiologia,* **23**:419-425.

Levi-Minzi, R., Riffaldi, R., and Saviozzi, A. 1986. Organic Matter and Nutrients in Fresh and Mature Farmyard Manure, *Agril. Wastes,* **16**:225–236.

Neuhauser, E.F., Loehr, R.C. and Malecki, M.R. 1988. The potential of earthworms for managing sewage sludge, In: Earthworms and Waste Management, (ed. C.A. Edwards and E. F. Neuhauser),), SPB Acad. Publ., The Netherlands, pp. 9-20.

Senesi, N. 1989. 'Composted Materials as Organic Fertilizers,' *The Sci. Total Environ.* **81/82**:521-524.

Stephenson, J. 1930. The Oligochaeta, Oxford University Press, U.K.

Suthar, S. 2006. Potential utilization of guargum industrial waste in vermi-compost production. *Bioresource Tech.,* **97**: 2474-2477.

Talashilkar. S.C. and Dosani. A.A.K. 2005. Earthworms in Agriculture, Agrobios (India), 21 p.

Tripathi, G. and Bhardwaj, P. 2004. 'Comparative Studies on Biomass Production, Life Cycles and Composting Efficiency of *Eisenia foetida* (Savigny) and *Lampito mauritii* (Kinberg),' *Biores Technol.* **92**:275-283.

Viel, M., Sayag, D., and Andre, L. 1987. 'Optimization of Agricultural, Industrial Waste Management through In-vessel Composting'. In: M. de Bertoldi, (ed.), *Compost: Production, Quality and Use. Elseiver Appl. Sci. Essex*, pp. 230-237.

34

New Record on Decomposition of Legume Biomass by Soil Insect in Cashew Ecosystem

D. Sundararaju

INTRODUCTION

In the west coast of India, cashew is grown in undulating terrain and during south west monsoon luxuriant growth of leguminous weed, *Mimosa pudica* L. is very common. When such green biomass of the weed is cut and heaped to the basins of cashew trees as mulch, gregarious feeding of litter feeding caterpillars was observed resulting in rapid decomposition of biomass similar to vermi-compost. It was interesting to note that those caterpillars did not feed directly on the live succulent foliage of the same species existing in the field. It was similar to the attack of *Simplicia robustalis* Guen. (Lepidoptera: Noctuidae : Herminiinae) on dry stacked fodder of ragi, jowar, etc., in south India (Ayyar, 1963).

In order to study the process of decomposition, the cut green biomass of common leguminous plants available in cashew plantation as well as in the borders of cashew plantation *viz., M. pudica, M. invisa* C. Martius ex Colla, *Gliricidia sepium* (Jacq.) Steud, *Peltophorum ferrugineum* Benth, *Erythrina* sp. and *Desmodium* sp. and *Crotalaria laevigata* Lam. were heaped during July- November period. Subsequently the symptom of gregarious feeding by first instar larvae were observed within three weeks of heaping. It gives an indication that the wilting of heaped leguminous foliage might be emitting a particular volatile which might have attracted female moth for egg laying. The young larvae are hyaline with black head and four black spots on the dorsal region of each segment. The grown up caterpillars are about five centimeter in length and dark brownish in colour. The caterpillars voraciously feed dry/decaying foliage and scrape the bark of stem cuttings and finally convert entire foliage as faecal pellets within a period of 45–60 days. The final decomposition was found to be similar to vermicast. In 30 × 30 cm area of the heap, 12–18 caterpillars were counted and the larval period varied from 26–35 days. Larval parasitism at a low level (below 5.0 %) by an unidentified tachinid parasitoid was also encountered. Pupation takes place within the cocoon of faecal pellets and 9-13 pupae/0.09 m^2 of the heap were observed. The emergence of moth was observed between 10–14 days after pupation. The moths are medium sized light to dark brown coloured with conspicuous two small white spots on each fore wing. On both fore and hind wings, two dark brown coloured continuous wavy bands are also seen. During July-September attraction of adult moths is sparingly seen at coastal Kerala and Karnataka and Malanad areas of Karnataka. The male and female moth has pectinate and setacious antennae respectively.

This species was identified as *Bocana manifestalis* Walker (Lepidoptera: Noctuidae). Information on occurrence of *B. manifestalis* was reported in Ceylon (Sri Lanka) along with other species of *Bocana*, viz., *B. jutalis* Wlk., *B.ophinsalis* Wlk., *B. vagalis* Wlk. and *B. turpatalis* Wlk. under family Herminiidae was reported (Tennent, 1861). Under latest taxonomical classification, Herminiidae is a sub-family designated as Herminiinae under family Noctuidae (Smetacek, 2008). Museum details on *B. manifestalis* are seen in the literature from Fiji, Papua New Guinea and Hawaii islands (CSIRO, 2006; Evenhuis, 2007; BM, 2008). In the literature, other than above taxonomic information, no further details on biology and host range of *B. manifestalis* were seen. Thus, the above observations on the decomposition of cut green biomass of legumes were new record in India as well as in the whole world.

The herbivore alteration of litter inputs may change litter decompositions and influence ecosystem nutrient cycling and augment nutrient acceleration in the soil (Chapman *et al.*, 2003). Thus, it provides scope for enhancement of nutrient recycling at a shorter period as a method of organic nutrient management. The green manuring in cashew orchards with glyricidia had enhanced the nitrogen content in the soil and yield at coastal Karnataka (Yadukumar *et al.*, 2007) and it may due to rapid recycling nutrients as a result of faster decomposition by *B. manifestalis* as its activities were observed repeatedly in the decomposition of cut green biomass of *G. sepium*. Thus, there is a wider scope for rapid enrichment of nutrients in cashew ecosystem by heaping the cut green biomass of legumes during the period of July-November as a method of organic nutrient management. As south west monsoon at west coast of India enhances luxuriant growth of various legumes (green manure plants or fence plants/trees) in cashew plantations, growth and utilization such legume needs to be encouraged for organic nutrient management.

CONCLUSION

The soil insect involved in the decomposition of cut green biomass of legumes was identified as *Bocana manifestalis* Walker (Lepidoptera: Noctuidae : Herminiinae). It acts as a mobile decomposer and the final decomposition is similar to vermicast. Therefore, legumes may be grown for producing maximum green biomass especially during south-west monsoon at west coast of India and added to basin of cashew tress between July and November as an organic nutrient management technique.

References

Ayyar, T.V.R. 1963. *Hand Book of Economic Entomology for South India.* Govt. Press, Government of Madras. pp. 360-361.

BM (Bishop Museum). 2008. ***Bocana manifestalis*** Walker, 1858. http://www2.bishopmuseum.org/natscidb/?w=PBIN&srch=b&pt=t&lst=o&cols=8 &rpp=500&pge=1&lID=-345088491&oID=716947466&tID=-1073389330

Chapman,.S.K, Hart, S.C., Cobb, N.S., Whitham, T.G and Koch, G.W. 2003. Insect herbivory increases litter quality and decomposition: an extension of the acceleration hypothesis. *Ecology,* **84**(11): 2867–76.

CSIRO (Commonwealth Scientific and Industrial Research Organisation). 2006. *Bocana manifestalis* Walker, 1859. Entoml D-PNG Specimen Database. http://anic.ento.csiro.au/entomid-png/taxon_details.asp?BiotaID=10171

Evenhuis, N.L. 2007. Checklist of Fijian Lepidoptera. HERMINIIDAE: *Bocana manifestalis* Walker, 1858: 171. *Bishop Museum Technical Report* **38**(13): 21. http://hbs.bishopmuseum.org/fiji/checklists/lepidoptera.html

Smetacek, P. 2008. Moths recorded from different elevations in Nainital district, KumaonHimalaya, India. *Bionotes* **10**(1):5-15.

Tennent, J.M. 1861. *Sketches of Natural History of Ceylon.* Natural History of Ceylon, p 437.

Yadukumar, N., Rejani, R. and Nandan S.L. 2007. Green manuring for cashew orchards. National Seminar on Research, Development and Marketing Cashew, Goa. *Souvenir and Extended Summaries.* pp. 59-60.

SECTION – VI
PLANT HEALTH MANAGEMENT

35

Identification of Indicator Plants for Early Detection of Basal Stem Rot (*Ganoderma* wilt) of Coconut

K.B. Palanna, T.B. Basavaraju, B. Boraiah and N.E. Thyagaraj

INTRODUCTION

Basal stem rot or Thanjavur wilt of coconut caused by *Ganoderma lucidum/ G applanatum* is one of the most destructive diseases affecting coconut production in India (Karunanithi *et al.*, 2005). The disease is soil borne and spreads through soil and irrigation water from palm to palm and garden to garden. The disease incidence was reported from almost all major coconut growing states of India (Bhaskaran *et al.*,1994). Basal stem rot disease incidence ranged from 6.06–36.15% in Arsikere Taluk of Karnataka (Naik *et al.*, 2000). Its incidence was observed maximum (up to 62.50%) in coconut gardens raised in sandy and red soils in coastal district of Andhra Pradesh (Srinivasulu *et al.*, 2003). Though several research workers (Bhaskaran *et al.,* 1994, and Srinivasalu *et al.,* 2001 and 2004) have reported different management practices for the basal stem rot disease, the results are inconsistent and giving variability in containing the disease depending on soil condition, age of the palm and time/stage of treatment imposition.

Besides coconut, the *Ganoderma* spp. has a wide host range attacking a variety of palms and several forest, avenue and fruit trees (Naidu *et al.,* 1966., Govindu *et al.*, 1983; Bhaskaran *et al.*, 1994). Basal stem rot of coconut can be contained by following integrated disease management practices, if the disease is detected in the early stages and it is very much essential for elimination of the pathogen at the earliest possible moment to minimize the disease incidence and subsequent spread. A few methods have been reported to be useful for early diagnosis of the disease. Transpiration rate (Vijayaraghavan *et al.*, 1987), Colorimetric method (Natarajan *et al.*, 1986), EDTA method (Natarajan *et al.*, 1986,

Vijayaragavan, *et al.*, 1987). Chemo diagnostic and Physological parameters (Karthikeyan and Bhaskaran, 2004). However, all these methods are costlier and not feasible to the farming community.

Attempt was made to identify basal stem rot at an early stage of infection by indicator plants to combat the disease effectively. Pulses, vegetables and a cereal (Table 35.1) were screened in pot culture as well as in field experiment.

Table 35.1. Plants tested against *Ganoderma* wilt disease of coconut

Sl. No.	*Crop*	*Scientific name*	*Family*
1.	Redgram/ Pigeon pea	*Cajanus cajan* (L.)Millsp.	Papilionaceae
2.	Soybean	*Glycine max.*Merl.	Leguminaceae
3.	Horse gram	*Dolicus uniflorus* Lam.	Papilionaceae
4.	Black gram	*Vigna mungo* Linn	Papilionaceae
5.	Green gram	*Vigna radiata* (L) Wilezek	Papilionaceae
6.	Cowpea	*Vigna unguculata*(L) Walp	Papilionaceae
7.	Bengal gram/chick pea	*Cicer arietinum*	Papilionaceae
8.	Avare	*Dolichos lablab.*Linn	Papilionaceae
9.	Ladies finger	*Abelmoschus esculentus*(L) Moench.	Malvaceae
10.	Tomato	*Lycopersicon escul*	*entum* Solanaceae
11.	Brinjal	*Solanum melongena.* Linn	Solanaceae
12.	Chilli	*Capsicum annum.* Linn	Solanaceae
13.	Maize	*Zea mays* Linn	Gramineae

Ganoderma isolates (G_{13} and G_{14}) were mass multiplied separately on sorghum grains in poly bags and were used for soil inoculation by mixing (100 g/pot) with pot mixture at the time of sowing. The treatments with out inoculum served as control. For field studies same set of plants were sown in the basins of diseased palms and healthy palm (control) for comparison of symptoms.

All the plants showed different types of symptoms for two isolates (G_{13} and G_{14}) tested (Table 35.2). In the case of red gram, seeds were infested with disease and failed to germinate and a few of the germinated plants showed wilting (60 %) symptoms after 40–50 days of sowing. In the case of soybean, the plants showed stunted growth and 70 % (G_{13})–95% (G14) plants showed bark splitting as against 20% in control. In Ladies finger, 67.5% (G_{13}) to 70% (G_{14}) of plants showed wilting as against 0% in control. Whereas, black gram, green gram and bengal gram showed 3.3 to 6.6, 6.6 and 53.4–60.0% wilting, respectively as against 0% in control and all these plants showed stunted growth. In the cases of black gram and green gram, wilting was observed after 30–40 days of sowing, whereas, in bengal gram wilting was observed at seedling stage as well as after 30–40 days of sowing. In horse gram, avare and tomato plants showed stunted growth compared to control. Cowpea, brinjal, chilli and maize did not exhibite any type of visual symptoms but germination was affected in case of chilli.

Table 35.2. Early detection of BSR (*Ganoderma* wilt) disease of coconut by indicator plants

Sl. No.	*Crop*	*Symptoms expressed by plants*				
		Pot culture			*Field*	
		G_{14}	G_{13}	*Control*	*Diseased palm*	*Healthy palm (Control)*
1.	Redgram	W (60%)	W (60%)	—	W& BS	—
2.	Soybean	BS(95%)	BS(70%)	BS(20%)	BS	BS
3.	Horse gram	SG	SG	—	—	—
4.	Black gram	W&SG(6.6%)	W&SG (3.3%)	—	—	—
5.	Green gram	W&SG (6.6%)	—	—	—	—
6.	Cowpea	—	—	—	—	—
7.	Bengal gram	W&SG (53.4%)	W&SG (60%)	—	W	—
8.	Hebbal avare	SG	SG	—	—	—
9.	Ladies finger	W(70%)	W(67.5%)	—	W	—
10.	Tomato	SG	SG	—	—	—
11.	Brinjal	—	—	—	—	—
12.	Chilli	—	—	—	—	—
13.	Maize	—	—	—	—	—

Note: (i) W – Wilting, BS – Bark splitting, SG – Stunted growth,
(ii) Figures in the parenthesis are % of plants wilted

When all the infested plants were subjected for re-isolation, *Ganoderma* was isolated from Red gram (infested seeds and wilted plant), ladies finger (wilted plant), soybean (bark splitted plant), Bengal gram (wilted seedling and plant), green gram (wilted plant), black gram (wilted plant) and horse gram (stunted plant). Thus, the pathogenicity was proved in red gram, soybean, bengal gram, green gram, black gram, horse gram and ladies finger and are found to be good indicator plants under pot culture studies.

In field conditions, among thirteen plant species tested, four plants showed specific symptoms. In the case of red gram, wilting of seedlings and bark splitting were observed at the base of the plants (60–70 DAS). In Ladies finger and Bengal gram, seedlings showed wilting symptoms. Bark splitting was observed in soybean (30–40 DAS). However, the percentage of wilting was less in the field conditions compared to pot culture studies. These disease symptoms were observed in sick soil, but soybean showed bark splitting symptoms both in sick soil and control. In pot culture studies, all the plants under study belonging to family papilionaceae expressed symptoms except cowpea. Whereas under field conditions among the 13 plant species belonging to 3 families (Table 35.1) tested the pathogenicity was proved only in bengal gram, pigeon pea and Ladies finger. Srinivasulu *et al.*(2006) reported that Pigeon pea (*Cajanus cajana.*L)- as an indicator plant to *Ganoderma* wilt disease of coconut. The present study revealed that, bengal gram, pigeon pea and Ladies finger are found to be indicator plants for *Ganoderma* wilt disease of coconut. Such information may helpful to combat disease effectively. However, this needs testing in different locations for further confirmation.

References

Bhaskaran, R., Rethinam, P. and Nambiar, K.K.N. 1994. *Ganoderma* wilt disease of coconut. In: *Advances in Horticulture, Plantation and Spice Crops*, Part 2, Vol. 10 (Eds. K.L.Chandha and P. Rethinam), pp. 899-920.

Govindu, H.C., Rao, A.N.S. and Kesavamurthy, K.V. 1983. Biology of *Ganoderma lucidum* (Leys.) Karst. and control of Anabe Roga of coconut. In Coconut Research and Development (Ed. N.M.Nayar). Wiley Eastern Ltd., pp. 325-329.

Karthikeyan A. and Bhaskaran R. 2004. Early diagnostic techniques for the management of basal stem rot disease of coconut. *Indian Coconut J.* 34 (12) 16-18.

Karunanithi, K., Sarala, L., Manickam, G., Rajarathinum, S. and Khan, H.H. 2005. Management of basal stem rot of coconut. *Indian Coconut J.* **35**(9) 10-11.

Naidu, G.V.B Kumar, S.N.S. and Sannamarappa, M. 1966. Anabe roga, *Ganoderma lucidum* (Leys.) Karst. On arecanut palm: a review and further observations. *J. Mysore Hort. Soc.*, **1**(3): 14-20.

Naik R.G., Palanimuthu, V., Hanumanthappa, M. and Indiresh, K.M. 2000. Prevalence and intensity of basal stem rot disease of coconut in Arsikere taluk of Karnataka. *Indian Coconut J.* **31**(1): 8-10.

Natarajan, S., Bhaskaran, R. and Shanmugham, N. 1986. Preliminary studies to develop techniques for early detection of Thanjavur wilt in coconut. *Indian Coconut J.* **17**(3): 3-6.

Srinivasulu B., Aruna. K. and Rao D.V.R., 2001. Biocontrol of *Ganoderma* wilt of coconut. *South Indian Horticulture.* **49**: 242-241.

Srinivasulu B., Aruna. K., Rao D.V.R., & Khan H.H. 2003. Epidemiology of basal stem rot (*Ganoderma* wilt) disease of coconut *J. Plantation Crops* **31**(1): 62-63.

Srinivasulu B., Vijaykrishnakumar K. , Aruna K., Lakshmi B.K.M. and Rao DVR, 2004. An eco friendly technology to manage BSR (*Ganoderma* wilt) disease of coconut. *The Andra Agric. J.* **50** (Spl.): 264-267.

Srinivasulu B.,Vijay Krishna kumar.K., Aruna. K. and Rao D.V.R. 2006. Pigean pea (*Cajanus cajana.* L) - an indicator plant to Basal Stem Rot (*Ganoderma* wilt) disease of coconut. AAPP.News letter **1**(4):3.

Vijayaraghavan, H., Ramadoss, N. and Bhaskaran, R. 1987. Approaches for early detection of Thanjavur wilt disease of coconut. *International Symposium on Ganoderma wilt diseases on Palms and other Perennial Crops.* Tamil Nadu Agrl. Univ., Coimbatore (Abst.). pp. 16-17.

36

Effect of Vermiwash on *Radopholus similis* on Coconut Seedlings

V.K. Sosamma

INTRODUCTION

The burrowing nematode, *Radopholus similis* (Cobb, 1893-Thorne, 1949) is the most important nematode parasite of coconut in Kerala. On infestation, it produces small, elongate, orange coloured lesions on tender orange to white roots (Fig. 36.1) and cause extensive rotting of the roots (Koshy *et al.*, 1975). *Eudrilus* sp is the only earthworm currently available in the country with the capacity to convert lignin rich (31%) coconut leaf biomass into humus rich granular odourless vermi-compost. In view of environmental and health concerns, a determined effort is necessary to reduce and rationalise use of nematicides and vermiwash was used to manage nematodes effectively.

One year old WCT coconut seedlings raised in methyl bromide fumigated soil in earthen pots were used for the study. Vermiwash was prepared from vermi-compost made in big earthen pots using coconut leaf waste and with earthworm *Eudrilus* sp. a native isolate (Fig. 36.2) collected from decomposing coconut biomass. Vermiwash was collected (Fig. 36.3) and diluted six times and one liter each of the diluted wash was introduced into each pot of the treatments wherever required.

For laboratory screening, different dilutions of vermiwash collected from coconut leaf compost viz. undiluted (UD), 1:5, 1:10 and 1:20 were used. Sterile water alone served as control. To study the nematicidal effect of vermiwash, suspensions of 100 active specimens of *R.similis* collected from coconut root was used. 2 ml of nematode suspension containing 100 nematodes was placed in 90 mm diameter Petri dish in which two ml of solution of double concentration of vermiwash was added so as to get a final dilution of

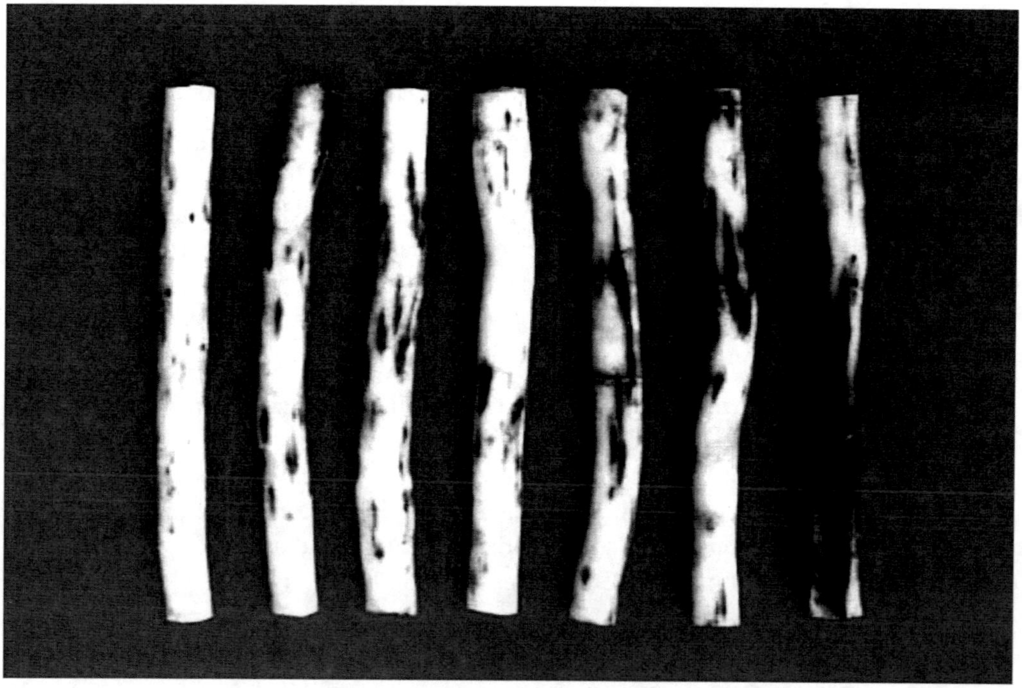

Fig. 36.1. *R similis* infested coconut roots showing lesions and rotting

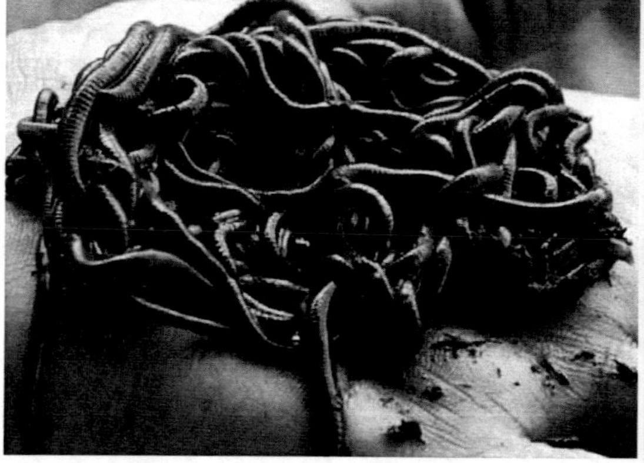

Fig. 36.2. Earthworm, *Eudrilus* sp dark in colour

Fig. 36.3 Vermiwash collection

1:5, 1:10 and 1:20. In undiluted vermiwash and picked nematodes were directly introduced. The number of dead and surviving nematodes were counted at 24 hr, 48 hr and 72 hr. The death of the nematode was ascertained after transferring them to water for one hour.

Vermiwash on *Radopholus similis* (burrowing nematode) of coconut seedlings was found to be very effective in suppressing its population (Table 36.1). Study conclusively demonstrated the toxic effect of vermiwash on *R.similis* to varying degrees. In the laboratory studies among the four dilutions, undiluted vermiwash caused maximum mortality followed by 1:5 dilution. No revival of nematodes was noticed in undiluted, 1:5, and 1:10 which clearly indicated the nematicidal property of vermiwash. No mortality was recorded in control where distilled water was used (Table 36.2).

Various decomposed organic amendments and oil cakes had been found toxic to nematodes (Badra *et al.*, 1979) and vermiwash collected from coconut leaf compost was reported to contain phenols (Murali Gopal *et al.*, 2005). Toxicity of phenols to various nematodes was reported. The biochemical analysis of fresh vermiwash produced from actively composting coconut leaves along with *Eudrilus* sp. had a pH of 7 to 8.5. It was found to have total phenol content of more than 10-14 μg/ml (Alka, personal communication). The observed mortality effect could be of significance in reducing nematode population in field by giving initial protection to crops in the field.

Table 36.1. Effect of vermiwash on *R. similis* infested coconut seedlings

Treatments	*Girth (cm)*	*Height (cm)*	*No. of leaves*	*Leaf length (cm)*	*Leaf breadth (cm)*	*Lesion Index*	*Rs. Popn.*
Control	10.25	167.0	7.25 (2.688)*	81.00	17.50	0.0	0.0(3.162)*
R.similis – 100	9.50	160.0	6.50 (2.540)	75.25	17.00	3.0	876.5(29.692)
R.similis – 100 + Vermiwash	10.60	172.75	7.50 (2.737)	82.75	17.50	2.125	259.0(16.345)
R.similis – 1000	9.0	151.25	5.50 (2.343)	69.00	15.25	4.00	1114.75(33.514)
R.similis – 1000 + Vermiwash	10.5	82.50	6.5 (2.548)	79.00	16.50	3.00	331.25(18.445)
Vermiwash alone	11.0	95.0	8.5 (2.914)	91.25	18.00	0.0	0.0(3.162)
Mean	10.14	171.42	2.63	79.71	16.96	2.02	430.25(17.39)
CD (P=0.5)	NS	15.85	0.24	5.27	1.38	0.43	1.88

*Figures in parentheses are transformed values (SQRT (x+10))
NS = Not Significant

Table 36.2. Effect of vermiwash on mortality of *R.similis*

Vermiwash dilution	*24 hr.*	*48 hr.*	*72 hr.*
UD	45	61	72
1:5	29	42	68
1:10	24	35	56
1:20	14	22	35
Sterile water	0	0	5

References

Badra, T., Saleh, M.A. & Oteifa, B.A. 1979. Nematicidal activity and composition of some organic fertilizers and amendments. *Revue de Nematologie* **2**:29-36.

Koshy, P.K., V.K. Sosamma and C.P.R.Nair 1975. Preliminary studies on *Radopholus similis* (Cobb, 1893) Thorme, 1949 infesting coconut and arecanut palms in South India. *Indian J. Nematol.* **5**:26-35.

Murali Gopal, Alka, G., Palaniswamy C.,Dhanapal, R. and G. V. Thomas 2005. Vermiwash from coconut leaf compost an ecotechnology for improving the crop production efficiency of soil. Paper presented in "*International conference on Human Centered Sustainable Development paradigm"* held at Chennai Trade Centre during 7–10 August 2005.

37

Role of *Pseudomonas fluorescens* in the Biological Control of Leaf Blight of Coconut

B. Meena and K. Rajamanickam

INTRODUCTION

Coconut (*Cocos nucifera* L.) is affected by many lethal and debilitating diseases. Among the foliar diseases of coconut, the leaf blight, *Lasiodiplodia theobromae* (Pat.) Griffon and Maubl. formerly known as *Botryodiplodia theobromae*. Kia is the most ubiquitous one. *L. theobromae* is a cosmopolitan fungus infecting the fruits of a wide range of host plants. With the introduction of coconut hybrids, the disease has assumed serious proportions. The leaf blight incidence decreased the coconut yield to the extent of 10-24 % (Karthikeyan, 1996). In adult coconut palms, the lesions spread to the entire frond, the leaflets get blighted, the leaves droop down and fall off. Leaf drooping and shedding are accompanied by shedding of buttons and nuts. Finally the crown topples down leading to the death of the tree (Nakkeeran *et al.*, 1998). The fungus also infects the buttons and nuts. Grey coloured lesions appear on the surface of the nuts which enlarge and surface of affected husk becomes rough and shreds. In advanced stage of infection the nut size is reduced (Bhaskaran *et al.*, 2007).

Biological control of plant diseases with antagonists is now considered as a promising alternative to the use of fungicides. Bio-pesticides are cheaper, ecofriendly and do not pose risk of the pathogen developing resistance.

The fungal leaf blight pathogen, *Lasiodiplodia theobromae* was isolated from infected leaves using potato dextrose agar medium. The effect of biocontrol agents *viz.*, *Trichoderma viride, T. harzianum* and *Pseudomonas fluorescens* in inhibiting the mycelial growth of *L. theobromae* was assessed by dual plate technique under *in vitro* conditions. On farm field experiments were conducted with biocontrol agent which was found effective in the laboratory screening for the management of leaf blight disease at Kambalapatti village of Pollachi Taluk, Coimbatore District in Tamil Nadu. *P. fluorescens* was applied to soil @ 50 g/palm/year along with neem cake @ 5 kg/palm/year. Root feeding of 100% culture filtrate of *P. fluorescens* was done at monthly interval, quarterly interval and annually. Pre-treatment and post-treatment observations on leaf blight disease were made and the difference between the two gives the reduction in disease severity.

In vitro screening of antagonistic organisms against leaf blight pathogen revealed that *P. fluorescens* significantly inhibited the mycelial growth of the pathogen (75.6% inhibition) followed by *T. viride* which inhibited the mycelial growth to 55.8 %. In the control, maximum mycelial growth of 8.6 cm was observed (Table 37.1). *P. fluorescens* significantly inhibited the mycelial growth of *Pestalotia palmarum* (Karthikeyan *et al.,* 1997). Talc-based powder formulation of the effective biocontrol agent, *P. fluorescens* was developed and evaluated for its efficacy in the management of leaf blight of coconut under field conditions.

Table 37.1. *In vitro* screening of biocontrol agents against *Lasiodiplodia theobromae*

S. No.	*Antagonistic organism*	*Mycelial growth * (cm)*	*% Inhibition over control*
1.	*T. viride*	3.2	62.7
2.	*T. harzianum*	3.8	55.8
3.	*P. fluorescens*	2.1	75.6
4.	Control	8.6	–
	CD(P=0.05)	2.4	

* Mean of five replications

The results of the on farm field experiment conducted at Kambalapatti revealed that combined application of talc-based powder formulation of *P. fluorescens* to soil (50 g/palm/year) along with neem cake (5 kg/palm/year) followed by root feeding of 100% culture filtrate of *P. fluorescens* at half yearly intervals was effective in managing the disease (14.2% disease reduction) (Table 37.2).

Control of fruit rot caused by *L. theobromae* in papaya has been reported earlier (Hunter and Buddenhagon, 1972; Saldana *et al.,* 1985). Soil application of neem cake alone was not effective in managing the leaf blight disease. In the control, maximum disease intensity of 35.4% was observed. In addition to disease reduction, the nut yield per palm was also found to be maximum in combined application of talc-based powder formulation of *P. fluorescens* to soil (50 g/palm/year) along with neem cake (5 kg/palm/year) followed by root feeding of 100 % culture filtrate of *P. fluorescens* at half yearly intervals (130) as against the lowest (74) in the control.

Table 37.2. Management of leaf blight disease

Treatments	Pre-treatment PDI	Post-treatment PDI	Reduction in disease severity (PDI)*	Mean nut yield/palm/ year
RF – 100% culture filtratePf1 @ 25 ml/palm at quarterly interval	25.8	17.2	8.6	102
RF – 100% culture filtratePf1 @ 25 ml/palm at half yearly interval	24.3	16.0	8.3	94
RF – 100% culture filtrate Pf1 @ 25 ml/palm per year	26.8	20.7	6.1	91
SA – Pf1 (50 g) + Neem cake – 5 kg/palm/year	23.7	13.1	10.6	104
SA – Pf1 (50 g) + Neem cake – 5 kg + RF – Pf1 @ 25 ml/palm at quarterly interval	25.8	11.2	14.6	130
SA – Pf1 (50 g) + Neem cake – 5 kg + RF – Pf1 @ 25 ml/palm at half yearly interval	22.4	8.2	14.2	128
SA – Pf1 (50 g) + Neem cake – 5 kg + RF – Pf1 @ 25 ml/palm per year	24.2	12.1	12.1	115
SA – Neem cake – 5 kg/palm/year	23.6	17.8	5.8	94
Control	27.1	35.4	-	74
CD (P = 0.05)			4.2	6.4

* Mean of five replications
SA – Soil application
RF – Root feeding at quarterly intervals

References

Abad, R.G. 1981. Evaluation of seven fungicides in the control of leaf spots in the nursery stage. *Philipp. Phytopathol.* **17:** 87-92.

Bhaskaran, R., Karthikeyan, A., Marimuthu, R., Ramanathan, A. and Sundaram, T. 2007. Leaf blight disease of coconut palm. *Indian Coconut Journal* **37**:2-5.

Chan, C.L. 1974. Chemical control of Curvularia leaf spot of coconut seedling. *MARDI Research Bulletin.* **2:**19-24.

Das, C.M. and Mahanta, I.C. 1985. Evaluation of some fungicides against *Pestalotia palmarum* incitant of grey blight of coconut. *Pesticides.* **19:**37-39.

Gunasekaran, M. and Srinivasan, N. 2000. Involvement of *Lasiodiplodia theobromae* (Pat.) Griffon and Maubl. in fruit rot of coconut in India. *Trop. Agric.* **77:**123-125.

Hunter, J.E. and Buddenhagen, I.W. 1972. Incidence, epidemiology and control of fruit disease of papaya in Hawaii. *Trop. Agri.* **49**:61-71.

Karthikeyan, A. 1996. Studies on pathophysiology, host resistance and management of leaf blight of coconut. Ph.D. Thesis, Tamil Nadu Agricultural University, Madurai, India. 163p.

Karthikeyan, A., Bhaskaran, R. And Giridharan, S. 1997. Management of leaf blight disease of coconut. *The Planter.* **73:**495-501.

Kudalkar, S.K., Joshi, M.S. and Pawar, D.R. 1991. Control of leaf blight of coconut caused by *Pestalotia palmarum. Indian Coconut Journal* **22**:18-19.

Menon, K.P.V. and Pandalai, K.M. 1958. The coconut palm, A Monograph, Indian Central Coconut Committee, Ernakulam, S. India, 384p.

Nakkeeran, S., Sridharan, C.S., Karivaratharaju, T.V., Valluvaparidasan, P. and Renuka Devi, P. 1998. Lethal leaf blight of coconut. *TNAU Newsletter* **28:**1.

Nene, Y.L. and Thapliyal, P.N. 1982. *Fungicides in Plant Disease control.* Oxford and IBH publishing House, New Delhi, 385p.

Ohler, J.G. 1999. Modern Coconut Management – Palm cultivation and products. Intermediate Technology Publicaitons Ltd., UK, 80p.

Papa Rao, A. and Govinda Rao, P. 1966. A survey of coconut disease in Andhra Pradesh. *Andhra Agric. J.* **13:**208-217.

Papa Rao, A., Laxminarayan, C. and Pandit, S.V. 1975. The nature of coconut leaf blight and its control. *Indian Phytopathology.* **28:**447-450.

Ram, C. 1989. Epidemiologia e control quimico da Queima-das-folhas (*Botryodiplodia theobromae*) do coqueiro (*Cocos nucifera*). *Fitopatologia Brasileira* (Brazil), **14:**215-220.

Saldana, M.I., Marquez, M. and Ruiz, P. 1985. Identification de enfermedades fungosas del cultivo de la papaya (*Carica papaya* L.) en el estado de Tabasco. *Revista Mexicana de Fitopatologia* **3:**14-17.

Warwick, D.R.N., Passos, E.E.M., Leal, M.L.S. and Bezerra, A.P.O. 1993. Influence of water stress on the severity of coconut leaf blight caused by *Lasiodiplodia theobromae. Oleagineux.* **48:**279-282.

38

Effect of Botanicals in Inhibiting Growth and Infection of *Lasiodiplodia theobromae*, the Causal Organism of Rotting and Immature Nut Fall of Coconut

S. Venugopal and R. Chandramohanan

INTRODUCTION

Recently immature nut fall of coconut due to rotting has become a serious problem in many localities, especially southern part of Kerala State with well-distributed rainfall. With wide spread occurrence of eriophyid mite (*Aceria guerreronis* Keifer), severe incidence of nutfall caused by *Lasiodiplodia theobromae* (Pat.). Griffon and Maub. has been observed throughout the year with varying intensities *L. theobromae* was found to be the predominant and major pathogen causing rotting of mite infested nuts (Venugopal and ChandraMohanan, 2006). Infection always started from the white and soft tissue beneath perianth, the tissue infested by mite. When nearly mature or mature nuts were infected by *L. theobromae*, the infection continued to spread internally without any external lesion or with a small lesion on the nut surface beneath the perianth. *L. theobromae* being a wound pathogen enters the tissue through the wound mainly caused by eriophyid mite in the soft tissue beneath the perianth. Studies on the diversity of the pathogen revealed marked variability within *L. throbromae* (Venugoapl *et al.*, 2008).

Considering the economic importance of the problem, detailed studies were undertaken to find out the botanicals which are effective in inhibiting growth of virulent isolate of *L. theobromae* as well as the infection caused by it. The extracts of 52 plants species belonging to 30 families and products of 10 plants belonging to 8 families were tested for fungitoxicity against *L.theobromae*. The details of the plants used in the study are given in Table 38.1. Most of the plants and plant products were selected based on the earlier reports on their efficacy in inhibiting growth and spore germination of other fungi as well as their availability in and around the place of study/local market. To study the effect of plant extracts/products, three virulent isolates of *L. theobromae* viz., dark gray (APA/L-172), grayish black (EKM/L-307) and white type (TPM/L-38) were selected. These three isolates were earlier identified from 3 main groups of *L. theobromae* isolates based on the detailed studies on diversity of *L. theobromae* causing nut rot and fall in different locations of disease endemic areas (Venugopal *et al.*, 2008).

All plants extracts were made out in 99% acetone. Fresh and healthy garlic and onion bulb of uniform maturity were obtained from local market and outer dry scales of bulb were removed. Fresh plant parts viz., leaves/bulbs/ rhizomes/ entire plant without roots as mentioned in the Table 38.1 were used for extrzction and the extract was tested for their effect on *L. theobromae* isolates by poisoned food technique (Nene, 1971).

Table 38.1. The details of plants used for the studies on the effect of plant extracts on *in vitro* mycelial growth of *L. theobromae* isolates

Sl. No.	*Botanical name*	*Vernacular/ Malayalam name*	*Family*	*Parts tested*
1.	*Acacia auriculiformis* A. Cunn	Acacia(E)	Mimosae	L
2.	*Adathoda vasica.* Nees	Adalodakam(M)	Acanthaceae	L
3.	*Adenocalymma allicea* L.	Garlic climber(E)	Bignoniaceae	L
4.	*Aegle marmelos* Corr	Koovalam(M)	Rutaceae	L
5.	*Aerva lanata* (Linn.) Juss.	Cherula(M)	Amaranthaecae	EP
6.	*Ageratum conyzoides* L	Mudiyumam(M)	Compositae	L
7.	*Allium cepa* L.	Onion(E)	Lilliaceae	B
8.	*Allium sativum.* L	Garlic(E)	Lilliaceae	B
9.	*Anacardium occidentale* L.	Cashew(E)	Anacardiaceae	L
10.	*Areca catechu* L.	Kamuku(M)	Palmae	L
11.	*Azadirachta indica* A.Juss.	Vep (M)	Meliaceae	L
12.	*Boerhavia diffusa* L.	Thazhuthama(M)	Nyctaginaceae	EP
13.	*Calotropis gigantea* R.Br.	Erukku (M)	Asclepiadaceae	L
14.	*Clerodendrum infortunatum* L.	Peruvelum(M)	Verbenaceae	L
15.	*Cinnamomum zeylanicum* Blume.	Cinnamon(E)	Lauraceae	L
16.	*Citrus aurantium* L .	Lemon(E)	Rutaceae	L
17.	*Citrus aurentifolia* Risso.	Orange(E)	Rutaceae	L
18.	*Coffea arabica* L.	Coffee(E)	Rutaceae	L
19.	*Coriandrum sativum* L.	Coriander(E)	Umbelliferae	EP
20.	*Costus* species	Insulin plant(E)	Zingiberaceae	L
21.	*Cocos nucifera* L.	Coconut(E)	Palmae	L
22.	*Curcuma longa* L.	Manjal(M)	Zingiberaceae	R
23.	*Elettaria cardamomum* (L).Maton.	Cardamum(E)	Zingiberaceae	L
24.	*Eucalyptus globulus* Labill.	Eucalyptus(E)	Myrtaceae	L
25.	*Eupatorium odoratum* L.	Communist pacha(M)	Asteraceae	L
26.	*Glycosmis pentaphylla* Corr.	Panal(M)	Rutaceae	L

Each plant extract at 1, 2 and 4 % concentration when added to medium was tested for the efficacy. The per cent inhibition of growth of *L. theobromae* in treatment was calculated by the equation given by Vincent (1927).

$$I = \frac{C - T}{C} \times 100$$

where,

I = Inhibition of fungal growth
C = Growth in control
T = Growth in treatment.

The nature of fungitoxicity was also tested in cases where no fungal growth was observed.

The most promising plant extracts were selected based on their efficacy in inhibiting the mycelial growth of *L. theobromae* isolates under *in vitro* condition. The extracts of *Allium sativum* (Garlic), *Syzygium aromaticum* (clove leaf), *Pimenta dioica* (All spice) at four per cent concentration completely inhibited the mycelial growth of *L.theobromae*. Therefore, these three plant extracts were prepared and evaluated.

The acetone extracts of plants which showed 70 % and more inhibition were selected (Table 38.2) and studied for their *in vitro* effect on growth of selected *L.theobromae* isolates viz., dark gray, grayish black and white types.

Table 38.2. The details of plant products used in *in vitro* studies against *L. theobromae* isolates

Sl. No	*Plant products*	*Botanical name of the source plant*	*Family*
1.	Clove oil	*Syzygium aromaticum* (L.) Merr.et Perry	Myristicaceae
2.	Camphor +Soap	*Cinnamomum camphora* L.	Lauraceae
3.	Eucalyptus oil	*Eucalyptus globulus*. Labill	Myristicaceae
4.	Eco neem plus (botanical Pesticide) 10000 ppm	*Azadirachta indica* A. Juss	Meliaceae
5.	Neem oil	*Azadirachta indica* A. Juss	Meliaceae
6.	Hydnocarpous oil +Soap	*Hydnocarpus wightiana* Gaertn.	Bixaceae
7.	Thespesia oil + Soap	*Thespesia populnea*.(L)	Malvaceae
8.	Neem oil + Garlic + Soap*	*Azadirachta indica* A. Juss, *Allium sativum* L.	Meliaceae Lillaceae
9.	Turmocure(Turmeric extract)	*Curcuma longa* L.	Zingiberaceae
10.	Turpentine	*Pinus* species	Pinaceae

*Natural turmeric extract manufactured by Narmada products, Bhavanishankar temple compound, Puttur, Karnataka

The effect of plant extracts and plant products on *L. theobromae* infection of detached coconut was studied using about three months old coconuts of WCT cultivar. Nuts at this stage were found to be convenient in making cavities with cork borer and to fill plant extract/ plant product solution.

The fungitoxicity of plant extracts against *L.theobromae* varied with the concentration. (Table 38.3) In general the response of dark grey (APA/L-172) and greyish black (EKM/L-307) isolates to different concentrations of plant extracts did not vary. The inhibition of mycelial growth of white type (TPM/L-36) isolate was more than the other two isolates in the presence of plant extract. All isolates exhibited a significant reduction in growth on first day after inoculation. The inhibition of growth is decreased with duration of incubation in most cases. The growth of the three isolates varied with plant extracts and their concentrations.

Among 52 plant extracts tested *in vitro*, the extracts of *A.sativum* (Garlic), *S. aromaticum* (Clove leaf) and *P. dioica* (All spice) at 4% concentration were found to be very effective in inhibiting the growth of selected isolates. The extract of garlic bulb at 1 and 2% concentration showed more than 70% of inhibition of all isolates on first day after inoculation. The growth of white type isolate was completely inhibited in the presence of *A. sativum* extract at2% concentration. The inhibition of growth of this isolate was more than 80% even at 1% concentration of *A. sativum* extract. There were significant variations in growth inhibition on 2nd day after inoculation. Dark grey isolate showed cent per cent inhibition with *A. sativum* 2% on 1st day after inoculation whereas greyish black isolate exhibited only 72.14% inhibition. Percentage of inhibition decreased with duration of incubation in these two isolates. The extract of *P. dioica* at 1 and 2% concentration was found significantly effective than *A. sativum.* Even at lower concentrations it effectively checked the growth of all the selected isolates. White type isolates showed 100% inhibition of growth in the presence of *P. dioica* extract (all concentrations) on first day after inoculation. The inhibition of growth decreased with duration of incubation. This isolate exhibited total inhibition of growth even at 2% concentration of *P. dioica* extract. Dark grey and grayish black isolates showed almost equal growth response with all concentrations of *P. dioica* extract. All the isolates exhibited total inhibition of growth in the presence of *P. dioica* extract at 4% concentration.

The extracts of *S. aromaticum* at2% concentration completely inhibited the growth of white type isolate. On the other hand, 2z% concentration of *S. aromaticum* extract exhibited only 61.46% and 72.96% inhibition of dark gray and grayish black isolates respectively. This extract at 4% concentration was found fungistatic to all the three isolates.

The extract of *G. pentaphylla* (all the concentrations) had stimulatory effects on growth of dark gray and grayish black isolates. The extract obtained from *A. occidentale* (all the concentrations) also promoted the growth of dark gray isolate. The extracts of *C. infortunatum* (at 1%) and *C. nucifera* (2 and 4 %) had stimulatory effect on the growth of grayish black isolate. *V. rosea* (1% concentration) and *M. pudica* (1 and 2% concentrations) had no effect on any of the three isolates.

Among 52 plants screened for fungitoxic properties against selected isolates of *L. theobromae*, the extracts (at 4% conc.) from *H. wightiana*, *L. inermis* and *S. nux-vomica* showed more than 70% of inhibition of growth of all the three isolates. The extracts of *A. auriculoformis, E. globules, A. marmelos, A. conyzoides, C. longa and Z. officinale* at 4% concentration showed 50% inhibition of growth of white isolate after 24 h of inoculation. Thus, among the 52 plant species screened for fungitoxic properties, extracts of 20 plant species exhibited 60% inhibition of dark grey isolate. There was 60% inhibition of growth of both greyish black and white isolates in the presence of extracts of 17 plant species.

Table 38.3. Percentage inhibition of growth of selected isolates of *L. theobromae* after 2 days of inoculation at different concentrations of plant extracts

Sl. No	Plant extracts	Conc.%	*Per cent inhibition after 2 days of inoculation *L. theobromae* isolates		
			Dark grey type	Greyish black type	White type
1.	*Acacia auriculiformis*	4	72.59 (58.44) -	61.85(51.83) +++	5.41(13.39) -
2.	*Aegle marmelos*	4	71.85(57.93) -	72.96(58.64) +++	34.90(36.19) -
3.	*Aerva lanata*	4	60.37(50.97) -	71.11(57.46) +++	36.86(37.29) -
4.	*Ageratum conyzoides*	4	40.00(39.20) -	72.96(58.64) +	57.09(49.05) -
5.	*Allium cepa*	4	7.40(12.97) -	30.37(33.41) +++	39.49(38.89) -
6.	*Allium sativum*	2	64.44(53.37) -	54.44(47.53) +	100.00(90.00) -
7.	*Calotropis gigantia*	4	100.00(90.00) -	100.00(90.00) -	100.00(90.00) -
8.	*Curcuma longa*	4	74.81(59.85) -	72.96(58.64) ++	55.01(47.85) -
9.	*Eucalyptus globulus*	4	70.74(57.23) +	71.85(57.93) +	71.02(57.42) -
10.	*Hydnocarpus* species	4	74.81(59.88) -	72.96(58.64) +++	61.75(51.77) -
11.	*Pimenta dioica*	4	71.85(57.94) -	85.18(67.34) ++	100.00(90.00) -
12.	*Piper nigrum.*	1	70.37(57.00)-	61.11(51.40)+	64.29(53.30) -
		2	83.33(65.93) -	81.48(64.49) -	100.00(90.00) -
		4	100.00(90.00) -	100.00(90.00) -	100.00(90.00) -
13.	*Psidium guajava*	4	75.18(60.10) -	67.78(55.39) +++	57.12(49.07) -
14.	*Syzygium aromaticum*	1	62.96(52.49) +++	58.15(49.67) +++	82.43(65.19) -
15.	*Strychnos nux- vomica*	2	68.89(56.07) +++	54.07(47.32) +++	82.43(65.19) -
		4	77.04(61.35) +++	39.26(38.76) +++	85.00(67.19) -
16.	*Tamarindus indicus*	1	100.00(90.00) -	100.00(90.00)-	100.00(90.00) -
		2	72.59(58.42) -	70.18(56.88) +++	76.21(60.79) -
17.	*Zingiber officinale*	4	76.66(61.10) -	74.07(59.38) +++	79.32(62.93) -
		4	81.48(64.56) -	75.37(60.24) +++	100.00(90.00) -
		4	70.74(57.24) -	45.18(42.21) ++	16.24(23.67) +
			70.74(57.23)	22.59(28.35)	62.20(52.07) -
	C.D (P = 0.05)		3.73	2.60	4.12

Results of only those plant extracts that gave higher percentage inhibition of growth of *L.theobrome* are given
*Mean of three replications; Figures in parenthesis are angular transformed values
Degree of sporulation after 14 days of inoculation +++ Abundant, ++ Moderate, + Poor, - Nil

When there was complete inhibition of growth, the nature of fungitoxicity was further tested. Extracts of *A. sativum, P. dioica* and *S. aromaticum* at 4% concentration were found fungistatic to all the three isolates. Therefore higher concentrations such as 5, 6, 8 and 10% of these extracts were tested. *A. sativum* at 10% and *P. dioica* and *S. aromaticum* each at 8% concentration were found fungicidal to all the three isolates of *L.theobromae*.

Among the extracts tested, 22 totally inhibited the sporulation in dark grey isolate. Different concentrations of remaining 30 plant extracts reduced the degree of sporulation to moderate level. Most of the extracts did not inhibit the sporulation of greyish black isolate. The extracts like *A. auriculoformis* (4%), *A. lanata* (4%), *A. sativum,* (1%) *C. gigantia* (2 and 4%), *C. zeylanicum* (2 and 4%), *C. arabica* (2 and 4%), *C. sativum* (1%), *C. longa,* (1%), *H. wightiana* (4%), *T. indica* (2 and 4%) and *Z. officinale* (1%) reduced the sporulation of greyish black isolate to moderate level. *Z. officinale* (2 and 4%), *S. aromaticum* (1%), *P. dioica* (1%), *C. longa* (2 and 4%), *Allium sativum* (2%), *and A. conyzoides* (4%) exhibited marked reduction in the degree of sporulation and thus there was only poor sporulation in the presence of these extracts. There was moderate sporulation in white type isolate in the presence of extracts of *A. allicea* (1%), *A. indica* (all conc.), and *M. koenigii* (all conc.). The extracts of 29 species completely inhibited the sporulation of white isolate even at 1% concentration.

Out of 52 plant species tested, the extracts of *A. sativum, S. aromaticum* and *P. dioeca* were the most promising in inhibiting the growth of all the three isolates. The sporulation of all the three isolates of *L.theobromae* was also completely inhibited by 4% concentation of *A. sativum* and 2% concentration of the other two plant extracts.

Fresh plant extracts of *A. sativum, S. aromaticum* and *P. dioica* without autoclaving were found to be the most promising in inhibiting the growth of *L.theobromae* under *in vitro* condition. The mycelial growth of *L.theobromae* in the presence of autoclaved extracts of *A. sativum* (4% conc.), *S. aromaticum* (4% conc.) and *P. dioica* (4% conc.) varied significantly among the three isolates as well as within the isolates with respect to different concentrations (Table 38.4).

The study also revealed that *S. aromaticum* was superior to all other extracts tested followed by *P. dioica*. Out of three extracts tested, *S. aromaticum* inhibited growth of dark gray and grayish black isolates one day after inoculation. The fungitoxicity varied with concentration of plant extracts and days of incubation. White isolate exhibited cent percent growth inhibition at 4% concentration of *S. aromaticum* and *P. dioica*. The autoclaved extracts of *A. sativum* inhibited the growth of white and greyish black isolates below 50 % level on first day whereas in dark grey it was 37.80%. When the extract of *A. sativum* at 4% concentration without autoclaving was incorporated into sterilized and cooled medium, it inhibited fungal growth completely. But autoclaving the extract decreased its antifungal property.

Table 38.4. Percentage of inhibition of growth of selected isolates of *L. theobromae* on 2nd day after inoculation at different concentrations of autoclaved plant extracts

Sl.No	*Plant extract*	*Conc.%*	**Percent inhibition on 2nd day after inoculation L.theobromae isolates*		
			Dark grey type	*Greyish black type*	*White type*
1.	*Allium sativum*	1	0.00(0.00)	0.00(0.00)	12.49(20.24)
		2	0.00(0.00)	0.00(0.00)	16.27(23.54)
		4	0.00(0.00)	0.00(0.00)	19.99(26.52)
2.	*Pimenta dioica*	1	24.07(29.34)	16.29(23.63)	39.73(39.05)
		2	55.18((47.95)	57.03(49.03)	80.27(63.62)
		4	78.52(62.38)	75.92(60.69)	100.00(90.00)
3.	*Syzygium aromaticum*	1	18.52(25.42)	29..26(32.72)	49.32(44.59)
		2	49.26(44.55)	54.44(47.53)	76.53(61.00)
		4	72.22(58.21)	83.33(66.16)	100.00(90.00)
	Gen. Mean		29.76	31.09	50.96
	C.D(P=0.05)		2.94	5.04	4.72

The acetone extracts of plants which showed 70% and above inhibition were selected and the aqueous extract of this plant species were tested to find out their inhibitory effects (Table 38.5). All isolates showed significant variation in growth in the presence of the selected plant extracts. Acetone extracts were found superior in inhibiting growth than aqueous extract.

Among the aqueous plant extract tested, *A. sativum* at 4% concentration showed above 50% of inhibition of dark grey and greyish black isolates on first and second day after inoculation. White isolate was found more sensitive and its growth was completely inhibited by aqueous extracts of *A. sativum* (4%) on first day. Aqueous extract of *S. aromaticum* at 4% concentration exhibited above 65% inhibition of growth of dark grey and white isolates. On the other hand there was only 55.34% inhibition of grayish black isolates in the presence of this extract on first day.

Among the three isolates screened, dark grey(APA/L-172) and greyish black (EKM/L-307) isolates showed almost similar response in growth in the presence of different plant products whereas isolate of white colony(TPM/L-36) group exhibited more response than other two groups (Table 38.6). Analysis of variance showed that there was significant difference among the plant products and their concentrations in inhibiting the mycelial growth. All isolates exhibited significant reduction in mycelial growth one day after inoculation on PDA incorporated with plant products when compared to control. The inhibition of growth decreased with duration of incubation in most cases.

Table 38.5. Percentage of inhibition of growth of selected isolates of *L. theobromae* on 2nd day after inoculation at different concentrations of aqueous plant extracts

Sl. No	Plant extract	Conc.%	*Percent inhibition on 2nd day after inoculation *L. theobromae* isolates		
			Dark grey type	Greyish black type	White type
1.	*Allium sativum*	1	5.18(13.09)	4.81(12.65)	45.11(42.18)
		2	39.63(38.99)	42.59(40.70)	69.39(56.39)
		4	56.29(48.59)	59.63(50.54)	79.95(63.37)
2.	*Hydnocarpus wightiana*	1	0.00(0.00)	45.55(42.42)	17.12(24.33)
		2	0.00(0.00)	37.77(37.89)	22.94(28.53)
		4	0.00(0.00)	52.22(46.25)	47.74(43.69)
3.	*Lawsonia inermis*	1	0.00(0.00)	21.48(27.48)	37.26(37.52)
		2	0.00(0.00)	32.96(35.02)	43.01(40.96)
		4	0.00(0.00)	32.22(34.56)	49.84(44.89)
4.	*Pimenta dioica*	1	0.00(0.00)	0.00(0.00)	0.00(0.00)
		2	0.00(0.00)	0.00(0.00)	0.00(0.00)
		4	0.00(0.00)	12.96(21.08)	0.00(0.00)
5.	*Strychynos nux- vomica*	1	0.00(0.00)	0.00(0.00)	0.00(0.00)
		2	0.00(0.00)	0.00(0.00)	0.00(0.00)
		4	0.00(0.00)	13.33(21.32)	17.12(24.33)
6.	*Syzygium aromaticum*	1	68.14(55.63)	0.00(0.00)	55.16(47.95)
		2	75.55(60.34)	0.00(0.00)	65.66(54.13)
		4	83.33(65.88)	17.03(23.98)	72.03(58.05)
	Gen. Mean		15.70	21.88	16.01
	C D (P = 0.05)		1.10	3.94	2.22

*Mean of three replications
Figures in parentheses are angular transformed values

Among 10 plant products screened in *vitro*, clove oil (2% and above), eucalyptus oil (1% and above), Eco-neem plus (commercial product) (1% and above) and neem oil + garlic + soap mixture (5% and above) were found very effective in inhibiting the growth of selected isolates. Eco neem at 8% concentration and neem oil + garlic + soap at 10% concentration completely inhibited the growth of all isolates and were found fungitoxic to all isolates. The response of dark grey and greyish black isolates to eco neem plus at 0.25 and 0.5% concentration was almost similar whereas white type isolate showed a significant reduction of growth at these concentrations. Neem oil (all con.), *Hydnocarpus* oil + soap (1 and 2%) and *Thespesia* oil + soap (1 and 2%) promoted the growth of dark grey isolate. All the three concentrations of turmocure (turmeric extract) significantly reduced the growth of all the three isolates. Dark grey and greyish black isolates at all the three concentrations of turpentine oil showed significant reduction in growth over the control. On the other hand, white types isolate showed total inhibition of growth even at 1% concentration of this plant product. The plant products significantly inhibited the sporulation in all isolates.

The effect of plant extracts and plant products which exhibited 100% inhibition of mycelial growth of selected isolate of *L.theobromae* (EKM/L-307) under *in vitro* condition was tested by inoculating detached coconuts with this isolate.

The bulb extract of *A. sativum* (10%) and leaf extract of *S. aromaticum* (8%) were found to be effective and showed a significant reduction in lesion size over control. The extracts of *A. sativum* (10%) and *S. aromaticum* (8%) inhibited the lesion development for 3 days after inoculation. The lesion size on 5th day was 12.66 mm × 10.33 mm and 11mm × 10.33 mm respectively (Table 38.7).

Among the plant products evaluated, Eco neem plus (botanical pesticide) at 8% concentration gave promising result. This plant product completely inhibited lesion development. Neem oil + garlic + soap mixture at 10% concentration produced lesion size of 10 mm × 10 mm on 5th day after inoculation. Eco neem at 8% concentration completely inhibited the lesion size.

The presence of antifungal compounds in higher plants has long been recognized as an important factor to disease resistance (Mahadevan, 1982). Such compounds, being biodegradable and selective in their toxicity, are considered as valuable in controlling some plant diseases (Singh and Dwivedi, 1987). The present study is the first attempt to find out the antifungal activity of plant extract against *L.theobromae*, the causal organism of fruit rot and immature nut fall in coconut. The *in vitro* evaluation of 52 plant extracts extracted in acetone revealed that extracts from bulb of *Allium sativum* (garlic), leaf extracts of *Syzygium aromaticum* (clove) and *Pimenta dioica* (all spice) were superior in inhibiting the mycelial growth of *L .theobromae*.

As observed in the present study the excellent performance of garlic in inhibiting several other fungi was also reported by other workers (Singh *et al.*, 1979; Singh and Singh, 1980). Aqueous and ethnol extracts of five Nigerian spices [*Allium sativum* (garlic), *Zingiber officinale, Afromomum melegueta, Xylopia aetheopica*, and *Monodora myristica*] inhibited the growth of fungi such as *Botryodiplodia theobromae, Aspergillus niger, Aspergillus flavus, Mucor* sp., *Rhizopus stolinifer, Penicillium* sp.and *Fusarium* sp (Ejechi *et al.*, 1997). Chakraborty *et al.* (2006) reported the effective management of dieback of bottle brush (*Callestemon citrinus*) caused by *Botryodiplodia theobromae* using biocides (extracts of *Azadirachta indica, Allium sativum, Allium cepa* and *Eucalyptus* species). The active principle in garlic viz., allicin had been reported to contain antimicrobial property (Cavellito and Bailey, 1944., Rao *et al.*, 1946 and Pereira *et al.*, 2006).

Leaf extracts of *S. aromaticum* at 2% concentration exhibited 61.46 and 72.96% inhibition of growth of dark grey and greyish black isolates respectively and total inhibition of growth of white isolate in the present study. *S. aromaticum* contains high percentage of eugenol, which has been identified as a compound exhibiting antifungal properties (Garg and Siddigui, 1992). Several workers observed that clove (*S. aromaticum*) promoted total inhibition of mycelial growth of *Penicillium roqueforti* (Pereira *et al.*, 2006; Azzous and Bullerman, 1992 and Farag *et al.*, 1989), *Rhizotonia solani* (Myeong *et al.*, 2003) and *L. theobromae* (Ranasinghe *et al.*, 2002).

The extract of *P. dioica* (all spice) at 1 and 2% concentration was more effective than *A. sativum.* Its lower concentrations effectively checked the growth of all the three isolates. Lima *et al.* (2006) reported eugenol as the active principle of the *Pimenta* sp. A recent study indicated that 1, 8 cineole and eugenol exhibited strong antifungal activity against *Fusarium moniliformae* (Garg and Siddigui, 1992). There are no reports in the literature on antifungal properties of *P. dioica* against *L. theobromae.*

Another interesting aspect in the present study was the effect of certain plant extract promoting the mycelial growth of *L.theobromae*. The leaf extracts of *Glycosmis pentaphylla* (at all the concentrations tested) had stimulatory effects on growth of dark grey and greyish black isolates. The extract obtained from *Anacardium occidentale* (all the concentrations tested) had stimulatory effect on growth of dark gray isolate. Leaf extracts of *C. infortunatum* (at 1%) and *C. nucifera* (2 and 4%) also stimulated the growth of greyish black isolate. Shekhawat and Prasada (1971) reported that the extract of *Ficus bengalensis* stimulated spore germination in *Aspergillus tenuis* and *Helmithsporium* species and the extract of *Xanthium straumarium* stimulated the growth of *Penicillium oxalicum* and *Aspergillus flavus.* The extracts of *Hyptis sauveoleens* had stimulating effect on the growth of *Aspergillus flavus* (Tiwari, 1987).

The antifungal activity of leaf extract of *L. inermis* was reported by various workers in many pathogenic fungi such as *Drechslera oryzae* (Natarajan and Lalitha Kumari, (1987a; 1987b), *C. gloeosporioides* (Babu and Reddy, 1986), and *Rhizotonia solani* (Kurucheve *et al.,* 1997). Muhamed *et al.* (1996) reported that thirty four other plants including *L. inermis, Mentha arvensis, Mimusops elangi, Ocimum sanctum, Phyllanthus niruri, Piper nigrum, Solanum nigrum* and *Tinospora tuberculata* showed selective antifungal activity against various fungi such as *Colletotrichum capsici, Fusarium palidoroseum, Botryodiplodia theobromae, Alternaria alternata, Penicillium citrinum, Phomopsis caricae-pepayae and Aspergillus niger.* But leaf extract of *L.inermis* at 4% conc. exhibited only 34.9%, 73.3 and 69.47% inhibition of growth of dark grey, greyish black and white type isolates of *L. theobromae* respectively.

Studies on the effect of seven aqueous plant extracts on the mycelial growth of *L. theobromae* revealed that acetone extract was superior to water extract. The result of the present findings is in accordance with the findings of Nwachukwu and Umechuruba (2001). They reported that crude leaf extracts of *A. indica, Vernonia amygdalina, Ocimum basilicum, Carica papaya* and *Cymbopogen citratus* were more effective than aqueous extracts in inhibiting mycelial growth and spore germination of seed borne fungi such as *F. moniliforme, B. theobromae* and *Aspergillus niger.*

Though the extract of *A. sativum* (garlic), *P. dioica* (all spice) and *S. aromaticum* (clove) inhibited the growth of the three selected *L. theobromae* isolates completely, all these extracts were only fungistatic at 4% concentration. The higher concentrations such as extract of *A. sativum* at 10% and extract of *P. dioica* and *S. aromaticum* each at 8% were found fungicidal. The extracts of *P. dioeca* (all spice) and *S. aromaticum* (clove) retained mycelial inhibitory effects even after autoclaving. This might be due to the thermostable nature of active principle present in these extracts. When garlic was incorporated into sterilized and cooled PDA medium, it inhibited growth of the isolates completely at 4%. But autoclaving of the extract reduced the antifungal property considerably and thus exhibited least inhibition of growth. Parkinson and Clarke (1964) reported the loss of inhibitory effect of onion bulb extract by boiling. They reported that though onion bulb extract totally inhibited the growth of *Pythium apanidermatum* boiling the extract destroyed the inhibitory effect.

It could be concluded that the leaf extracts having fungicidal properties, markedly differed in their fungitoxicity. The difference was also significant between the isolates of *L.theobromae* selected for the study. Thus, the plant extracts in general and garlic and clove leaf extracts in particular, have great potential in inhibiting the growth and sporulation of *L. theobromae* isolates.

Striking differences were also noticed in the response of selected isolates of *L. theobromae* to plant products tested. All the isolates exhibited significant reduction in mycelial growth one day after inoculation compared to control. The rate of growth increased with duration of incubation in most cases. Among 10 plants products screened *in vitro*, eco neem plus at 8% and neem oil + garlic + soap mixture at 10% were found fungicidal to all three isolates. When fresh leaf extracts of *A. indica* was used there was least inhibition of growth of isolates. Clove oil was found effective in inhibiting the growth of various fungal pathogens viz., *Penicillium roqueforti* (Farag *et al.*, 1989; Azzous and Bullerman, 1992; Pereira *et al.,* 2006), *Rhizotonia solani* (Myeong *et al.*, 2003) and *L. theobromae* (Ranasinghe *et al.,* 2002).

When garlic extract at 4% concentration completely inhibited the growth of the isolates, a mixture of neem oil + garlic + soap at 4% did not inhibit the growth completely. Spraying of Eco neem plus and neem oil + garlic + soap at 2% are recommended for effective management of eriophyid mite. But the recommended dose (2%) for mite management did not completely inhibit the growth of *L.theobromae* isolates. Spraying a mixture of neem oil + garlic + soap at 10 % concentration can successfully check mite as well as *L. theobromae* infection.

The effect of plant extract and plant products which exhibited 100% inhibition of growth of the three selected isolates under *in vitro* studies was tested on detached coconut inoculated with the most virulent isolate of *L.theobromae* (greyish black isolate). Eco neem plus (8%) was found to be the most effective treatment when tested on detached nut. There was no lesion development when treated with this plant product. Neem oil + garlic + soap (10%) significantly inhibited the lesion development when compared to control. Garlic bulb extract (10% concentration) and clove leaf extract (8% concentration) inhibited lesion development for 3 days from the day of inoculation. Hence, Eco neem plus (8%), neem oil + garlic + soap mixture (10%), garlic bulb extract (10%) and clove leaf extract (8%) can be

Table 38.6. Percentage inhibition of growth of selected isolates of *L. theobromae* on 2nd day after inoculation at different concentrations of plant products

Sl. No	*Plant products*	*Conc.%*	**Per cent of inhibition on 2nd day after inoculation L. theobromae isolates*		
			Dark grey type	*Greyish black type*	*White type*
1.	Clove oil	1 %	83.33(65.88) +	88.15(69.84) -	100.00(90.00) -
		2 %	100.00(90.00) -	100.00(90.00) -	100.00(90.00) -
		4%	100.00(90.00) -	100.00(90.00) -	100.00(90.00) -
2.	Camphor + soap	1 %	78.51(62.44) -	85.55(63.81) +	100.00(90.00) -
		2 %	80.37(63.71) -	100.00(90.00) -	73.80(59.19) -
		4 %	100.00(90.00)	100.00(90.00) -	100.00(90.00) -
3.	Eucalyptus oil	0.25 %	-61.48(51.64) -	69.26(56.40)++	100.00(90.00) -
		0.5 %	100.00(90.00) -	100.00(90.00)++	100.00(90.00) -
		1 %	100.00(90.00) -	00.00(90.00) -	100.00(90.00) -
		2 %	100.00(90.00) -	100.00(90.00) -	100.00(90.00) -
		4 %	100.00(90.00) -	100.00(90.00) -	100.00(90.00) -
		0.25 %	5.55(11.17) -	25.92(30.56) +	100.00(90.00) -
4.	Eco neem plus	0.5 %	4.44(7.13) -	38.14(38.11) -	100.00(90.00) -
	(Botanical Pesticides)	1 %	100.00(90.00) -	100.00(90.00) -	100.00(90.00) -
		2%	100.00(90.00) -	100.00(90.00) -	100.00(90.00) -
		4 %	100.00(90.00) -	100.00(90.00) -	100.00(90.00) -
		6 %	100.00(90.00) -	100.00(90.00) -	100.00(90.00) -
		8 %	100.00(90.00) -	100.00(90.00) -	100.00(90.00) -
		1 %	0.00(0.00) -	0.00(0.00) -	100.00(90.00) -
5.	Hydnocarpous oil +	2 %	0.00(0.00) -	0.00(0.00) -	0.00(0.00) -
	soap Neem oil	4 %	0.00(0.00) -	0.00(0.00) -	24.08(29.16) -
		1 %	0.00(0.00) +	2.59(7.39) ++	29.80(32.92) -
6.	Thespesia oil+ soap	2 %	0.00(0.00) +	17.78(24.57) ++	3.63(8.98) -
		4 %	23.70(29.08) -	125.92(30.57) ++	14.14(22.05) -
		1 %	0.00(0.00) -	6.67(14.92) ++	27.25(31.42) -
7.	Neem oil+ garlic +soap	2 %	0.00(0.00) -	12.59(20.70) ++	23.56(28.99) -
		4 %	32.22(33.82) -	23.33(28.84) +	17.25(24.43) -
8.	Turmeric extract	2%	0.00(0.00) +	0.00(0.00) +	29.80(32.92) -
		4 %	14.81(22.53) -	27.78(31.79) -	63.33(52.71) -
		5%	100.00(90.00) -	100.00(90.00) -	77.97(62.02) -
		6 %	100.00(90.00) -	100.00(90.00) -	100.00(90.00) -
		8 %	100.00(90.00) -	100.00(90.00) -	100.00(90.00) -
9.	Turpentine	10%	64.07(53.16) -	100.00(90.00) -	100.00(90.00) -
		1%	67.78(55.39) -	70.00(56.79) ++	100.00(90.00) -
		2%	79.26(62.88) -	75.37(60.22) -	69.62(56.53) -
		4 %	88.89(70.51) -	82.40(65.17) -	74.88(59.09) -
10.		1 %	90.00(71.53) -	88.52(70.17) -	78.53(62.37) -
		2 %	100.00(90.00) -	89.63(71.19) -	100.00(90.00) -
		4 %			100.00(90.00) -
					100.00(90.00) -
	Gen. Mean		54.47	58.95	66.42
	C.D (P = 0.05)		5.74	3.07	3.53

*Mean of three replications

Figure in parenthesis are angular transformed values

Degree of sporulation after 14 days of inoculation ++ Moderate, + Poor, - Nil selected for the field trial to find out effective control measure under conditions of natural incidence of the disease. Thus, this method of testing the plant extracts and plant products on detached nuts before field trial was found to be very helpful in selecting the effective concentrations for field trial.

Table 38.7. Effect of plant extracts and plant products on *L. theobromae* infection of detached coconut

Sl.No.	Plant products /Plant extracts	Conc.(%)	No. of days taken for appearance of lesion	Mean lesion diameter (in mm) on 5th day of inoculation	
				length (mm)	breadth (mm)
1.	Plant products Neem oil + garlic + soap	6	3	23.33	17.00
		8	3	14.66	12.00
		10	5	10.00	10.00
2.	Eco neem plus (Botanical Pesticide) Plant extract	4	3	27.66	13.66
		6	3	14.33	12.33
		8	-	0.00	0.00
3.	*Allium sativum* (garlic bulb extract)	6	3	28.00	22.00
		8	3	17.00	15.66
		10	5	12.66	10.33
4.	*Syzygium aromaticum* (clove leaf extract)	4	3	31.33	25.33
		6	3	18.66	16.33
		8	5	11.00	10.33
5.	Control		3	42.66	39.33

References

Azzous, M.A. And Bullerman, L.R .1992. Comparative antimycotic effects of selected herbs, spices, plant components and commercial antifungal agents. *J. Food Prot. Ames.* **45**:1298–1301.

Babu, K.J. and Reddy, S.M. 1986. Efficacy of some indigenous plant extracts in the control of lemon rot by two pathogenic fungi. *Nat. Acad. Sci. Lett.* **9**:133-134.

Chakraborty, M.R., Dutta. S., Ojha And Chatterjee N. C. 2006. Management of dieback of bottle brush (*Callistemon citrinus*) – An integrated approach. *Intenat. J. Agric Sci.* **2**:33-36.

Cavellito, C.J. and Bailey. J.H. 1944. Allitin, the antibacterial principle of *Allium sativum* - 1. Isolation, physical properties and antibacterial action. *J. Amer. Chem. Soc.* **66**:1950-1951

Ejechi, B.O., Ojeata, A. and Oyeleke, S.B. 1997. The effects of extracts of some Nigerian spices on biodeterioration of Okro (*Abelmoschus*(L) Moench) by fungi. *Journal of Phytopathology.* **145**:469-479.

Farag R.S, Daw, Z.Y. and Abo-Raya S.H .1989. Influence of some spice essencial oils on *Aspergillus parasiticus* growth and production of aflatoxinas in a synthetic medium. *J. Food Sci. Chicago.* **54**:54-74.

Garg, S.C. and Siddigui, N. 1992. Antifungal activity of some essential oil isolates. *Pharmazie.* **47**:467-468.

Kurucheve, V., Gerard Ezhilan and Jayaraj. 1997. Screening of higher plants for fungitoxicity against *Rhizotonia solani in vitro. Indian Phytopath.* **50**:235-241.

Lima ,M.E.L., Caordiero.I., Young,M.C.M., Marcos.E.G., Sobra., Paulo Roberto, Moreno, 2006. Antimicrobial activity of the essential oil from two specimens of *Pimenta pseudocaryophyllus* (Gomes) L.R. Landrum. (Myratace) native from Saopaulo state -Brazil. *Pharmacologyonline* **3**:589-593.

Mahadevan, A.1982. Biochemical aspects of plant disease resistance Part 1. Preformed inhibitory substances-Prohibitions. Today and tomorrow's Printers and Publishers, New Delhi, India, 425p.

Muhamed, S., Sizama Saka., Elsharkawy, S.H., Ali, A.M and Sepiah Mavid. 1996. Antomycotic screening of 58 Malaysian plants against plant pathogens. *Pesticide Science.* **47:**259-264.

Myeong, L.S., Jo HyunJin., Kim Dong Soo., Kim Do Wan., Leek Dong Woon., Choo Ho Yul and Lee Chongkyu. 2003. Effect of clove extract on antifungal activity against *Rhizotonia solani* and plant growth. *KFRI Journal of Forest Science* **66**:105-111.

Natarajan, M.R. and LalithaKumari, D. 1987a. Antifungal activity of the leaf extract of *Lawsonia inermis* on *Drechslera oryzae. Indian Phytopath.* **40**:393-395.

Natarajan, M.R. and LalithaKumari, D. 1987b. Leaf extracts of *Lawsonia inermis* as antifungal agent. *Curr. Sci.* **56**:1021-1022.

Nene, Y.L. 1971. *Fungicides in plant disease control.* Oxford and IBH publishing Co., New Delhi. 385p.

Nwachukwu, E.O. and Umechuruba, C. I. 2001. Antifungal activities of some leaf extracts on seed-borne fungi of African yam bean seed, seed germination and seedling emergence *J. Applied Sciences and Enviromental Management.* **5:** 29-32.

Parkinson, D. And Clark, J.H. 1964. Studies on the fungi in the root region. III. Root surface fungi of three species of *Allium. Plant and Soil.***20**:166-170.

Pereira, M.C., Chalfoun, S.M., Pimenta., Angelico C.L. and Maciel. W.P. 2006. Spices, fungi mycelial development and Ochratoxin A production. *Scientific Research and Essay* **1**: 38-42.

Rao, R.R., Rao . S.S. and Venkataraman. P.R. 1946. Investigations on plant antibiotics. 1.Studies on allitin, the antibacterial principle of *Allium sativum* (garlic). *J. Sci. Ind. Res. India* **5**: 31-55.

Ranasinghe, L., Jayawardena, B. and Abeywickrama, K. 2002. Fungicidal activity of essential oils of *Cinnamomum zeylanicum* (L.) and *Syzygium aromaticum* (L.) Merr et L.M.Perry against crown rot and anthracnose pathogens isolated from banana. *Letters in Applied Microbiology* **35**: 208-211.

Singh, U.P., Pathak, K.K., Khare, M.N. and Singh, R. B.1979. Effects of leaf extract of garlic on *Fusarium oxysporum* f.sp.ciceri and *Sclerotinia sclerotiorum* on gram seeds. *Mycologia* **71**: 556-564.

Singh, H.B. and Singh, U.P. 1980. Inhibition of growth and sclerotium formation in *Rhizotonia solani* by garlic oil. *Mycologia* **72**:1022-1025.

Singh, R.K. and Dwivedi, R.S. 1987. Fungitoxicity of different plant species against *Sclerotium rolfsii* Sacc. a foot rot pathogen of barley. *Nat. Acad. Sci. Letters India* **10**: 89 - 91.

Tiwari. R. 1987. Fungitoxicity of volatile constituents of some higher plants against predominant storage fungi. *Indian Phytopath.***50**:548-551.

Venugopal, S. and Chandramohanan, R. 2006. Role of fungi in fruit rot and immature nutfall of coconut. *Cord* **22**:33-40.

Venugopal, S. Chandramohanan, R. and Nampoothiri, C.K. 2008. Variability in cultural and morphological characters within *Lasiodiplodia theobromae* isolates causing rotting and immature nut fall of eriophyid mite infested coconut. *Plant Achives* **8**:803-811.

Vincent, J.M. 1927. Distortion of fungal hyphae in presence of certain inhibitors. *Nature.* **159**:850.

39

Evaluation of Antagonistic Potential of *Trichoderma* Isolates Against *Pythium aphanidermatum* (Edson) Fitz. Causing Rhizome Rot in Turmeric (*Curcuma longa* L.)

K. Anoop and R. Suseela Bhai

INTRODUCTION

In turmeric, rhizome rot is the most destructive disease that causes economic damage to the crop. Survey conducted in various turmeric growing tracts of South India revealed that the disease is caused predominantly by *Pythium aphanidermatum* (Edson.) Fitz. (Anoop *et al. unpublished)*. Fungicides like Metalaxyl and various organic amendments were reported to manage the disease (Setty *et al.*, 1996). The indiscriminate use and undesirable side effects of fungicides have increased the importance of alternative methods of disease management like biological control. Biological control appears to be the best strategy for the long-term sustainability and effective management of soil borne diseases. *Trichoderma* strains have long been recognized as biological agents for the control of plant disease and for their ability to increase root growth, crop productivity and uptake and use of nutrients. The most common targeted pathogens are species of *Pythium, Fusarium, Phytophthora* and *Rhizoctonia* reflecting their world wide importance. The basic mechanisms responsible for their ability to control pathogens have included antibiosis, mycoparasitism and competition for nutrients. Several species of this *Trichoderma* have been reported for their ability to produce antifungal metabolites (Sivasitamparam and Ghisalberti, 1998) and lytic enzymes (Haran *et al.,* 1996). There are numerous reports of compounds derived from *Trichoderma* species with volatile and non volatile activities (Sivasitambaram and Gisalberti, 1998). The synergistic effect of these compounds with enzymes appears to be important in the management of the disease (Sivasithamparam, 1998). Many species in this genus can be characterized as opportunistic avirulent plant symbionts (Harman *et al.,* 2004). The pathogenicity tests conducted do not suggest that these organisms pose a threat to human or animal health (Nevalainen and Neethling, 1998). They have the ability to promote growth and induce resistance in plants (Monte, 2001) and are highly resistant to a range of toxicants (Harman, 2004). *Trichoderma* spp. is known to be the most commonly used antagonists against *P. aphanidermatum* (Chet *et al.*, 1981, Hazarika *et al.*, 2000 and Manoranjitham 2001). *Trichoderma* spp. and VAM have proved to be effective against the rhizome rot of turmeric which was then reported to be caused by *Fusarium solani* (Reddy, 2003).

Antagonistic potential of *Trichoderma* isolates, obtained from rhizosphere soil of 32 major turmeric growing locations of South India, against *P. aphanidermatum*, the predominant species causing rhizome rot in Turmeric was assessed during the survey conducted in 2005-07 seasons.

The rhizosphere soil samples were subjected to dilution plating using *Trichoderma* Selective Medium (Elad and Chet, 1983). The sieved soil sample (10 g) was shaken in 90 ml of sterile distilled water (SDW) for 10 min. The resulting soil suspension was diluted to 10^{-4}. Aliquot of 1 ml was added to Petri dishes containing 20 ml of TSM. The *Trichoderma* isolates obtained were purified and maintained on PDA slants at 4°C for further studies.

The potential of *Trichoderma* spp. was evaluated *in vitro* using dual culture technique (Morton and Stroube, 1955). The percentage inhibition was calculated by the formula:

$$PI = \frac{C - T}{C} \times 100$$

PI = Percentage inhibition

C = Radial growth of the pathogen in control plate (mm)

T = Radial growth of the pathogen in Dual culture (mm)

The activity of non volatile metabolites of *Trichoderma* isolates was studied using the culture filtrate of the antagonist (Dennis and Webster, 1971) and the effect of volatile metabolites was studied by the method followed by Dennis and Webster (1971).

A total of 22 isolates of *Trichoderma* (Table 39.1) were obtained including four isolates from Andhra Pradesh, four from Karnataka, six from Kerala and eight from Tamil Nadu representing nineteen locations of south India. All the isolates were tested against *P. aphanidermatum* by dual culture technique. The isolates showed inhibition ranging from 32.37–72.15% (Table 39.2). All the isolates covered entire plate within seven days of inoculation. The highest inhibition (71.78%) was shown by the isolate CLT102 collected from Guntur Dist. of Andhra Pradesh followed by CLT 102 collected from Wayanad Dist. of Kerala. Isolate CLT 115 had comparatively slow growth.

Table 39.1. Details of isolates collected during 2005-07

Isolate No.	*Location*	*State*
CLT 101	Calicut	Kerala
CLT 102	Wayanad	Kerala
CLT 103	Wayanad	Kerala
CLT 104	Kodumudi	Tamil Nadu
CLT 105	Thondamuthoor	Tamil Nadu
CLT 106	Thondamuthoor	Tamil Nadu
CLT 107	Kunjanapalle	Andhra Pradesh
CLT 108	Vallabapuram	Andhra Pradesh
CLT 109	Cheraooru	Andhra Pradesh
CLT 110	Chittoor	Kerala
CLT 111	Kalpetta	Kerala
CLT 112	Nathamkuni	Kerala
CLT 113	Jagtial	Andhra Pradesh
CLT 114	Bidar	Karnataka
CLT 115	Annur	Tamil Nadu
CLT 116	Kolathoor	Tamil Nadu
CLT 117	Mettoor	Tamil Nadu
CLT 118	Settiputhoor	Tamil Nadu
CLT 119	Settiputhoor	Tamil Nadu
CLT 120	Ainoli	Karnataka
CLT 121	Chincholi	Karnataka
CLT 122	Bidar	Karnataka

Smix isolates namely CLT 102, CLT 107, CLT 110, CLT 114, CLT 118 and CLT 121 showed inhibition over 70%. These isolates also showed fast growth rate in comparison with others showing their potential to overgrow the pathogen. The growth of the antagonist over pathogen was scored using modified Bell's Scale (Bell *et al.* 1982) (Table 39.3). CLT 114 showed the cytoplasmic coagulation of *P. aphanidermatum* at the point of meeting between the antagonist and pathogen showing the property of antibiosis on disease suppression.

The isolates of *Trichoderma* varied in the production of non volatile metabolites. The inhibition of the isolates ranged from 0.00 to –46.30%. The highest inhibition was shown by CLT 110 isolated from Palakkad Dist., Kerala followed by CLT 114 (42.22%) isolated from Bidar Dist, Karnataka. CLT 103, CLT 108, CLT 116, CLT 117 and CLT 122 showed no inhibition against pathogen. This may be due to the low production of the metabolites.

Table 39.2. *In vitro* inhibition of *P. aphanidermatum* by *Trichoderma* isolates

Isolate No.	*Inhibition (%)*		
	Dual	*Volatiles*	*Non-volatiles*
CLT 101	47.78 (43.68)	27.41(31.56)	12.22(20.44)
CLT 102	71.78 (57.86)	60.00 (50.77)	21.11 (27.35)
CLT 103	45.33 (42.3)	0.00(0.00)	37.78(37.88)
CLT 104	43.56 (41.27)	5.74 (13.81)	16.67 (24.04)
CLT 105	59.11 (50.18)	17.41 (24.65)	21.11 (27.35)
CLT 106	54.89 (47.75)	4.81 (12.66)	5.56 (13.56)
CLT 107	72.15 (58.12)	41.11 (39.87)	38.52 (38.35)
CLT 108	49.27 (44.14)	0.00(0.00)	18.14 (25.18)
CLT 109	38.89 (38.53)	5.56 (13.56)	0.00(0.00)
CLT 110	71.26 (57.48)	49.63 (44.77)	46.3 (42.88)
CLT 111	45.93 (42.65)	24.04 (29.33)	24.07 (29.33)
CLT 112	38.44 (38.29)	13.33 (21.39)	27.78 (31.76)
CLT 113	44.00 (41.55)	10.74 (19.09)	29.63 (32.96)
CLT 114	70.98 (57.35)	38.15 (38.12)	42.22 (40.51)
CLT 115	32.37 (34.63)	29.63 (32.96)	0.00(0.00)
CLT 116	35.64 (36.63)	0.00(0.00)	0.00(0.00)
CLT 117	40.00 (39.23)	0.00(0.00)	25.56(31.33)
CLT 118	70.37 (57.04)	84.82 (67.05)	34.44 (35.97)
CLT 119	45.93 (42.65)	12.22 (20.44)	12.22 (20.44)
CLT 120	47.04(43.28)	8.89(17.6)	0.00(0.00)
CLT 121	70.37(56.98)	82.22(65.05)	20.00(26.56)
CLT 122	46.96(43.22)	0.00(0.00)	0.00(0.00)
LSD	0.053	0.120	0.246

Table 39.3. Ratings of selected isolates of *Trichoderma* spp. on the inhibition *P. aphanidermatum* (Bell's Scale, 1981)

Isolate No.	*Days*			
	3	*4*	*5*	*6*
CLT 101	S3	S3-S2	S2	S1
CLT 102	S1			
CLT 103	S3	S2	S1	
CLT 104	S2	S2-S1	S1	
CLT 105	S3	S3-S2	S2-S1	S1
CLT 106	S3	S3-S2	S2-S1	S1

(*Contd.*)

Isolate No.	Days			
	3	4	5	6
CLT 107	S1			
CLT 108	S3	S3-S2	S2	S1
CLT 109	S3	S3-S2	S2	S1
CLT 110	S1			
CLT 111	S3	S3-S2	S1	
CLT 112	S3	S3-S2	S2-S1	S1
CLT 113	S2	S2-S1	S1	
CLT 114	S1			
CLT 115	S3	S3-S2	S2	S1
CLT 116	S2	S2-S1	S1	
CLT 117	S2	S2-S1	S1	
CLT 118	S2	S1		
CLT 119	S3	S2	S1	
CLT 120	S3	S2	S1	
CLT 121	S2	S1		
CLT 122	S3	S2	S1	

*S1 – 100% overgrowth, S2 – 75% overgrowth, S3 – 50% overgrowth, S4 – locked at the point of contact.

However, some isolates reduced the profuse growth of pathogens in comparison with its control showing the presence of metabolites that can reduce the growth of *P. aphanidermatum.* These results show the ability of certain *Trichoderma* isolates to control *P. aphanidermatum* by way of non-volatile metabolite production and thus support the mechanism involved in the biocontrol (Sivan *et a*l., 1984, Shanmukham and Varma, 1999, Hazarika *et al.*, 2000).

The production volatile metabolites varied with the isolates. The inhibition due to volatile metabolites ranged from 0.00–84.82%. The highest inhibition was shown by the isolate CLT 118 (84.82%) collected from Coimbatore Dist., Tamil Nadu followed by CLT 121 (82.22%) from Gulbarga Dist., Karnataka. These isolates are comparatively fast growing and are showing high inhibition in dual culture showing their potential as an antagonist to *P. aphanidermatum.*

The twenty two isolates of *Trichoderma* showed varying ability in their production of volatile and non volatile metabolites. Similarly the mechanism of inhibition such as antibiosis, mycoparasitism, competition etc. also varies with isolates. The mechanisms for inhibition such as antibiosis, mycoparasitism and competition have already been explained by several authors (Howell, 1998) and are supporting the present investigation. The fast growing isolates causing more inhibition of pathogen may be due to mycoparasitism and competition for nutrients. Hyphal parasitism of *Pythium* spp. by *Trichoderma* was also observed *in vitro* by many workers (Dennis and Webster, 1971, Chet *et al.*, 1981). Considering the results from the different methods of screening to study the potentialies of *Trichoderma* spp. isolated from rhizosphere of turmeric, it is found that six isolates have the potential for being exploited for the biological control rhizome rot of turmeric. Hence, these isolates were short listed for their evaluation in the field for the management of the disease. All the six isolates have shown over 70% inhibition in dual culture. The highest volatile activity showed by the isolates CLT 118 and CLT 121 have taken into consideration rather than their non-volatile inhibition. The slow growing isolates showed poor inhibition by non-volatile metabolites probably because they may not be producing sufficient metabolites to inhibit the pathogen. Moreover, the slow growing isolates cannot be selected for biological control, since fast growing nature is one of the criteria for the selection of antagonists (Otis, 1993). The six isolates of *Trichoderma* shortlisted are capable of inhibiting the pathogen by any one of the mechanism involved in the biological control. They have proved their potential to use as a biological control agent *in vitro*. The performances of these short listed isolates in the field are further to be studied in the conducive soil to develop them as a biocontrol agent against rhizome rot of turmeric.

References

Bell, D.K., Wells, H.D. and Markham, C.R. 1982. *In vitro* antagonism of *Trichoderma* species against six fungal plant pathogens. *Phytopathology* **72:**379–382.

Chet, I., Harman, G.E., Baker, R. 1981. *Trichoderma harzianum*: its hyphal interaction with *R.solani* and *Pythium* spp. *Microbial Ecology* **7**:834-839.

Dennis, C. and Webster, J. 1971. Antagonistic properties of species-group of *Trichoderma*. I. Production of nonvolatile antibiotics. *Transaction of the British Mycological Society.* **57**:25-39.

Elad, Y., Chet, I. and Henis, Y. 1982. Degradation of plant pathogenic fungi by *Trichoderma harzianum. Can. J. Microbiol.* **28:**719-725.

Elad, Y. and Chet, I. 1983. Improved selective media for isolation of *Trichoderma* spp. or *Fusarium* spp. Phytoparasitica **11**:55-58.

Haran, S., Schickler, H. and Chet, I., 1996. Molecular mechanisms of lytic enzymes involved in the biocontrol activity of *T. harzianum. Microbiology* **142**:2321-2331.

Harman, G.E. 2006. "Overview of mechanisms and uses of *Trichoderma* spp.". *Phytopathology* **96**:190-194.

Hazarika, D.K., Sarmah, R., Paramanik, T., Hazarika, K. and Phookan, A.K. 2000. Biological management of tomato damping off caused by *P. aphanidermatum. Indian J. Plantpathology* **54**:59-61.

Lewis, J.A. and Papavizas, G.C. 1984. A new approach to stimulate population proliferation of *Trichoderma* species and other potential biocontrol fungi introduced into natural soils. *Phytopathology* **74:**1240–1244.

Lifsbitz. R., Windham, M.T. and Baker, R. 1986. Mechanism of Biocontrol of pre-emergence damping off of pea by seed treatment with Trichoderma spp. *Phytopathology* **76**:720-725.

Manoranjitham, S. K., Prakasan, V. and Rajappan, K. 2001. Biocontrol of Damping of Tomato caused by *P. aphanidermatum. Indian Phytopathology* **54:**59-61.

Monte. E. 2001. Understanding *Trichoderma*: between agricultural biotechnology and microbial ecology. *International Microbiology* **4:**1-4.

Morton, D.T. and Stroube, W. H. 1955. Antagonistic and stimulatory effects of microorganism upon *Sclerotium rolfsii. Phytopatholgy* **45:**419-420.

Nevalainen, H. and Neethling, D. 1998. The safety of *Trichoderma* and *Gliocladium.* In *Trichoderma and Gliocladium. Vol. I Basic Biology, Taxonomy and Genetics* (Eds. C. P. Kubicek & G. E. Harman): 193-200. Taylor and Francis, London.

Otis, C. M. 1993. *Plant Disease Control. Principles and Practice.* John Wiley & Sons INC pp: 235-250.

Reddy, M.N., Charitha Devi, M. and Sreedevi N.V. 2003. Biological control of rhizome rot of turmeric (*Curcuma longa* L.) caused by *Fusarium solani. Journal of Biological Control* **17**(2):193-195.

Sreerama Setty, T. A., Mohan, E., Herle, P. S. and Parameswar, N. S. 1996. Effect of seed dressing fungicides and organic amendments on rhizome rot of turmeric. *Journal of Plantation Crops* **24**:240-242.

Shanmugam,.V. and Varma,A.S. 1999. Effect of native antagonists against *P.aphanidermatum*, the causal organism of rhizome rot of Ginger. *J. Mycol. Pl. Pathol.* **29**:375-379.

Sivan, A., Elad Y., and Chet, T. 1984. Biological control effect of new isolate of *Trichoderma harzianum* on *P.aphanidermatum.* Phytopathology **74**:498-501.

Sivasitamparam, K. and Ghisalberti, E.L. 1998. Secondary metabolism in Trichoderma and Gliocladium. In Trichoderma and Gliocladium. Vol. I Basic Biology, Taxonomy and Genetics (Eds. C. P. Kubicek & G. E. Harman). pp.139-191. Taylor and Francis, London.

40

Inhibition of Polypathogenic Fungi by Leaf Extract of Matty (*Ailanthus triphysa*, Mill.) and Guava (*Psidium guajava* L.)

J. Jacob, V. Sinoby, N.S. Saritha, U. Sheeba and J. Shirmila

INTRODUCTION

Diseases in crops are being managed through cultural practices, use of resistant varieties and application of fungicides. However, non-judicious fungicide use created problems such as development of resistant strains of fungi, and harmful effects on human health and environment. The botanicals containing secondary plant metabolites also control the diseases caused by various fungi, bacteria and virus in plants (Fawcet and Spencer (1970); Mishra and Tewari (1993); Naqvi *et al.*, (1991) and are superior to synthetic pesticides because of low mammalian toxicity, target specificity and biodegradability and may replace present synthetic and non-biodegradable fungicides, i.e., exploitation of allelopathy for plant protection through botanicals. The simple phenolic acids present in the plants impart resistance to attack by plant pathogens (Putnam and Duke, 1978). matty (*Ailanthus triphysa* Mill.) and guava (*Psidium guajava* L.) are trees commonly planted in the home gardens of coastal ecosystem of Kerala and tropical regions. The allelopathic effect of *matty* and guava on crops has been reported (Jacob, 1998).

The study was undertaken at the Cropping Systems Research Centre, Karamana, Thiruvananthapuram during 2007. Six test fungi viz. *Phytophthora infestans, Colletotrichum gloeosporioides, Alternaria alternata*, *Fusarium oxysporum*, *Rhizoctonia solani and Sclerotium rolfsii* were isolated from the infected parts of various crops (*P. infestans* from black pepper, *C. gloeosporioides* from mango, *A. alternata* from water hyacinth, *F. oxysporum* from cowpea, *R. solani* from rice *and S. rolfsii* from elephant foot yam). The test fungi cause serious diseases in many crop plants, i.e., they are polypathogenic and cause much economic loss to farmers. *P. infestans* causes foot rot in black pepper, *C. gloeosporioides* anthracnose in mango, *A. alternata* leaf spot in water hyacinth, *F. oxysporum* wilt disease in cowpea, *R. solani* sheath blight in rice *and S. rolfsii* collar rot/ black spot in yams.

Plant part showing early infection was cut into small pieces and surface sterilized with 0.1% (w/v) mercuric chloride for 1 min. followed by three washings in sterile distilled water. Sterilized infected tissue bits were plated on Potato Dextrose Agar (PDA) medium containing the antibiotic sporidex (i.e., Cephalexin) and incubated in laboratory condition at room temperature of 25 ± 3 °C. Pure culture of the pathogens were obtained by single spore/mycelium isolation method. The pathogenicity of all the above fungi was tested on respective plant parts under laboratory condition. The respective fungi were re-isolated from the infected tissues to prove Koch's Postulates. The pure cultures of fungi were maintained in PDA medium supplemented with sporidex antibiotic under laboratory condition.

Fresh extract of *matty* and guava leaf 100% (w/v) was prepared in distilled water. For this mature leaves of *matty* and guava were collected. Leaves were thoroughly washed under running tap water and shade dried for 1 h to remove their wetness. Five hundred gram of leaves were weighed out, chopped to 0.5 cm size and homogenized in a mixer after adding 200 ml of distilled water. The homogenate was initially filtered through a layer of absorbent cotton and subsequently through three layers of muslin cloth to remove all plant debris. The filtrate was collected and the residue was again homogenized twice in a mixer, each time with 100 ml of distilled water and filtered through absorbent cotton and three layers of muslin cloth to collect the excess extract. The final volume of the extract was made to 500 ml with distilled water to form 100 % aqueous extract. PDA media containing 5, 10, 15, 25 and 50% (v/v) leaf extracts of *matty* and guava were prepared. To solidify the media, variable amounts of agar were added and sterilized in autoclave at 120 °C and 15 pounds pressure for 20 min.

The sterilized PDA containing extracts were poured into sterilized Petri dishes (9 cm dia.) after cooling to luke warm temperature of 40 °C. The media was inoculated with 9 mm discs taken from growing edges of 5 day old cultures of test fungi and incubated at laboratory condition (25 ± 3 °C). The growth of fungi on different concentrations of various extracts was measured as the diameter of mycelial growth of fungi (cm) at 3, 5 and 7 days after inoculation (DAI). The PDA medium alone served as the control. The percent inhibition/ stimulation was calculated using the formula:

$$\text{Inhibition/stimulation (\%)} = [(T\text{-}C)/C] \times 100$$

where C and T are mycelial growth (cm) of test fungi in control and treatment respectively. All treatments were replicated thrice. The data were subjected to analysis of variance for completely randomized design.

The aqueous leaf extract of *matty* significantly affected the mycelial growth of the different test fungi (Table 40.1). The growth of *P. infestans* was significantly influenced at 3, 5, and 7 DAI by all treatments. At 3 DAI, the inhibition by 50 % extracts was very high (53.1%) which was on par with 10, 15 and 25% extracts. At 5 DAI, 50% extract caused inhibition as high as 67.5%, which was on par with 15 and 25% concentrations. At 7 DAI, all the concentrations significantly inhibited the fungal growth.

Inhibition of *C. gloeosporioides* was significant at all concentrations. At 3 DAI, significantly high inhibition was observed at 50% concentration (48.9%) while the lowest was at 5% extract (27.8%). At 5 DAI, greatest inhibition was by 50% extract (40.9%) which was on par with 25%. At 7 DAI, 24.7% inhibition was brought about by 50% extract. However, all the other concentrations (5-25%) were on par with control.

Leaf extract suppressed the growth of *A. alternata*. No significant inhibition was observed at 3 and 7 DAI. At 5 DAI, 50% extract caused highest inhibition (23.1%) and, was on par with 25% extract, while 5, 10 and 15% extracts were on par with control.

Table 40.1. Effect of *matty* leaf extract on growth (cm) of polypathogenic fungi

Leaf extract	*Phytophthora infestans*			*Colletotrichum gloeosporoides*			*Alternaria alternata*		
(%)	*3 DAI*	*5 DAI*	*7 DAI*	*3 DAI*	*5 DAI*	*7 DAI*	*3 DAI*	*5 DAI*	*7 DAI*
0	*2.13(0)*	*3.83(0)*	*4.30(0)*	*3.00(0)*	*5.30(0)*	*5.80(0)*	*2.00(0)*	*4.03(0)*	*4.77(0)*
5	1.20(-43.7)	2.70(-29.5)	3.10(-27.9)	2.17(-27.8)	4.77(-10.1)	5.43(-6.3)	1.97(-1.7)	3.80(-5.7)	5.03(5.5)
10	1.03(-51.5)	1.67(-56.5)	2.13(-50.4)	2.03(-32.2)	4.27(-19.5)	5.67(-2.3)	2.03(1.7)	3.70(-8.2)	5.03(5.5)
15	1.00(-53.1)	1.50(-60.8)	2.17(-49.6)	2.03(-32.2)	3.83(-27.7)	5.47(-5.7)	2.07(3.3)	3.90(-3.2)	5.07(6.2)
25	1.00(-53.1)	1.40(-63.5)	2.10(-51.2)	1.73(-42.2)	3.47(-34.6)	5.27(-9.2)	1.97(-1.7)	3.20(-20.6)	4.10(-14.6)
50	1.00(-53.1)	1.23(-67.5)	2.10(-51.2)	1.53(-48.9)	3.13(-40.9)	4.37(-24.7)	1.97(-1.7)	3.10(-23.1)	4.30(-9.9)
CD (P = 0.05)	0.094	0.317	0.375	0.187	0.367	0.710	NS	0.540	NS

Leaf extract	*Fusarium oxysporum*			*Rhizoctonia solani*			*Sclerotium rolfsii*		
(%)	*3DAI*	*5DAI*	*7DAI*	*3DAI*	*5DAI*	*7DAI*	*3DAI*	*5DAI*	*7DAI*
0	2.80(0)	5.87(0)	6.13(0)	2.80(0)	5.87(0)	6.13(0)	2.80(0)	5.87(0)	6.13(0)
5	1.97(-29.8)	4.20(-28.5)	5.70(-7.0)	1.97(-29.8)	4.20(-28.5)	5.70(-7.0)	1.97(-29.8)	4.20(-28.5)	5.70(-7.0)
10	1.93(-31.0)	4.00(-31.9)	5.77(-5.9)	1.93(-31.0)	4.00(-31.9)	5.77(-5.9)	1.93(-31.0)	4.00(-31.9)	5.77(-5.9)
15	1.63(-41.7)	3.73(-36.4)	5.67(-7.6)	1.63(-41.7)	3.73(-36.4)	5.67(-7.6)	1.63(-41.7)	3.73(-36.4)	5.67(-7.6)
25	1.83(-34.5)	2.07(-64.9)	4.43(-27.7)	1.83(-34.5)	2.07(-64.9)	4.43(-27.7)	1.83(-34.5)	2.07(-64.9)	4.43(-27.7)
50	1.63(-41.7)	2.67(-54.6)	4.17(-32.0)	1.63(-41.7)	2.67(-54.6)	4.17(-32.0)	1.63(-41.7)	2.67(-54.6)	4.17(-32.0)
CD (P = 0.05)	0.552	0.592	0.952	0.552	0.592	0.952	0.552	0.592	0.952

DAI = Days after inoculation
NS = Not Significant

Leaf extract, at all concentrations, were effective in restricting the fungal growth *F. oxysporum* at 3 DAI and the inhibition ranged from 29.8-41.7%. At 5 DAI, the highest inhibition was observed at 25% concentration (64.9%). At 7 DAI, the greatest inhibition was observed at 50% (32%) which was on par with 25% extract (27.7%). All other concentrations were on par with control.

R. solani was considerably inhibited by 50% extract (34.2) at 3 DAI. At lower concentration, no inhibition was noticed. At 5 DAI, no significant inhibition was observed. The fast growing nature of fungus made it difficult to assess the impact at 7 DAI.

The growth of *S. rolfsii* was suppressed at 3 DAI. The highest inhibition was by 50% extract (41.7%) and lowest by 5% extract. It is to be noted that these fungi are fast growing causing the mycelium to cover the petri plate at 5 and 7 DAI. Therefore, in these fungi, the inhibitory effect observed at 3 DAI can be considered.

The aqueous leaf extract of guava significantly affected the growth of different test fungi (Table 40.2). The growth of *P. infestans* was significantly influenced at 3, 5 and 7 DAI. At 3 DAI, 10, 15, 25 and 50% extract significantly suppressed the fungal growth. The highest inhibition was at 25 and 50% concentration (24.7%). The effect of 5% extracts was on par with control. At 5 DAI, except 5% extract, growth was suppressed considerably by all the other concentrations. At 7 DAI, treatments significantly suppressed the fungal growth. The 50% extract resulted in inhibition as high as 46.1%, which was on par with 10, 15 and 25% concentrations.

Table 40.2. Effect of guava leaf extract on growth (cm) of polypathogenic fungi

Leaf extract	*Phytophthora infestans*			*Colletotrichum gloeosporoides*			*Alternaria alternata*		
(%)	*3 DAI*	*5 DAI*	*7 DAI*	*3 DAI*	*5 DAI*	*7 DAI*	*3 DAI*	*5 DAI*	*7 DAI*
0	2.17(0)	3.80(0)	5.57(0)	3.33(0)	5.40(0)	7.63(0)	2.07(0)	4.40(0)	6.27(0)
5	2.30(6.0)	4.03(6.1)	5.03(-9.6)	2.67(-19.9)	4.97(-8.0)	7.37(-3.5)	2.03(-1.8)	4.17(-5.3)	6.60(5.3)
10	2.10(-3.2)	2.50(-34.2)	3.20(-42.5)	2.23(-32.9)	4.90(-9.3)	7.30(-4.3)	2.03(-1.8)	4.23(-3.8)	6.47(3.1)
15	1.40(-35.5)	2.43(-36.0)	3.33(-40.2)	2.40(-27.9)	5.03(-6.8)	7.30(-4.3)	2.13(3.1)	4.10(-6.8)	5.40(-13.9)
25	1.63(-24.7)	2.40(-36.8)	3.23(-42.0)	1.97(-40.9)	4.33(-19.8)	5.03(-34.0)	2.03(-1.8)	4.03(-8.3)	5.50(-12.3)
50	1.63(-24.7)	2.40(-36.8)	3.00(-46.1)	1.97(-40.9)	4.30(-20.4)	5.10(-33.2)	1.63(-21.1)	3.43(-22.0)	4.43(-29.3)
CD (P = 0.05)	0.241	0.507	0.432	0.668	0.356	0.697	0.126	0.302	0.548

Leaf extract	*Fusarium oxysporum*			*Rhizoctonia solani*			*Sclerotium rolfsii*		
(%)	*3DAI*	*5DAI*	*7DAI*	*3DAI*	*5DAI*	*7DAI*	*3DAI*	*5DAI*	*7DAI*
0	1.77(0)	6.03(0)	7.13(0)	5.90(0)	9.00(0)	9.00(0)	5.63(0)	9.00(0)	9.00(0)
5	2.23(26.2)	4.53(-24.8)	6.67(-6.5)	6.93(17.5)	9.00(0)	9.00(0)	6.00(6.6)	9.00(0)	9.00(0)
10	2.17(22.4)	4.20(-30.3)	6.27(-12.1)	6.07(2.8)	9.00(0)	9.00(0)	6.03(7.2)	9.00(0)	9.00(0)
15	2.17(22.4)	4.60(-23.7)	6.23(-12.6)	6.20(5.1)	9.00(0)	9.00(0)	5.63(0.1)	9.00(0)	9.00(0)
25	2.07(16.8)	3.97(-34.2)	5.87(-17.7)	5.80(-1.7)	8.67(-3.7)	9.00(0)	5.17(-8.2)	9.00(0)	9.00(0)
50	2.03(14.9)	3.77(-37.5)	4.67(-34.5)	4.97(-15.8)	8.67(-3.7)	9.00(0)	4.87(-13.6)	9.00(0)	9.00(0)
CD (P = 0.05)	0.262	0.870	0.370	0.625	NS	NS	0.542	NS	NS

DAI = Days after inoculation NS = Not Significant

Leaf extracts suppressed growth of *C. gloeosporioides*. At 3 DAI, the highest inhibition was caused by 25 and 50% extracts (40.9%) while 5% extracts was on par with control. At 5 DAI, the highest inhibition was at 50% extract. At 7 DAI, 25 and 50% extracts inhibited most (33-34%) while other concentrations were on par with control.

The growth inhibition of *A. alternata* was significant at higher concentrations. At 3 DAI, significant inhibition was noticed at 50% concentration (21.1%) only. At 5 DAI, greatest inhibition was observed at 50% (22%) while the 5-15% extracts were on par with control. At 7 DAI, the inhibition was significant at all concentrations and the highest inhibition was caused by 50% extract (29.3%).

In *F. oxysporum*, at 3 DAI, all concentrations significantly stimulated fungal growth. However, at 5 and 7 DAI growth was inhibited by all concentrations. At 5 DAI, 50% extract resulted in 37.5% suppression, which was on par with other concentrations. At 7 DAI, growth was inhibited at all concentrations and degree of inhibition was more at 50% (34.5%).

The inhibition of *R. solani* could be assessed at 3 DAI only. Due to the fast growing nature of fungus, the mycelium covered the petriplate at 5 and 7 DAI. The 50% extract resulted in 15.8% suppression, which was significantly higher than that observed with other concentrations.

The growth of *S. rolfsii* was suppressed at 3 DAI. The effect of 15 and 25% extracts were on par with control while 5 and 10% had a slight stimulatory effect. However, 50% extract caused significant inhibition (13.6%). The fast growing nature of fungus made it difficult to assess the impact at 5 and 7 DAI.

The inhibition of mycelial growth and spore germination of various phytopathogenic fungi by water extracts of many plants has been reported (Bindu *et al.*, 1998; Charu, *et al.*, 2003; Jain and Pathak, 1970; 1963, Kishore *et al.*, 1982).

Certain allelochemicals viz., 1-methoxy canthin-6-benzoquinone, Dimethoxy-p-benzoquinone, Ailanthone, beta-sitosterol, Canthin-6- one, gallic acid, lignin, quercetin, Scopoletin and tannin were identified in *matty* leaf extract. Ascorbic acid, benzaldehyde, beta-bisabolene, beta-ionon, beta-sitosterol, citral, citric acid, linoleic acid, myricetin, myristic acid, oleanolic acid., pectin, quercetin, tannin and ursolic acid were identified in guava (Duke,1992). It is likely that any of these chemicals may be responsible for inhibition of the test fungi.

The varied effect of leaf extract on growth of the different test fungi may be due to the differential response of leaf extract to different fungi, i.e., certain plants having toxic impact against specific pathogens may be stimulatory to other pathogens (Dixit and Tripathi, 1975; Mishra and Tewari, 1993).

The results suggest that the selective principles influencing the fungal growth have a slightly stimulatory effect at lower concentrations whereas at higher concentration, the quantity of these principles is high, thereby causing inhibition. Thus, the same selective principles have differential functions at low and high concentrations. Mishra and Tewari (1993) reported the stimulatory properties of plant extracts on mycelial growth of pathogenic fungi.

CONCLUSION

The study reveals that *matty* and guava leaf causes varying degrees of stimulatory and inhibitory effect on test fungi. The study highlights the need to explore the potential of using *matty* and guava leaf extract for developing plant-based natural fungicides/botanicals to control seed, soil and air borne phytopathogenic fungi causing diseases in multiple crops under homestead farming situation but only after its further careful evaluation *in vivo*. Also, detailed field experiments are necessary before any conclusion on antifungal effects of *matty* and guava for which the results of this preliminary investigation will form a base.

References

Bindu, T.K., Shafi, P.M., Rajan, P.P. and Sarma, Y.R. 1998. Antifungal activity of *Uvaria narum* extracts. *Allelopathy Journal* **5**:67-74.

Charu Arora., Kaushik, R.D., Kumar, A. and Garg, G, K. 2003. Fungicidal potential of Kumaon and Tarai region plants against mushroom fungal pathogens. *Allelopathy Journal* **11**: 63-70.

Dixit, S.N. and Tripathi, S.C.1975. Fungistatic properties of some seedling extracts. *Current Science* **44**:279-280.

Duke, J. A. 1992. *Handbook of Phytochemical Constituents of Grass Herbs and Other Economic Plants.* CRC Press, Boca Raton, FL, USA

Fawcet, C.H. and Spencer, D.M. 1970. Plant chemotherapy with natural products. *Annual Review of Phytopathology* **8**:403-416.

Jacob, J. 1998. Allelopathic effect of leaf leachates of multipurpose trees. *Allelopathy Journal* **6**:81-85.

Jain, J.P. and Pathak, V.N. 1970. Antifungal activity in leaf extracts of certain plants. *Journal of Science and Technology* **8**:58.

Kishore, N., Dubey, N.K. Tripathi, R.D. and Singh, S.K. 1982. Fungitoxic activity of leaves of some higher plants. *National Academy of Science Letters* **5**:9-10

Mishra, M. and Tewari, S.N. 1993. *In vitro* studies on the mycitoxic properties of *Acacia arabica*, *Adhatoda vasica* and *Pongamia glabra*. In: *Proceedings of the National Symposium on Botanical Pesticides in Integrated Pest Management* (Eds. M.S. Chari and G. Ramaprasad). pp. 413-416. Central Tobacco Research Institute, Rajahmundry, India: Indian Society of Tobacco Science.

Naqvi, S.A.H., Khan, M.S.Y. and Vohora, S.B. 1991. Antibacterial, antifungal and antihelmintic investigations of Indian medicinal plants. *Fitoterepia* **62**:221-228.

Putnam, A.R. and Duke, W.B. 1978. Allelopathy in agroecosystems. *Annual Review of Phytopathology* **16**:431-451.

41

Antagonistic Effect of *Trichoderma* Against Soil-Borne Plant Pathogens

R. Poornima, Priya Nair, R.S. Misra and M.L. Jeeva

INTRODUCTION

Trichoderma is one of the most potent bio-control agents known (Blakeman and Fokkema, 1982). They are free living fungi that are common in soil and root ecosystem. Trichoderma species are frequently isolated from forest or agricultural soils at all latitudes. *Trichoderma* strains are known to induce resistance in plants (Ranasingh *et al.*, 2006). The effectiveness of *Trichoderma* in suppressing plant pathogens have been studied by various scientists (Elad *et al.*, 1983; Reghuchander *et al.*, 1998; Sharma and Sain, 2004. and Sen, 2000.) Cultures are typically fast growing at 25-30 °C, typically not growing at 35 °C. *Trichoderma* cultures were isolated from the rhizosphere soil collected from major tuber crop growing areas. The isolates were tested for their antagonistic effect on *Sclerotium rolfsii*, *Phytophthora colocasiae* and *Phytophthora palmivora* by the dual culture technique. Percentage inhibition of mycelial growth was calculated using the formula :

$$I = 100 \frac{(C - T)}{C}$$ (Dickson and Skidmore).

where I = Percentage inhibition of mycelial growth
C = Growth of the pathogen in the control plates (cm).
T = Growth of the pathogen in dual culture (cm).

A total of seventeen *Trichoderma* isolates were tested against the pathogens. Among the isolates tested, TCF 26 isolated from Western Ghats showed maximum percentage inhibition of the pathogens (75, 76 and 81, respectively) under *in vitro* condition (Table 41.1 and Plate 41.1). Isolate TCF 239 was found to be the least effective.

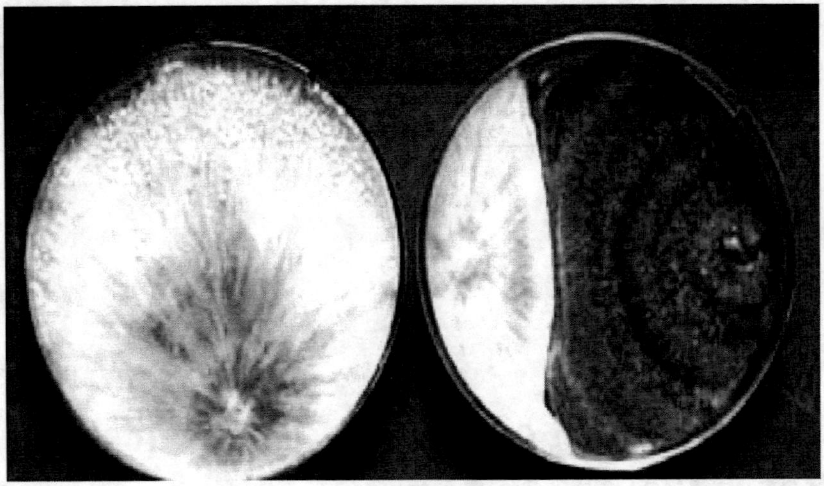

Plate 41.1. TCF 26 against *Sclerotium rolfsii*

Table 41.1. Effect of *Trichoderma* isolates against *Sclerotium rolfsii, Phytophthora colocasia* and *P. palmivora*

Sl.. No.	*Isolate No.*	*Percentage suppression against*		
		Sclerotium rolfsii	*Phytophthora colocasia*	*P. palmivora*
1.	TCF 2	41.11	67.77	75.55
2.	TCF 6	57.77	67.22	83.33
3.	TCF 10	–	66.66	76.66
4.	TCF 16	47.22	64.44	65.00
5.	TCF 24	–	–	40.00
6.	TCF 26	75.00	76.11	80.55
7.	TCF 28	–	70.55	76.66
8.	TCF 229	70.00	71.11	63.88
9.	TCF 232	50.55	65.00	77.22
10.	TCF 239	–	–	–
11.	TCF 311	33.88	46.66	68.33
12.	TCF 320	22.22	57.77	44.44
13.	TCF 324	58.88	52.77	58.88
14.	TCF 430	45.00	62.22	65.00
15.	TCF 457(b)	–	41.66	41.66
16.	TCF 459	39.44	67.22	63.33
17.	TCF 460	74.00	67.22	65.00

References

Blakeman, J.P. and Fokkema, N.J. 1982. Potential for biologial control of plant diseases on the phylloplane. *Rev. Phytopath.* **20**:167-192.

Dickson, C.H. and Skidmore, A.M. 1976. Colony interactions and hyphal interference between *Septoria nodorum* and phylloplane fungi. *Trans. Br. Mycol. Soc.* **66:**57-64.

Elad, Y., Chet, I. and Kalan, J. 1983. *Trichoderma harzianum*: A biocontrol agent effective against *Sclerotium rolfsii* and *Rhizoctonia solani. Phytopathology.* **70:**109-121.

Ranasingh N., Surabh A. and Nedunchezheyan M. 2006. Use of *Trichoderma* in disease management. *Orissa Review,* October 2006 68-70.

Reghuchander, T. Rajappan, K. and Samiyappan, R. 1998. Influence of biocontrol agents and organic amendments on soyabean root rot *Int. J. Tropical Agric.* **16:**247-252.

Sharma P. and Sain, S.K. 2004. Induction of systemic resistance in tomato and cauliflower by *Trichoderma* spp. against stalk rot pathogen, *Sclerotinia sclerotiorum* (Lib) de Bary. *J. Biol. Control,* **18(1)**:21-28.

Sen, B. 2000. Biological control: A success story. *Indian Phytopath.* **53:**243-249.

42

Large-Scale Field Validation of Biocontrol Strategy for Management of Coconut Leaf Eating Caterpillar *Opisina arenosella* Walker Utilizing Indigenous Parasitoids

Chandrika Mohan, C.P.R. Nair and P. Rajan

INTRODUCTION

Opisina arenosella Walker (Lepidoptera: Oecophoridae)) is a major pest of coconut palm in India (Nirula, 1956). This pest periodically assumes epidemic proportions in coastal and backwater areas and in the vicinity of water bodies in the interior parts of peninsular India. The caterpillars of *O. arenosella* construct galleries of silken webs reinforced with excreta and scrapes of leaf bits. Hiding in these galleries they feed on the chlorophyll containing parenchymatous tissues. Damage results in drying of the outer and middle whorl of leaves which affects the photosynthetic efficiency and yield potential of the palm. The damage to leaves renders them unsuitable for thatching and other purposes also.

A bio-intensive pest management strategy has been developed for the management of this pest when it is in an epidemic form (Sathiamma, 1993; Nair *et al.*, 1997). Pest incidence can be checked by the biocontrol methods utilizing indigenous parasitoids. Among the parasitoids recorded from India (Pillai and Nair, 1993), the larval parasitoid *Goniozus nephantidis* Mues. (Bethylidae), the pre-pupal parasitoid *Elasmus nephantidis* Rohw. (Elasmidae) and the pupal parasitoid *Brachymeria nosatoi* Habu. (Chalcididae) are the most promising ones. *Bracon brevicornis* Wesmael (Braconidae) also parasitizes the late larval stage of the pest. Sathiamma *et al.* (1987) fixed the norms for field release of these effective parasitoids. Field evaluation of these parasitoids effected significant reduction in the pest population in the field (Sathiamma *et al.,* 1996).

Large-scale field validation programme on biointensive IPM for the management of coconut leaf eating caterpillar was implemented by Central Plantation Crops Research Institute in different geographic locations in coastal Karnataka (Ullal and Jeppinamogru) and Coastal Kerala (Purakkad and Ayiramthengu) in a total area of 1400 ha. All the four locations were the endemic areas of pest incidence. The total area covered was 500 ha each in Ullal (D. Kannada, Dist., Karnataka) and Jeppinamogru (D. Kannada, Dist. Karnataka) and 200 ha each in Purakkad (Alappuzha Dist. Kerala) and Ayiramthengu (Kollam Dist., Kerala). The plots were selected with the help of extension officials of the State Department of Agriculture, Kerala and Department of Horticulture, Karnataka. The operational area was divided into clusters of 25 ha for assessing incidence of the target pest. In each 25 ha cluster, central area of 5 ha was selected for sampling palms and release of parasitoids.

Contact programmes were arranged in the field during the initial period for creating awareness among the beneficiaries. In each cluster, group discussions and demonstrations were conducted on pest identification, sampling, parasitoid identification, parasitoid release, etc.

Studies were initiated during October 1999 and observations continued till March 2002. Pre treatment pest incidence was recorded during October 1999. Ten percent palms in 5 ha of each cluster was selected and marked as sample palms. The data was recorded on intensity of infestation based on percentage of affected leaves assessed from ground observation. For recording detail count of pest stages, 80 leaflets at random were collected from the sample palms in each cluster. Sample leaves were collected from the lower or middle whorl of leaves. The pest stages *viz.,* larvae, pre pupae and pupae were counted. The sampling formula developed by George *et al.* (1982) was used for estimation of pest population in the affected palms. At the initiation of experiment, fully dried one or two outer leaves were removed from highly pest infested palms and burned.

Larval parasitoids (*Goniozus nephantidis* and *Bracon brevicornis*), Prepupal parasitoid (*Elasmus nephantidis*) and Pupal parasitoid (*Brachymeria nosatoi*) were multiplied in the Entomology laboratory of CPCRI, Kasaragod and CPCRI, Regional Station, Kayamkulam. The parasitoids were released in the field after assessing the population of the target stage of the pest at fortnightly/monthly intervals as per the norms. *G. nephantidis*/*B. brevicornis* were released @ 20% of the estimated larval population, *E. nephantidis* @ 49% of the pre pupal population and *B. nosatoi* @ 32% of the pupal population. Whenever multistage population of the pest was observed, parasitoids were released @ 40% of the pest stages estimated based on sampling. The releases of parasitoids were carried out in the initial period at fortnightly intervals for a period of 5-6 months depending on the pest population to effect an inundative release in each location. In the second year, pest monitoring was done every month and the parasitoid releases were done during initial stages of pest build up after the post-monsoon period and also during the peak period of pest population. Details of parasitoid released during the experimental period are given in Table 42.1.

Table 42.1. Details of parasitoids released in the demonstration plots

Parasitoid		*Locations*			
		Ullal	*J. mogru*	*Purakkad*	*Ayiramthengu*
G. nephantidis	1999	10565	8385	6900	9660
	2000	4500	3730	1200	1380
	2001	560	1110	1080	–
B.brevicornis	1999	31944	25270	2902	4060
	2000	12720	11400	500	580
	2001	1736	3336	456	–
E.nephantidis	1999	400	380	2100	2990
	2000	150	140	350	430
	2001	100	80	330	–
B.nosatoi	1999	220	68	370	52
	2000	40	52	64	74
	2001	20	–	58	–

A total of 8–12 release of parasitoids were made in the first year and need based releases during the subsequent years.

During pre-treatment period, pest infestation in coastal Karnataka was very high with 96% (Ullal) and 86 % (J. Mogru) compared to Coastal Kerala with 33% (Purakkad) and 21% (Ayiramthegu). Pest incidence during the post-release period could be brought down to 4–24% in the first year (with 75–78% reduction) and nil to 6.0% incidence in the 2nd year (93–100% reduction). In the third year, all the plots were free of pest incidence. Incidence of the pest during the observation period is presented in Fig 42.1.

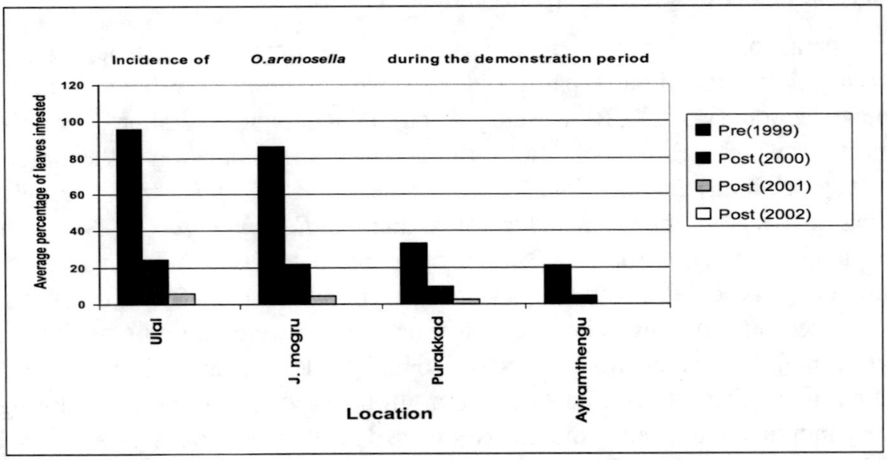

Fig. 42.1. Incidence of *O. arenosella* in the demonstration plots

Population count of *Opisina* in the sample palms at pre-release and post-release period are presented in Table 42.2. The pest population showed a sharp decline during the post-parasitoid release period. The pest suppression was very much evident from the end of the first year with a decline of 60–85% in various locations and the pest incidence become very negligible by the second year with 84–100% reduction in pest population. In all the locations the infested palms were free of pest in the third year. Thus, the impact of release of bioagents in the suppression of the pest was clearly visible.

Table 42.2. Population of *O. arenosella* in the demonstration plots

Place	*Pre-release 1999*	*Post release*			*T value*	
		2000	*2001*	*2002*	*1999-2000*	*1999-2001*
Ullal	790.8 ± 96.2	314.6 ± 55.1	43.9 ± 6.8	0.0	25.5*	35.3*
J.mogru	621.7 ± 92.3	280.4 ± 28.1	82.2 ± 14.8	0.0	17.3*	26.1*
Purakkad	70.8 ± 10.3	12.8 ± 2.2	10.5 ± 1.7	0.0	14.8*	15.3*
Ayiramthengu	98.0 ± 10.6	14.7 ± 1.9	0.0	0.0	24.7*	26.0*

* significant at 1%

There was a build up of released parasitoids in the field. This was evidenced by the observations during post-release period (Fig 42.2).

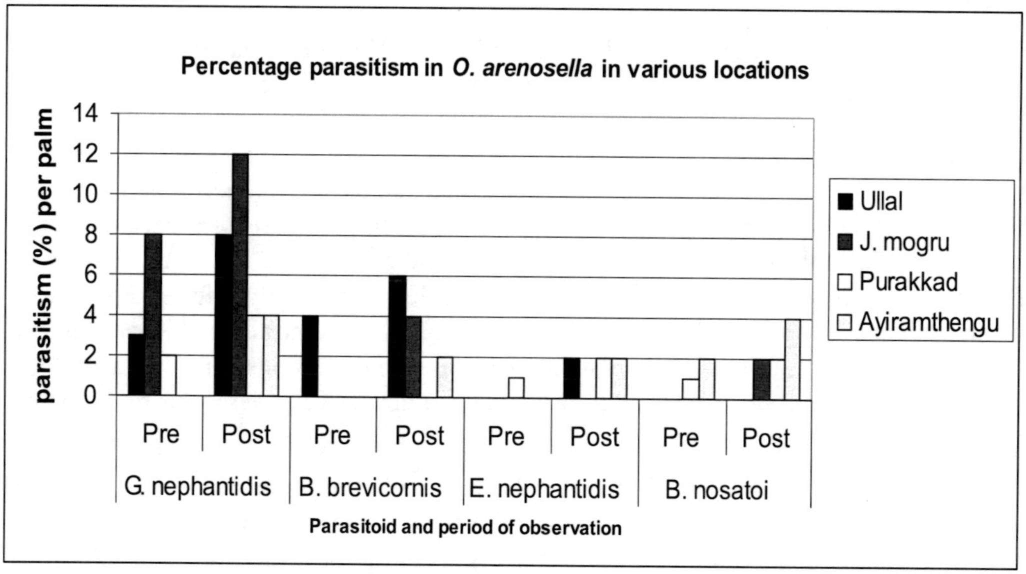

Fig. 42.2. Percentage parasitism on *O. arenosella* during pre and post release period

In the field, natural parasitism by the released parasitoids could be observed to be more frequent (Table 42.3). In addition to the released parasitoids, *Apanteles taragamae* an early instar parasitoid was consistently observed in Ullal. Presence of predators *viz*., anthocoreids, *Parena nigrolineata* and various species of spiders was noticed in the field in all the locations. The higher frequency of occurrence of *G. nephantidis* in all the locations during post-release period indicated the suitability of this larval parasitoid to adapt in various locations in the coastal belt of Kerala and Karnataka. Similarly the predominance of *Bracon brevicornis*, another larval parasitoid in Coastal Karnataka also highlights the importance of releasing this parasitoid.

Table 42.3. Relative abundance of parasitoids present during the experimental period

Location	*G. nephantidis*		*B. brevicornis*		*E. nephantidis*		*B. nosatoi*		*Other parasitoid and predators*
	A	*B*	*A*	*B*	*A*	*B*	*A*	*B*	
Ullal	++	+++	+	++	–	+	–	+	*A.taragamae* Anthocoreids
J. mogru	++	+++	–	++	–	–	–	+	Anthocoreids
Purakkad	+	++	–	–	+	+	+	+	Spiders
Ayiramthengu	–	++	–	+	–	+	++	+++	Spiders, Parena

Time of observation: April-May; A – pre-release period (1999); B – post-release period (2001), – Absent or not observed; + present; ++ available more frequently; +++ always available.

The results of this large area field release of parasitoids for management of coconut leaf eating caterpillar indicated that the pest population could be brought down to very negligible level by the timely release of stage specific parasitoids. These results are in accordance with the findings of Sathiamma *et al.*(1996) who had reported 94 % reduction in pest population by the release of stage specific parasitoids at fixed norms and intervals in a heavily infested coconut garden of 2.8 ha (Thodiyoor, Kerala) for a period of four years. Chandrika Mohan ad Nair (2002) reported 52.6 and 94.7% reduction in pest population after one and two years, respectively of parasitoid release in heavily infested homestead coconut gardens in an area of 3 ha in Neendakara (Kerala). For large area field management of the pest in an epidemic out-break, inundative releases of parasitoids in the initial periods at fortnightly intervals and further monitoring and augmentative releases during probable periods of pest build up is required.

The population fluctuation of the pest shows distribution throughout the year with peak in summer months. In all the locations after the pest suppression in the first year of parasitoid release, slight build up of the pest begins at the post monsoon period during October–November. Proper monitoring for the build up of pest population is the most important step to be adopted in endemic areas. Clipping of the affected leaflets and cutting and burning one or two highly affected lower whorl of leaves, if they do not support the bunch, is helpful to prevent the pest from attaining severe proportion. During early stage of pest build up, one or two inoculative release of parasitoids is essential to maintain the pest/ parasitoid ratio.

The present study shows more light on the importance of regular surveillance and monitoring on the pest parasite buildup and is suggestive of triggering augmentative release of appropriate parasitoids at the early stage of pest build up. This technology of non pesticidal IPM for field management of *O. arenosella* can be well integrated in the organic farming technology for better crop protection.

References

Chandrika Mohan and Nair, C.P.R. 2002. A case study on demonstration of Bio-intensive IPM for the control of coconut leaf eating caterpillar, *Opisina arenosella* Walk. Paper presented in "*Symposium on biological control of lepidopteran pests*- July 17-18, 2002, Bangalore, India, Abstract of papers. 64p.

George, M.V., Sathiamma, B. and Vijaya Kumar, K. 1982. A technique to estimate field populations of coconut leaf eating caterpillar. pp. 638 - 643. In: *Proceedings PLACROSYM - V, (Eds.* Bavappa, K.V.A. and Iyer R.D.), CPCRI, Kasaragod, Kerala, India.

Nair, C.P.R., Daniel, M. and Ponnamma, K.N. 1997. Integrated Pest management in Palms. Coconut Development Board, Cochin. 30p.

Nirula, K.K. 1956. Investigations on the pests of coconut palm. Part III. *Nephantis serinopa* Meyrick. *Indian Coconut J.* **9**: 101-131.

Pillai, G.B. and Nair, K.R. 1993. A checklist of parasitoids and predators of *Opisina arenosella* Wlk. on coconut. *Indian Coconut J.* **23**(9): 2-9.

Sathiamma, B. 1993. *Opisina arenosella* Wlk., the leaf eating caterpillar of coconut palm. *Tech. Bull.* 10; Central Plantation Crops Research Institute, Kasaragod, 12 p.

Sathiamma, B., Pillai, G.B., Abraham Jose, Bhat, S.K., Jayapal, S.P. and Nair, K.R. 1987. Norms for release of larval, pre-pupal and pupal parasitoids of *Opisina arenosella* Wlk. the leaf eating caterpillar of the coconut palm. *J. Plantn Crops* **15** (2):118-122.

Sathiamma, B., Sabu, A.S. and Pillai, G.B. 1996. Field evaluation of the promising species of indigenous parasitoids in the biological suppression of *Opisina arenosella* Wlk. the coconut leaf eating caterpillar. *J. Plantn Crops* **24**:9-15.

43

Management of Rhinoceros Beetle *Oryctes rhinoceros* L. in Coconut Gardens by Biocontrol Method

C.P.R. Nair, P. Rajan, Chandrika Mohan and Murali Gopal

INTRODUCTION

The rhinoceros beetle, *Oryctes rhinoceros* L. (Coleoptera: Scarabaeidae) is a major pest of coconut with ubiquitous distribution. It infests coconut palm at all stages. The adult beetle bores into the unopened spear leaf and also the inflorescence in severe cases of pest attack. Damage by the pest causes reduction in yield to the extent of nearly 6%. The injury by this pest leads to subsequent infestation by red palm weevil and invasion by fungal pathogens which are fatal to the palm.

Biological control using the fungal pathogen *Metarhizium anisopliae* and viral pathogen *Oryctes rhinoceros* virus (OrV) (earlier known as *Oryctes* baculovirus) is the most important component in the IPM of this pest (Nair *et al.*,1997). Mohan *et al.(*1983) reported 54% natural virus infection in adult beetles for the first time in India. The biological suppression by OrV has been successfully field demonstrated in many parts of the world (Purrini, 1989; Mohan *et al.*, 1989; Mohan and Pillai, 1993; Jacob,1996). OrV is a self perpetuating pathogen and is mainly transmitted through the contaminated food. Release of OrV in an already infected contiguous area in Kerala also showed significant reduction in pest population in 3 years. (Biju *et al.,*1995). *Metarhizium anisopliae,* also known as the green muscardine fungus is pathogenic to all stages of *Oryctes rhinoceros* (Nirula *et al.*,1956). Methods for mass production of this fungal pathogen using cheaper substrates like cassava chips rice bran mixture (Mohan and Pillai, 1982) and coconut water (Danger *et al.,* 1991) were developed at CPCRI by earlier investigations.

Large scale field demonstration of biocontrol of *Oryctes rhinoceros* using the fungal and viral pathogens was carried out in an area of 1,000 ha homestead coconut gardens in Krishnapuram village (Alappuzha Dist., Kerala) and 1,400 ha in Mogral Puthoor (Kasaragod Dist., Kerala) during 1999-2002. The operational areas were divided into clusters of 100 ha, and in each cluster 25 ha area in the centre was selected for recording observations.

M. anisopliae was mass multiplied in coconut water as per method developed by Danger *et al.* (1991). All the breeding sites in the operational areas were treated with *Metarhizium* spore suspension @ 5×10^{11} spores/m^3 in June-July during the demonstration period. *Oryctes rhinoceros* virus culture maintained in the biocontrol laboratory of CPCRI, Regional station, Kayamkulam was used for mass multiplication of the virus for field release. The viral culture was mass produced in *Oryctes* grubs. Viral inoculum for infecting adults for field release was prepared by homogenizing mid gut of infected grubs (1g of infected mid gut homogenized in 100 ml of phosphate buffer). *Oryctes* adults were infected by oral feeding the inoculum with a syringe or allowing the adults to wade through the inoculum for half an hour. The beetles were then kept in plastic boxes under starvation for 12-24 hours and released in the field preferably at dusk. The microbial agents were released only during the first and second year of the demonstration. In each cluster a total of 600 infected beetles were released over a period of first two years in four representative release sites so as to effect uniform spread of the viral inoculum in the breeding sites/native beetle population.

Pre treatment observation on damage by rhinoceros beetle in various clusters in Krishnapuram reveled that leaf damage due to the pest ranged from 17.0 to 28.0% with an average incidence of 21.7 % and spindle damage ranged between 14.0 to 25.0% with an average of 18% (Fig 43.1 and 43.2 and Table 43.1). Observations on leaf and spindle damage during the post treatment period showed steady decreasing trend.

Table 43.1. Average percentage leaf and spindle damage percentage at Krishnapuram (1999-2002)

Period	*Damage by rhinoceros beetle (%)(n = 10)*	
	Leaf	*Spindle*
Pre-treatment (1999)	21.66 ± 3.6	18.00 ± 3.74
Post–treatment (2002)	5.40 ± 1.35	3.70 ± 1.42
Percent reduction over pre treatment	75.1%	79.44%
t statistic (P = 0.05)	14.07*	15.90*

*significant at P = 0.05(table t value 2.262)

There was a reduction of 75.1% leaf damage and 79.4% spindle damage in a period of three years in the demonstration area at Krishnapuram.

At Mogral Puthur, the pre treatment observations on pest incidence recorded an average of 36.0% leaf damage, 12.0% spindle damage and 16.0% spathe damage (Table 43.2). Post-treatment observations on the above parameters recorded from the sample clusters

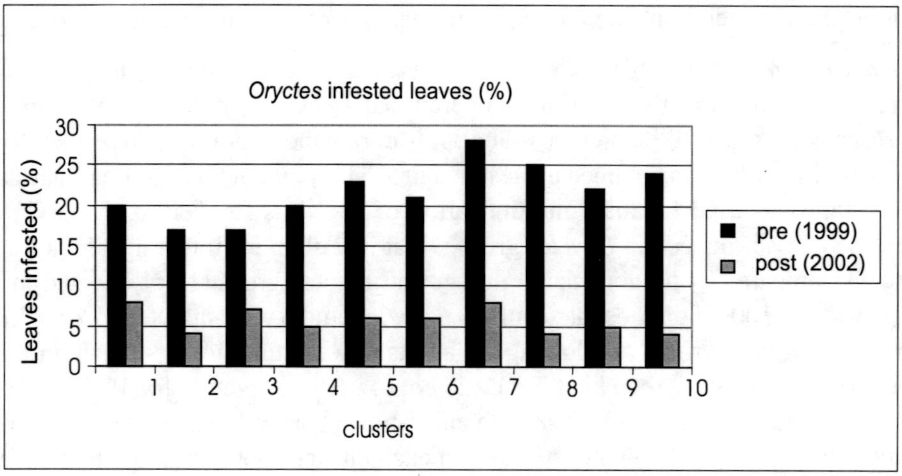

Fig. 43.1. Leaf damage by *Oryctes* at demonstration area (Krishnapuram)

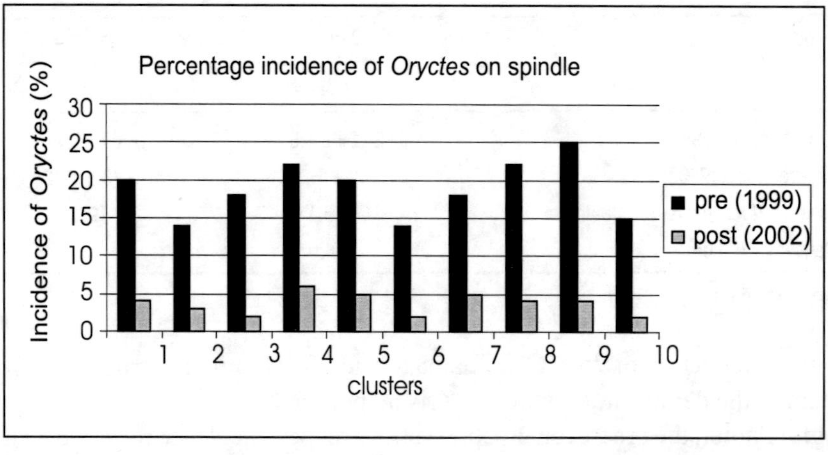

Fig. 43.2. Spindle damage by *Oryctes* at demonstration area (Krishnapuram)

after three years showed high reduction in pest incidence. On an average there was 66.6% reduction in leaf damage, 95.8% reduction in spindle damage and 62.5% reduction in spathe damage (Fig. 43.3, 43.4, 43.5 and Table 43.2).

Table 43.2. Incidence of rhinoceros beetle in Mogral Puthoor (1999-2002)

Period	*Damage by rhinoceros beetle (%) (n = 14)*		
	leaf	*spindle*	*spathe*
Pre-treatemnt (1999)	36.0±10.03	12.0 ±3.38	16.0 ±5.08
Post-treatment (2002)	12.0 ±3.24	0.58 ± 0.2	6.0 ±1.42
Percent reduction over pretreatment	66.6	95.8	62.5
t statistic (P = 0.05)	8.05*	12.74*	8.03*

·significant at P=0.05 (table t value 2.16)

The impact of re-release of *Oryctes rhinoceros* virus in the mainland of India had been demonstrated in smaller areas earlier in Thrissur District (Biju *et al* ,1995). Release of the viral pathogen of the pest over large and contiguous areas has not been so far done in the mainland. The demonstration has resulted in significant reduction in the pest damage by the combined use of two potential biocontrol agents of rhinoceros beetle. The results suggest necessity for similar demonstrations concurrently over contiguous areas in major coconut growing states of India so that effective management of the pest is achieved without disrupting the ecosystem. Large-scale adoption and popularization of the technology is viable and advantageous in the wake of organic farming getting popularized in coconut plantations that accommodate a variety of intercrops and animal components serving the food requirements of millions in the world.

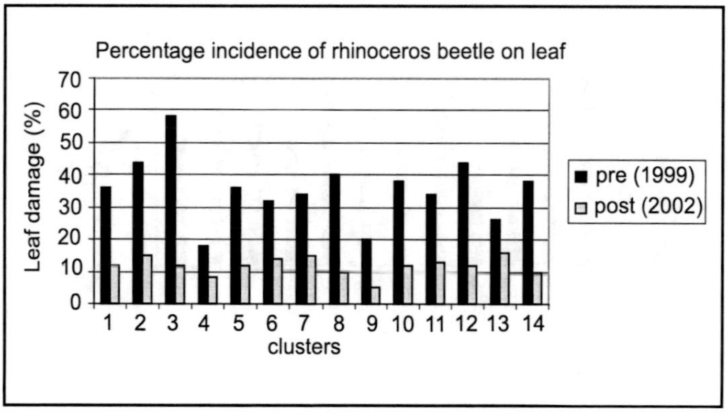

Fig. 43.3. Percentage leaf damage by rhinoceros beetle at demonstration area (Mogral Puthoor)

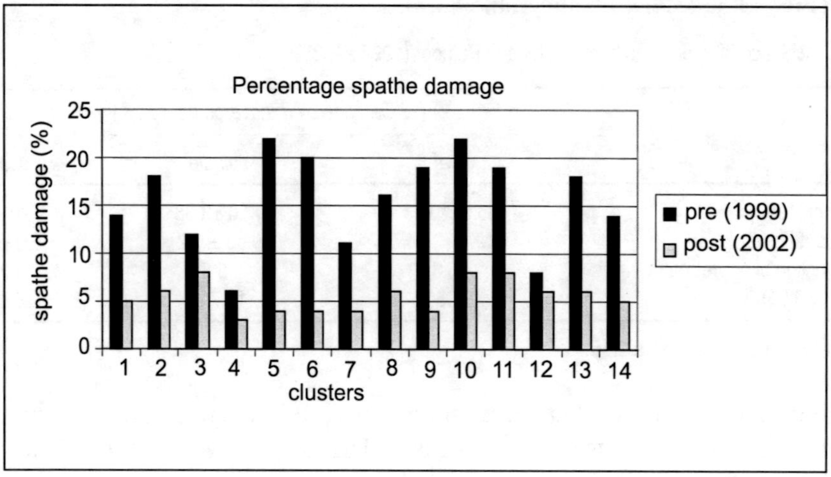

Fig. 43.4. Percentage of spathe damage by rhinoceros beetle at the demonstration area (Mogral Puthur)

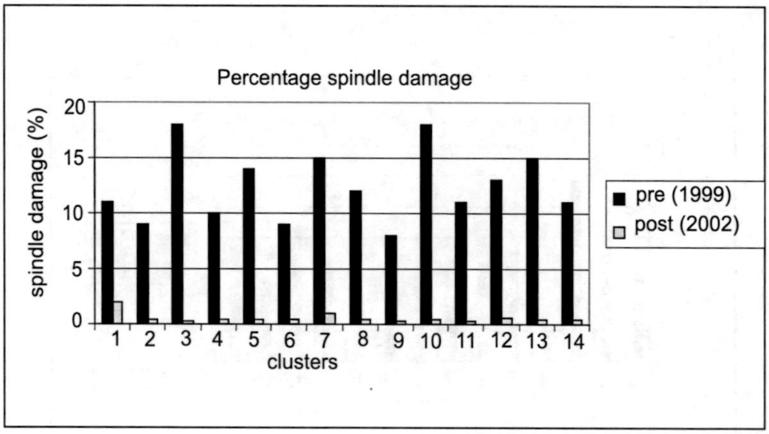

Fig. 43.5. Percentage incidence on spindle by rhinoceros beetle at the demonstration area (Mogral Puthur)

References

Biju, B., Sudhadevi, K., Danger, T. K. and Sathiamma, B. 1995. Biological suppression of *Oryctes rhinoceros* by re-release of *Baculovirus oryctes* in an infected contiguous area. *J. Plantn Crops* **23**:62-63.

Dangar, T.K., Geetha, L., Jayapal, S.P. and Pillai, G.B. 1991. Mass production of entomopathogen *Metarhizium anisopliae* in coconut water wasted from copra making industry. *J. Plantn Crops* **19**(1):54-69.

Jacob, T.K. 1996. Introduction and establishment of Baculovirus for the control of rhinoceros beetle *Oryctes rhinoceros* (Coleoptera: Scarabaeidae) in the Andaman Islands (India). *Bull. Ent. Res.* **86**:257-262.

Mohan, K.S. and Pillai, G.B. 1982. A method for laboratory scale mass cultivation of *Metarhizium anisopliae. Folia Microbiol.* **27**:281-283.

Mohan, K.S. and Pillai, G.B. 1993. Biological control of *Oryctes rhinoceros* (L.) using an Indian isolate of *Oryctes* baculovirus. *Insect Sci. Applic.* **14** (5-6):551-558.

Mohan, K.S., Jayapal, S.P. and Pillai, G.B. 1983. Baculovirus disease in *Oryctes rhinoceros* population in Kerala. *J. Plantn Crops* **11** (2):154-161.

Mohan, K.S., Jayapal, S.P. and Pillai, G.B. 1989. Biological suppression of coconut rhinoceros beetle *Oryctes rhinoceros* (L.) in Minicoy, Lakshadweep by *Oryctes* baculovirus - impact on pest population and damage. *J. Plantn Crops* **16** (Suppl.):163-170

Nair, C.P.R., Daniel, M. and Ponnamma, K.N. 1997. Integrated Pest management in Palms. Coconut Development Board, Cochin. 30 p.

Nirula, K.K., Radha, K. and Menon, K.P.V. 1956. The green muscardine disease of *Oryctes rhinoceros* L. II. The causal organism. *Indian Coconut* Journal **9**(2):83-89.

Purrini, K. 1989. Baculovirus *oryctes* release into *Oryctes monoceros* population in Tanzania, with special reference to the interaction of virus isolates used in our laboratory infection experiments. *J. Invertebrate Pathology* **53**(3): 285-300.

44

Evaluation of Organic Pest Management Techniques Against Tea Mosquito Bug in Cashew

D. Sundararaju, P. Shivarama Bhat and T.N. Raviprasad

INTRODUCTION

Mirid bugs belonging to genus *Helopeltis* called tea mosquito bugs (TMB) are the most serious foliage and fruit pests of cashew in India. Under west coast condition, three species of *Helopeltis* (*H. antonii* Sign. *H. theivora* Waterh and *H. bradyi* Waterh.) infest cashew, whereas in other regions, only one species (*H. antonii*) exists on cashew. However, in all regions, *H. antonii* is the dominant species causing severe damage. Its feeding injury leads to necrosis which is attributed as one of the causes of infection and manifestation of blossom blight and dieback diseases due to various fungal pathogens resulting in maximum yield loss. On cashew, its build up commences during October–November synchronising with the emergence of new flushes, after cessation of the monsoon. The population reaches a peak during January, when the trees are in full bloom. The pest prevails predominantly in the plantation till May and subsequently exists in negligible number during monsoon period (June–September) especially in older plantations (Sundararaju and Sundararbabu, 1999a). Under severe out-break situation, only insecticidal application rescues the cashew crop from ravage of TMB. As the major cashew importing countries (USA and EU) have already imposed restrictions on cashew kernels having pesticidal residues, research on non-chemical pesticides techniques viz., host plant resistance, botanical pesticides and biological control were done at National Research Centre for Cashew (NRCC) since 1987. The information available at NRCC on above aspects was compiled and presented in this paper.

Host Plant Resistance

A total of 76 cashew accessions consisting of released varieties, promising accessions and tolerant accessions located in TMB hot spot area (endemic area), germplasm and progeny evaluation trial were screened during 1988-92 under laboratory through cage screening technique in order to confirm genetic basis of resistance. The results revealed that all accessions were susceptible (Sundararaju and John, 1993 and Sundararaju, 1999). Subsequently, 392 cashew accessions available in the National Cashew Field Gene Bank (NCFGB) were also screened for resistance to TMB and no resistance could be detected.

Bhaskara variety, a promising selection was collected from severe TMB infested hot spot area of Goa state as a tolerant type to TMB. In order to confirm its tolerance to TMB under field condition, thirty cashew grafts were planted in a plot with a spacing of 6 × 6m during 1989 and the whole plot was maintained without any insecticidal application. In each year, the overall TMB damage of each tree was also assessed qualitatively as; very low (<10%), low (10–25%), moderate (26–50%) and severe (>50%). Tree-wise-yield data was collected from 1992-93–2001-02 cropping season (i.e., from 2nd to 11th harvest). The damage of TMB was in the range of very low to severe in the initial years (upto six years of age) and subsequently fluctuated between very low to moderate and moderate to severe). The varied level of TMB damage during different years was mainly due to clumping/ patchy distribution of TMB. Since, Bhaskara variety has the mid-season flowering habit, clear cut delay of one to two months in buildup of population was seen over early flowering varieties especially in older plantations and thus, it has got potential to escape from TMB attack as a phenomenon of pseudo resistance when compared to early varieties (Sundararaju, 2005).

Table 44.1. Yield and TMB (tea mosquito damage) in various ages of Bhaskara cashew variety under unprotected condition

Year (Planted during 1989)	*Range of TMB damage*	*Yield in kg*		
		Per tree		*Per ha***
		*Mean**	*Range*	
2nd harvest, 1992-93	Very low – Severe	1.1	0.0–4.6(3)	306
3rd harvest, 1993-94	Very low – Severe	0.8	0.0–4.1(4)	222
4th harvest, 1994-95	Very low – Severe	4.4	0.5–7.1	1223
5th harvest, 1995-96	Very low – Moderate	3.6	1.9–5.9	1001
6th harvest, 1996-97	Very low – Moderate	7.2	3.2 11.5	2002
7th harvest, 1997-98	Very low – Moderate	6.2	3.1–13.3	1724
8th harvest, 1998-99	Moderate – Severe	4.3	2.4–6.4	1195
9th harvest, 1999-00	Very low – Moderate	9.4	5.3–17.2	2613
10th harvest, 2000-01	Moderate – Severe	6.6	2.1–12.9	1835
11th harvest, 2001-02	Very low – Moderate	10.7	5.0–19.0	2975

* Mean of 21 trees; **Planted in a spacing of 6 x 6m.
Figures in brackets indicate no. of trees with complete loss in yield due to severe TMB attack.

From the yield data of sixth to eleventh harvest, it was found that whenever the TMB damage was low to moderate, the yield level was not affected drastically when compared to damage level of moderate to severe. Further, this variety has also yielded more than 1 t/ha from fourth harvest onwards and the highest yield of 2.975 t/ha (10.7 kg/tree) was obtained during 11th harvest (Table 44.1) (Sundararaju *et al.,* 2006).

Evaluation of Plant Products (botanicals)

A number of plant products *viz.*, Neem oil, Cotton seed oil, *Butea frondosa* L. leaf extract (LE), *Adathoda vasic* L., LE, *Pongamia* oil suspension (OS), *Pongamia* OS in lime solution, *Pongamia* solvent extract (in n- hexane), *Calophyllum* oil, *Vitex negundo* L., LE, *Thevatia neriifolia* Juss. LE, *Strychnous nux vomica* L., LE and fruit extract, tobacco decoction, *Annona* seed extract (commercial formulation from M/s Godrej Ltd.), Palmrosa oil (RAZE®), Neem Commercial formulations (Nimbicidine, RD-9 Repllin, Godrej Achook, Limanool,), Dillapiole extract, Garlic extract (GB+) and Phalada III C [extracts of Seeds of *Saphora japonica and Annona reticulata* L. + leaf extracts of *Adathoda vasica, Sapindus trifoliatus* L. and *Pongamia pinnata* (L.)] were evaluated at different concentrations ranging from 0.1 to 6.0 % (Table 44.2) against TMB during 1992 -2006 and none were found to be effective (NRCC, 2002; Raviprasad *et al*., 2002; Sundararaju, 2004a; Bhat and Raviprasad, 2007). Even smoking of the canopy in the early morning with dried leaves of cashew or *V. negundo* neither caused any mortality of TMB nor reduced its damage on cashew (NRCC, 2000). Though neem contains repellent and insecticidal properties due to secondary metabolites (limonoids/tetratriterpenoids), TMB causes severe damage on neem and overcomes those phytochemicals through its salivary mechanism. It is possible as its salivary glands secrete various detoxifying salivary enzymes (catechol oxidase, peroxidase and catalase) and free amino acids (Sundararaju and Sundarababu, 1996 and 1999b). Because of this phenomenon, all the plant products tested against TMB were found to be ineffective.

Table 44.2. List of plant products (botanicals) evaluated against tea mosquito bug at National Research Centre for Cashew, Puttur.

Sl. No.	*Name of plant products evaluated*	*Concentration tested (%)/ dosage*
1.	Neem oil	5.0
2.	Cotton seed oil	5.0
3.	*Butea frondosa* L. leaf extract	5.0
4.	*Adathoda vasica* Nees. leaf extract	5.0
5.	*Pongamia* oil suspension	5.0 & 6.0
	Pongamia oil suspension in lime solution	2.0
	Pongamia oil suspension in lime solution	4.0
	Pongamia solvent extract (in n- hexane)	5.0
6.	*Calophyllum* oil	5.0

(*Contd.*)

Sl. No.	*Name of plant products evaluated*	*Concentration tested (%)/ dosage*
7.	*Vitex negundo* L. leaf extract	5.0
8.	*Thevatia neriifolia* Juss. leaf extract	5.0
9.	*Strychnous nux vomica* L. leaf extract	5.0
	S. nux vomica Fruit extract	5.0
10.	Tobacco decoction	100g/3 litre water
11.	*Annona* seed extract (commercial formulation from M/s Godrej Ltd.)	5.0
12.	Palmrosa oil (RAZE®)	0.5
13.	Neem Commercial formulations	
	Nimbicidine	1.0
	RD-9 Repllin	1.0
	Godrej Achook	1.0
	Limanool	1.0
14.	Dillapiole extract	1.0
15.	Garlic extract (GB+)	0.4
16.	Phalada III C [extracts of Seeds of *Saphora japonica and Annona reticulata* L. + leaf extracts of *Adathoda vasica, Sapindus trifoliatus* L. and *Pongamia pinnata* (L.)]	0.1
17.	Fumigation of *V. negundo* / cashew leaves under cashew canopy	30 minutes twice/ week

Biological Control

This pest was found to be regulated under natural condition mainly through its egg parasitoids (*Telenomus* sp., *Ufens* sp. and *Chaetostricha* sp.). Among them *Telenomus* sp. and *Ufens* sp. are dominant in west coast of India and coastal Tamil Nadu with highest natural parasitism upto 70.8 and 47.4%, respectively. These parasitoids are specialist type and not amenable for mass multiplication under conventional laboratory rearing methods (Sundararaju and Sundarababu, 1999a and 1999c). Therefore, they could not be exploited under inundative release methods. However, their activities are naturally enhanced under favourable weather condition (increased minimum temperature and relative humidity) during vulnerable period (November-February) (Sundararaju, 2003). During the outbreak of TMB, among the cashew trees which had nests and intensive activities of *Oecophylla smaragdina* Fabr., the infestation of TMB was in the range of low to moderate; whereas remaining trees had severe TMB infestation. But under natural condition, *O. smaragdina* did not establish on majority of cashew trees in spite of cashew plantation being never sprayed with insecticides or in cashew plantation having prolonged withdrawal of insecticide application. When the impact of various arthropod predators on incidence of TMB was studied under unsprayed situations, the results indicated that among total arthropod predatory population, spiders accounted maximum upto two third of total population and appeared to be the dominant predators in cashew ecosystem. Therefore, their role in reducing the TMB damage was critically analysed. Actually, the mean spider population varied from 22-31 spiders per 100 cashew panicles and in spite of this, out break of TMB had occurred. No significant relationship between TMB damage and spiders was observed (Sundararaju, 2004b).

Sex Pheromone Behaviour

The information collected in cashew plantation clearly indicated that the male population of TMB can be detected within 15 minutes and highest population of 18 males were collected on single bait cage containing virgin females. In the entire trapping period of 90 minutes, 132 males were collected (Sundararaju and Sundarababu, 1999d). Thus it provides substantial scope in future for exploiting this sex pheromone behaviour for mass trapping in the plantations of organic cashew.

CONCLUSION

Even though research on organic pest management techniques (non-chemical pesticides techniques) viz., host plant resistance, botanical pesticides and biological control were done at National Research Centre for Cashew, Puttur, Karnataka for the management of tea mosquito bug on cashew since 1987, under host plant resistance only, some lead could be obtained. 'Bhaskara' cashew variety appeared to be a promising escape (pseudo resistant) variety and high yielding and it can be popularised in future organic farming propositions.

References

Bhat, P.S. and Raviprasad, T.N. 2007. Managementt of tea mosquito bug on cashew *Helopeltis antonii* Sign. with newer insecticides/products. *Extended Abstract. pp.80-81.* National Seminar on Research, Development and Marketing of Cashew. ICAR Research Complex, Goa, India.

NRCC (National Research Centre for Cashew.), 2000. Annual report for 1999-2000, Puttur, Karnataka, 70p.

NRCC (National Research Centre for Cashew.), 2002. Annual report for 2001-02, Puttur, Karnataka, pp. 40-42.

Raviprasad, T.N., Sundararaju, D. and Bhat, P.S. 2002. Efficacy of botanicals against *Helopeltis antonii* Sign., (Miridae: Hemiptera) infesting cashew. *In: Plantation Crops Research and Development in the New Millennium* (Eds. Rethinam, P., Khan, H.H., Reddy, V.M., Mandal, P.K. and Suresh, K.) pp.485-488. Coconut Development Board, Kochi, India.

Sundararaju, D. 1999. Screening of cashew accessions to tea mosquito bug *Helopeltis antonii* Sign. (Heteroptera : Miridae). *The Cashew*, **13**(4):20-26.

Sundararaju, D. 2003. Behaviour of egg parasitoids of *Helopeltis antonii* Sign. (Heteroptera : Miridae) in relation to weather parameters. *Journal of Pest Management in Horticultural Ecosystem* **9**:113-116.

Sundararaju, D. 2004a. Evaluation of certain insecticides against tea mosquito bug on cashew. *The Cashew* **18(2)**:22-26.

Sundararaju, D. 2004b. Influence of spiders and insect predators on incidence of tea mosquito bug in cashew. *The Cashew* **18 (1)**:9-13.

Sundararaju, D. 2005. Seasonal abundance of tea mosquito bug and extent of damage on cashew. *J. Plantn Crops* **33**:53-58.

Sundararaju, D. and John, N.J. 1993. Susceptibility of cashew accessions to *Helopeltis antonii* Sign. (Heteroptera : Miridae) in the pre-flowering phase. *J. Plantn Crops,* **21**:50-53.

Sundararaju, D. and Sundara Babu, P.C. 1996. Neem pest not a mystery. *Nature,* 381:108.

Sundararaju, D. and Sundara Babu, P.C. 1999a. *Helopeltis* spp. (Heteroptera : Miridae) and their management in plantation and horticultural crops of India. *J. Plantn. Crops,* **27**:155-174.

Sundararaju, D. and Sundara Babu, P.C. 1999b. Morphology and contents of salivary glands of *Helopeltis antonii* Sign. (Heteroptera : Miridae). *J. Ent. Res.* **23**:41-46.

Sundararaju, D. and Sundara Babu, P.C. 1999c. Survey and tritrophic interactions between host plants, *Helopeltis antonii* Sign. (Heteroptera : Miridae) and its egg parasitoids. *J. Biol. Control,* **13**:19-23.

Sundararaju, D. and Sundara Babu, P.C. 1999d. Trapping male tea mosquito bug *Helopeltis antonii* Sign. using virgin females. *Pest Management in Horticultural Ecosystem,* **5**:102-106.

Sundararaju, D Yadukumar, N., Bhat, P.S., Raviprasad, T.N., Venkatakumar, R. and Sreenath Dixit. 2006. Yield performance of 'Bhaskara' cashew variety in coastal Karnataka. *J. Plantn Crops,* **34**:216-219.

45

Field Evaluation of Methyl Eugenol Traps for Controlling Mango Fruit Flies [*Bactrocera dorsalis* (Hendel)] in Homesteads

T. Sivakumar and P. Anithakumari

INTRODUCTION

Mango (*Mangifera indica*) is the most common popular fruit tree in homesteads of Kerala especially Alleppey district. The State owns certain traditional varieties with good aroma and taste. The fruit crop is infested by more than forty five insect pests (Nair, 1999). Fruit fly, [*Bactrocera dorsalis* (Hendel)] is the major pest, attacking fully matured and ready to ripe fruits. 20–30% loss to produce was recorded due to fruit fly attack (Kalloo, 2005). The adult female flies oviposit their eggs just underneath the skin of fruits and developing maggots will feed on the sweet pulp remaining within. Fully matured maggots pupate in soil. Spraying chemicals on trees, standing near to residence and water bodies is difficult as it will cause direct contamination. The botanical method of management was also found difficult as raw material availability is very scarce during the flowering/fruiting period.

Testing of different fruit fly control measures was done during 2005-06 in Chingoli Panchayath through a farmer participatory approach under technical supervision of KVK-Alleppey. A preliminary study was conducted to record the present status of pest attack, damage and other socio-economic status of the participants. Fifteen farm families were selected in a ward comprising of 5 ha land and different management packages were followed. Spraying the trees with Malathion 0.1% + 2% sugar/jiggery, setting up of Ocimum and Methyl eugenol traps (ME were adopted. ME trap was prepared by impregnating plywood blocks of size 5 × 5 × 0.9 cm in a solution of ethyl alcohol, methyl alcohol and malathion in a ratio of 6:4:1 for two weeks and shade dried for another two weeks The treated blocks were hung inside a plastic bottle with 5 × 5 cm size windows on opposite sides and one trap was set for 3-4 trees. 20 g *Ocimum* leaves were crushed to extract juice and the juice + waste were taken in a coconut shell to which 3 g carbofuran 3 G and 10 g jaggery/sugar were added. Four traps were tried per fruiting tree.

The study showed that 'Kilichundan' variety recorded 75.25% damage by fruit fly followed by Neelum (67.67) and Moovandan (60.88) (Table45.1). The local traditional variety ('Nadan') characterized by peculiar aroma, taste, high fibre content and thick fruit rind showed least damage percentage by the pest (25.5). Eleven different varieties of mangoes were selected from the study area, each having its own taste and aroma. The variation in damage on different varieties may be due to sweetness, fruit rind thickness and fibre content of fruit.

Table 45.1. Fruit fly damage on different varieties recorded in participatory appraisal

Sl. No.	*Variety*	*Average percentage of damage*
1.	Vellari	63.0
2.	Moovanadan	60.88
3.	Pandi	30.0
4.	Nadan	25.5
5.	Neelum	67.67
6.	Kilichundan	75.25
7.	Appies	30.0
8.	Karpooram	50.0
9.	Perakka	25.0
10.	Salem	35.0
11.	Kappa	40.0

It was found that ME trap caught maximum number of flies (26.55) compared to *Ocimum* trap (6.45) (Table 45.2). ME trap persisted in the field for 90 days where as *Ocimum* traps lasted for only 4.5 days. ME traps lasted for one mango season and hence replenishing of traps could be avoided.

Plots with ME trap recorded 51.42% fruit damage before the treatment which got reduced to 25.46% after the treatment (Table 45.3). ME traps showed 50.27 % reduction in attack (Fig. 45.1). *Ocimum* trap plots recorded 52.8 and 47.65% damage before and after the treatments respectively, while control plots showed an increase in damage (1.4%).

Table 45.2. Average fruit fly catch in different traps

Treatments	*1MAS*	*2MAS*	*3MAS*	*Average fly/trap/week*
M.E. trap	6613	2502	1400	26.55
Ocimum trap	1808	1330	343	6.45

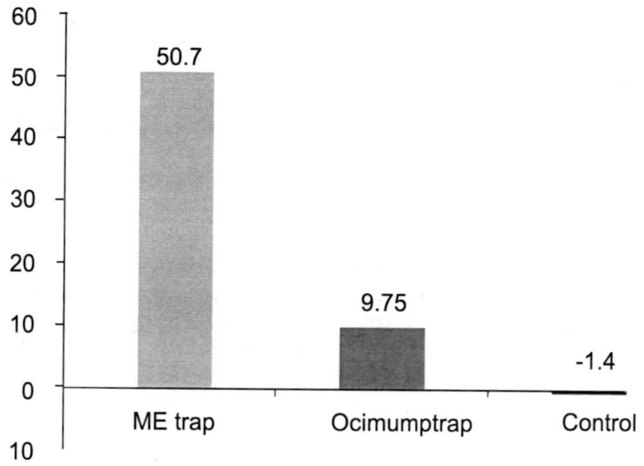

Fig. 45.1. Percentage reduction in fruit damage after the trial

Table 45.3. Effect of different treatments on fruit damage by fruit fly, *Bactrocera dorsalis* in mango

Treatments	*Percentage of fruit damage*	
	Before	*After*
ME trap	51.42	25.46
Ocimum trap	52.80	47.65
Control	57.30	58.10

The reduction in fruit damage was more with ME traps when compared to *Ocimum* traps. It may be due to the strong male annihilation by the ME traps and its persistency over the other traps tested. Similar observations were recorded by Thomas *et al.,* (2005) and Patel *et al.,* (2005).

The farmers were given a chance to appraise the effect and various aspects of the test traps. They ranked various aspects like the cost effectiveness, availability, efficacy, etc. based on their experience (Table 45.4). Farmers gave top rank for ME trap in their catching ability, followed by the efficacy of ME traps. The last rank was given to the cost of traps and availability of ME traps.

CONCLUSIONS

The study on fruit fly management using ME traps showed that use of one ME trap in homesteads with 3-4 mango trees is sufficient in reducing fruit damage. Festival season coupled with summer which coincides with flowering season of mango cuts the availability of *Ocimum* leaves. Use of ME traps along with other cultural measures of pest control such as raking up of soil and proper sanitation/disposal of affected fruits will lead to a non chemical alternative for fruit fly management in mango. The use of traps was found to be a farmer friendly, eco-safe method which suitably fit into the eco-friendly pest management practice against fruit fly.

References

Kallo, G. 2005. Fruit fly research in India. *Pest Management in Horticultural Ecosystems.* 11(2).

Nair, M.R.G.K., 1999. *Monograph on crop pests of Kerala and their management.* Kerala Agricultural University, Thrissur, Kerala.

Patel, R.K., Verghese, A., Patel, V.M., Joshi, B.K., Stonehouse, J.M. and Mumford, J.D. 2005 Bait, lure and cultural IPM of fruit flies in mango in Gujarat. *Pest Management in Horticultural Ecosystems.* **11**(2):155-158.

Thomas, J., Vidya, C.V., Verghese, A., Stonehouse, J.M. and Mumford, J.D. 2005 Bait and lure fruit fly control in mangoes in Kerala using manufactured and home made materials. *Pest Management in Horticultural Ecosystems.* **11**(2):153-154.

46

Entomopathogenic Fungi for the Biological Control of Important Pests of Vegetable Crops

T. Jiji, R. Praveena, A. Naseema and G. Suja

INTRODUCTION

Insect pests form a major threat to vegetable cultivation. Use of chemical pesticides for pest management leads to many environmental problems like residue hazards, disruption of natural control, etc., Vegetables are often consumed in non-processed form, making the problem of pesticide residues more prominent. In this context, biological control methods, which impart minimum harmful impact on the environment and non target organisms, are gaining importance.

Entomopathogenic fungi are reported from most of the insects. Under natural condition, fungi cause epizootics among the insect pests. More than 700 species of entomopathogenic fungi under 90 genera have been recorded (Charnely, 1989). Fungi, being able to invade directly through the cuticle, offer an advantage over the bacterial and viral pathogen. These fungi have wide host range, including many important pests of vegetable crops. *Beauveria* spp. were reported as entomopathogenic fungi as early as 1836 for the management of chinch bugs (Bassi, 1836). The earliest report of using this fungus in the management of vegetable pests was by Klochko (1969) who found *B. tenella* and *B. bassiana* as efficient pathogens of *Epilachna vigintioctopunctata* .

The paecilotoxin produced by *Paecilomyces lilacinus* was found to suppress the growth of insects (Milcami *et al*., 1989). The efficacy of *P. fumosoroseus* for the management of white flies was reported by Osborne and Landa (1992). Jiji *et al*. (2003) reported the occurrence of *Aspergillus candidus* from *B. dorsalis* infesting guava. Infection of scale insect by *A. candidus* was reported by Samsinakova and Kalalova (2006).

Though commercial formulations of fungi are available in the market, it fails to manage same pests worldwide. The local strains of the fungi are found to be pathogenic and efficient in pest management.

Pathogenicity tests of local isolates of entomopathogenic fungi *P. lilacinus* isolated from *B. cucurbitae*, *Aspergillus candidus* isolated from *B. dorsalis* and *B. bassiana* isolated from Okara leaf roller *S.derogata* were conducted for vegetable pest management was conducted in the laboratory of College of Agriculture, Vellayani. Different concentrations of test fungi were sprayed on target pests. For preparing the spore concentrations of the test fungus, seven mm fungal discs from well sporulating fungal cultures grown in petridishes were shaken well in 10 ml water blank. Second instar larvae were sprayed with different spore concentrations of the fungus using an atomizer. Two days old pupae and adults were taken for the study in the case of fruit flies. Based on insect mortality LC_{50} values of the fungus were worked out (Table 46.1).

Table 46.1. Cumulative percent mortality of test insects on the fifth day of application of *B.bassiana* in the laboratory

Test insect	*LC_{50}(spores/ml)*	*Mortality (%)*
Sylepta derogate	2.89 X 10	93.33
A. peponis	8.63 x10	66.67
D. indica	1.1 X10	83.33
Psara bipunctalis	1.41 X10	86.67
Henosepilachna vigintioctopunctata	3.8 X 10	63.33

Field trials were conducted at the Instructional Farm of the college. Okara and snake gourds were raised separately in field following the Package of Practices Recommendations of Kerala Agricultural University, without any pesticide application. The plants were naturally infested with *S. derogata, A. peponis*, *H. vigintioctopunctata* and *D. indica*. Pest count on okara and snake gourd plants were taken before spraying. Concentrations standardised under *in vitro* study was used for testing in the field condition. The talc based formulations developed also were tested in field.

Local isolates of fungi *Beauveria bassiana*, *Paecilomycess lilacinus* and *Aspergillus candidus* obtained were identified and deposited as ITCC 6063, ITCC 6064 and ITCC 5428 by the Indian Type culture collection, IARI, New Delhi. *B. bassiana* was found to infect pests of bhindi (*Sylepta derogata*), cucurbits (*Anadevidia peponis, Diaphania indica, Henosepilachna vigintioctopunctata*), Amaranthus (*Psara basalis*) and brinjal (*Psara bipunctalis* , *H. vigintioctopunctata*)

Larvae of lepidopterna pests showed various symptoms when sprayed with the spore suspension of *B. bassiana.* During the initial stages, the infected larvae were sluggish in their movement, became cream in colour and died from the next day onwards. The infected larvae failed to moult and early pupation also was observed. Malformed adults were emerged from the pupae, developed from infected larvae. In some cases adults did not emerge from the pupae. These pupae were black in colour, compared to brown colour of normal pupae and white and puffy mycelial growth of *B. bassiana* appeared on pupae. Mean mortality percentages of 93.33, 86.67, 83.33 and 66.67 were recorded for the larvae of *S. derogata, P. bipunctalis, D. indica* and *A. peponis*, respectively (Table 46.1) (Plates 46.1 to 46.5).

Plate 46.1. Infected larvae of *Sylepta derogate*

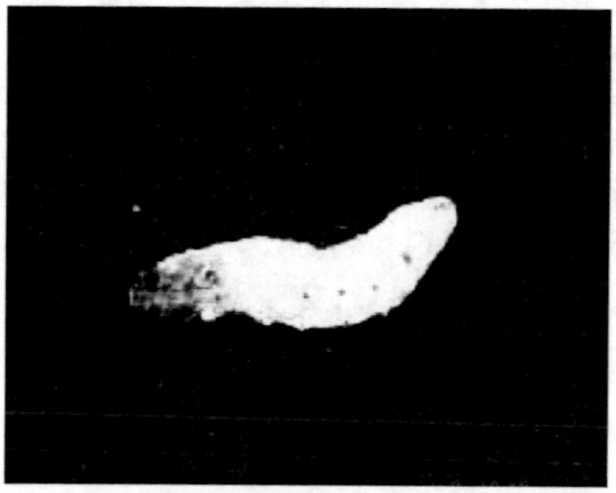

Plate 46.2. Infected larvae of *Psara bipunctalis*

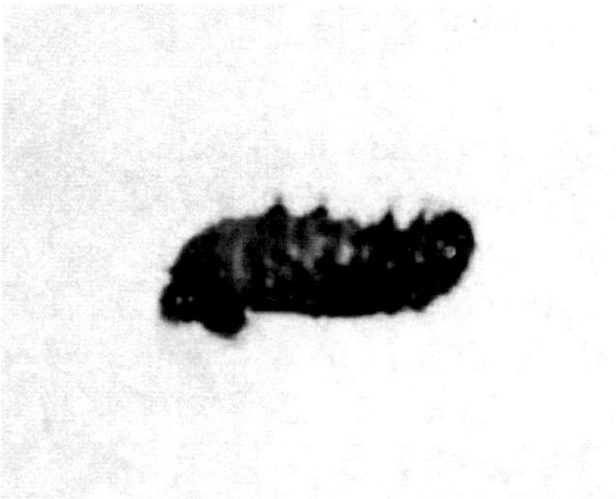

Plate 46.3. Infected larvae of *Anadevidia peponis*

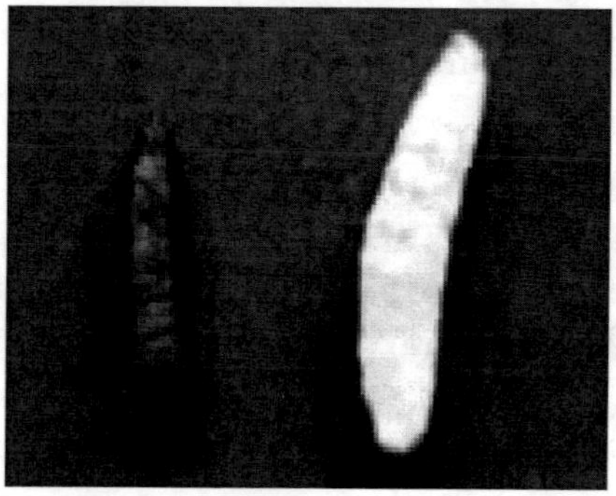

Plate 46.4. Infected larvae of *Diaphania indica*

Plate 46.5. Infected grubs of *Henosepilachna vigintioctopunctata*

Epilachna beetle (*H. vigintioctopunctata*), a voracious feeder of cucurbits and brinjal was found to be susceptible to the local strain of *B. bassiana*. Infected grubs showed less appetite. Later the colour of the spines changed to brownish black and finally whole grub became black in colour and dead. On dead grubs, white mycelial growth of the fungus appeared. Adult beetles when sprayed *with B. bassiana* did not develop any disease symptoms. A mean mortality of 63.33% was recorded on epilachna beetles.

Adult fruit flies sprayed with *P. lilacinus* died within 2-3 days. Pinkish violet puffy mycelial growth of the fungus appeared on dead insects. *P. lilacinus* @ 1.0×10^9 spores/ml caused 96.67% and 100% cumulative mortality in fruit flies on second and third day respectively. Fungus @ 1.0×10^7 was found to cause more than 50% cumulative mortality on second day LC_{50} values of *P. lilacinus* on *B. cucurbitae* were 5.0×10^6, 8.0×10^5, 7.0×10^5 spores/ml on 2nd, 3rd and 4th day, respectively. *P. lilacinus* treated pupae were black in colour and failed to emerge (Fig. 46.1).

Local strain of *B. bassiana* developed disease symptoms on *B. cucurbitae* and *B. dorsalis*. Adult fruit flies sprayed with *B. bassiana* died within 4-5 days. White puffy mycelial growth of the fungus covered these dead flies. *B. bassiana* @ 1×10^9 was found to cause 70%, 80% and 90% mortality on fourth, fifth and sixth day respectively. LC_{50} values of *B. bassiana* on *B. dorsalis* were 7.0×10^8, 2.0×10^7, 5.0×10^6 spores/ml on 3rd, 4th and 5th day, respectively (Fig. 46.2).

Adult fruit flies sprayed with *A. candidus* were found to die within three to four days. On the dead insect dull yellowish coloured fungal growth appeared within two days with sporulating heads of the fungus. Pupae treated within *A .candidus* also showed yellowish powdery growth of the fungus. *A. candidus* @ 1.25 x 10^9 spores/ ml caused 63.33%, 83.33% and 100% cumulative mortality in fruit flies on third, fourth and fifth day respectively (Fig. 3). LC_{50} values of *A.candidus* on *B.cucurbitae* were 1.29×10^8, 1.22×10^7, 2.27×10^6 spores/ml on third, fourth and fifth day respectively (Fig. 46.3).

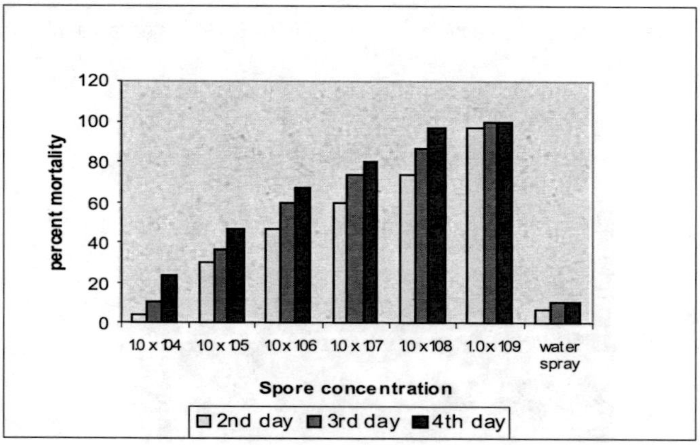

Fig. 46.1. Cumulative per cent mortality of *Bactrocera cucurbitae* treated with *Paecilomyces lilacinus*

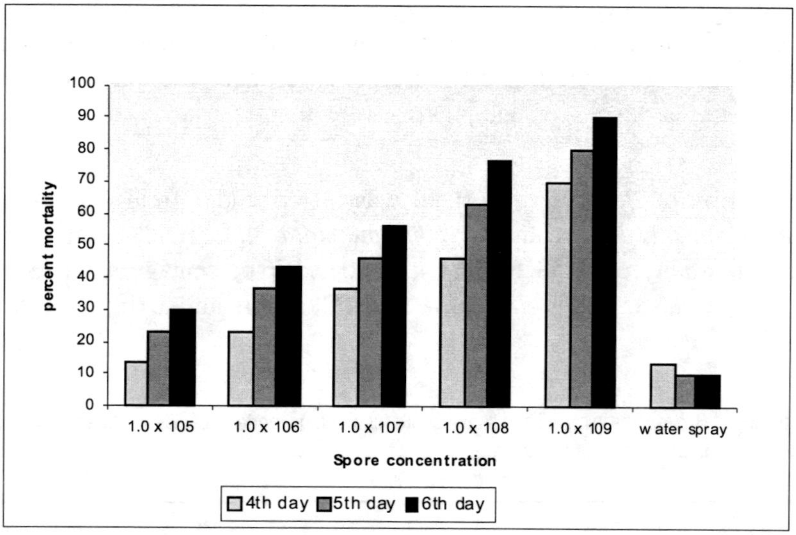

Fig. 46.2. Cumulative per cent mortality of *Bactrocera dorsalistreated* with *Seau veria*

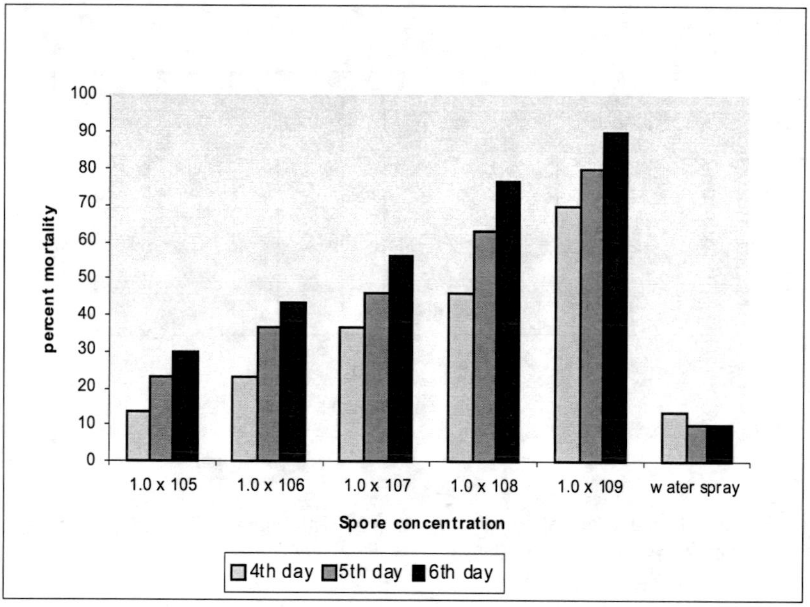

Fig 46.3 Cumulative percentage mortality of *B. curcurbitae* treated with *Aspergillus candidus*

The field studies revealed that the fungi *B.bassiana* and *P.lilacinus* were effective in field condition against major pests of vegetables. Among the treatments, spore suspension with wetting agent recorded the highest mortality when compared to other treatments. Spore suspension in starch solution and jaggery were also superior to control (Tables 46.2 and 46.3).

The formulation of *B. bassiana* (10%) was tested under field condition against *S. derogata, A. peponis, D. indica,* and *E.vigintiopunctata.* under field condition the fungus recorded a mean mortality of 69.58, 65.27, 64.36, 60 and 76 percentage against *S. derogata, A. peponis, D. indica,* and *E.vigintiopunctata* (Cucurbit and Brinjal), respectively (Fig. 46.4-7).

Table 46.2. Field testing of *B. bassiana* in bhindi and snake gourd (percentage mortality of insects) (mean of five replications)

Treatments	*S. derogata in bhindi*	*A. peponis*	*Epilachna*	*D.indica*
Spore suspension in water	23.081	29.349	23.081	29.719
Spore suspension in starch solution	30.79	41.482	30.79	29.349
Spore suspension in 10% jiggery	17.884	30.263	17.884	19.545
Spore suspension in 1% wetting agent (Teepol)	51.551	61.835	51.551	52.394
Water spray	0	0	0	0

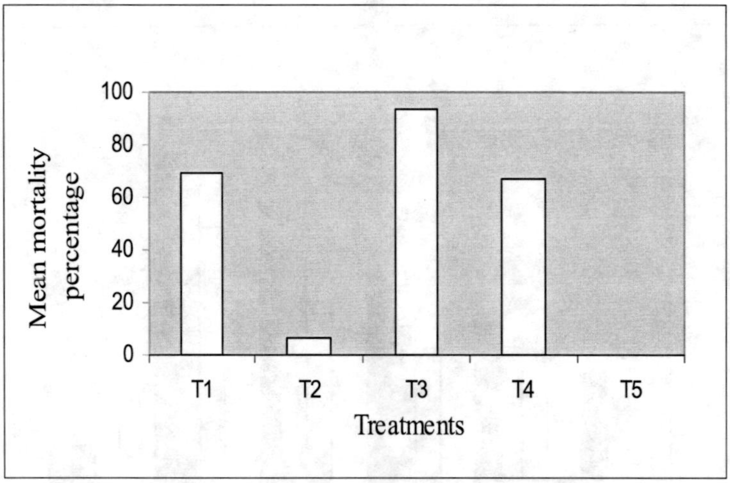

T1 – spraying 10% WP formulation of *B. bassiana;* T2 – Talc spray; T3 – POP (Malathion spray); T4 – 5% soap; T5 – Water spray

Fig. 46.4. Field evaluation of *Beauveria bassiana* against *Sylepta derogata*

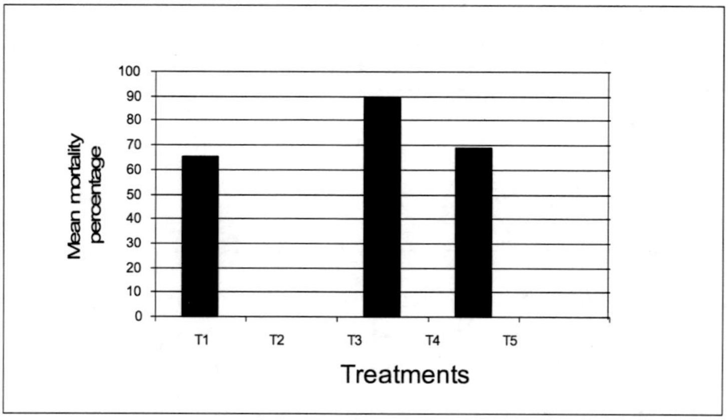

Fig. 46.5. Field evaluation of *B. basiana* against *A. peponis*

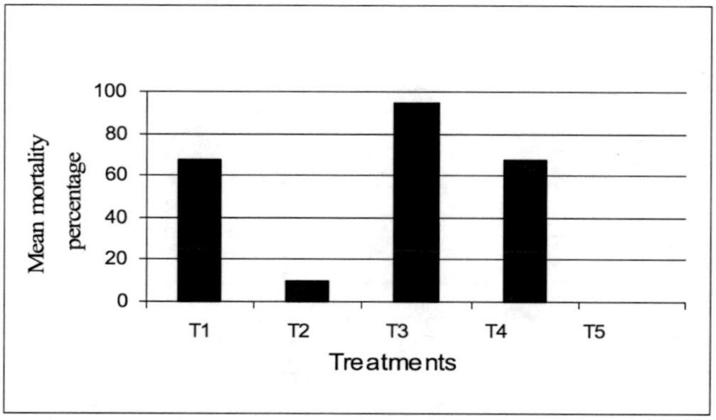

Fig. 46.6. Field evaluation of *Beauveria bassiana* against *Diaphania indica*

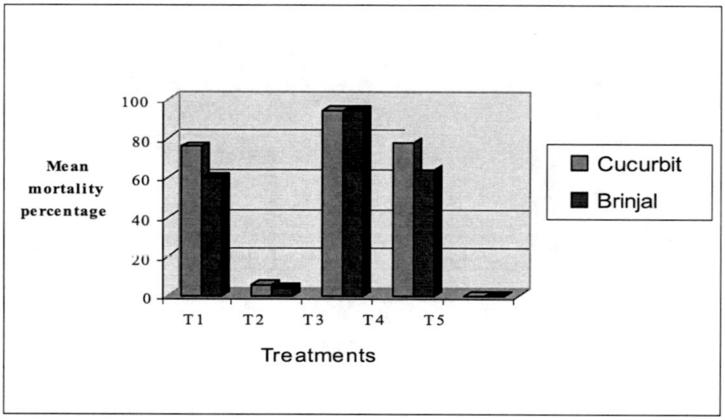

Fig. 46.7. Field evaluation of *Beauveria bassiana* against in cucurbit and brinjal

Table 46.3. Field testing of *P.lilacinus* in snake gourd (Percentage mortality of insects (mean of five replications)

Treatments	*A. peponis*	*Epilachna*	*D. indica*
Spore suspension starch solution	20.605	26.25	16.25
Spore suspension in 10% jiggery	41.665	26.665	35
Spore suspension in 1% wetting agent(Teepol)	25	51.668	7.5
Water spray	0	0	'0

The formulation of *P. lilacinus* (5%) was tested under field condition against *S. derogata, A. peponis, D. indica,* and *B.cucurbitae* pupae. under field condition and the fungus recorded a mean mortality % of 60.3, 58.66 and 62 against, *D. indica A. peponis*, and *E.vigintiopunctata,* respectively (Fig. 46.8-11). Pathogenicity of these fungi on target pests in field condition was already reported by Ramesh *et al. (*1999) and Wraight *et al.* (1998) and Jiji *et al,.* (2005).

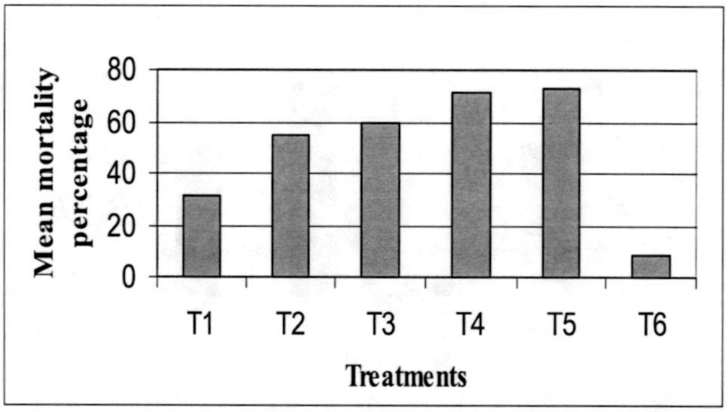

Fig. 46.8. Field evaluation of *Paecilomyces lilacinus* formulation *Bactrocera dorsalis* pupa

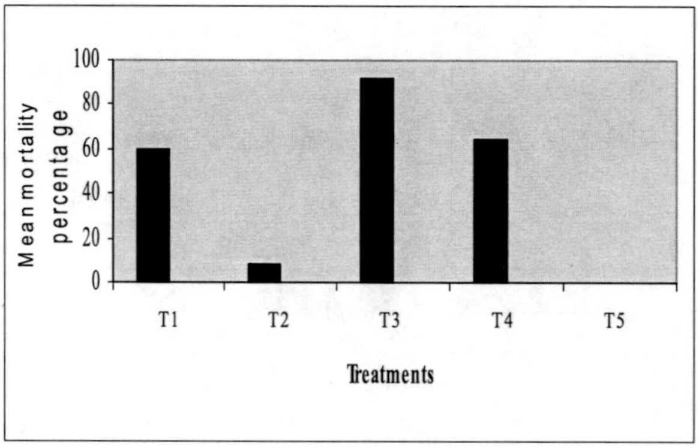

Fig. 46.9. Field evaluation of *Paecilomyces lilacinus* formulation against *Diphania indica*

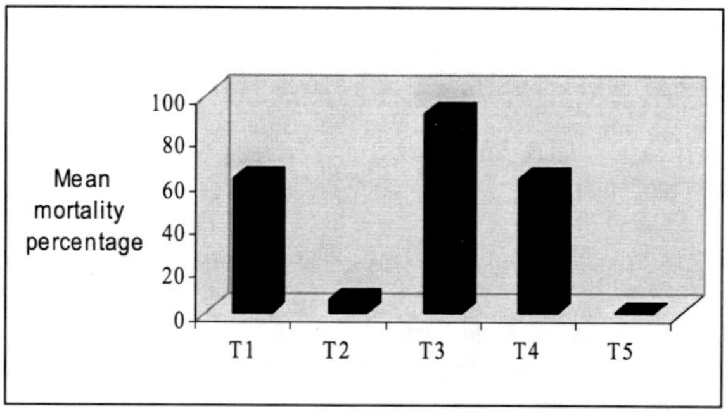

Fig.46.10. Field evaluation of *Paecilomyces lilacinus* against *Epilachna vigintiopunctata*

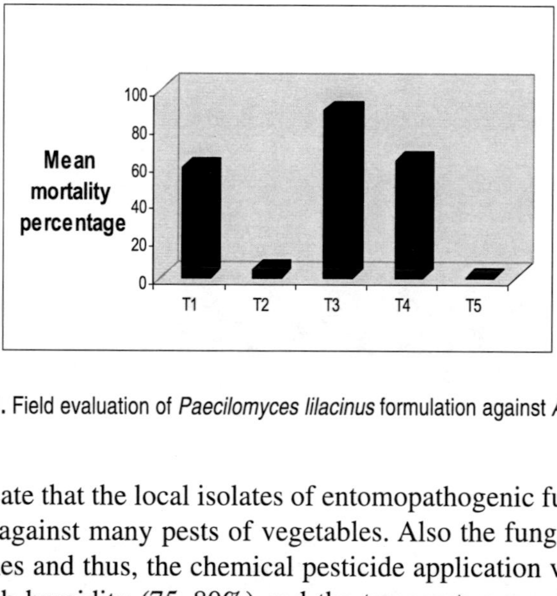

Fig.46.11. Field evaluation of *Paecilomyces lilacinus* formulation against *Anadevidia*

The results indicate that the local isolates of entomopathogenic fungi can be used an effective biocontrol against many pests of vegetables. Also the fungus can be integrated with IPM programmes and thus, the chemical pesticide application vegetable cultivation can be reduced. High humidity (75–80%) and the temperature condition (20–35 °C) in Kerala render a congenial atmosphere for the natural occurrence and multiplication of entomopathogenic fungus. Use of biopesticide is an eco-friendly practice and may help develop organic farming package in vegetables.

References

Bassi, A. Del Mal del Segno and Calcinaccio O Moscardino. 1836. Parte Seconda, Practica, Lodi, Orcesi, P. 1-67.

Charnley, A. K. 1989. .Mycoinsecticides : Present use and future prospects. Progress and prospects in insect control. *British Crop Protection Council monograph* **43:**165-181.

Jiji, T. Ann Napolean, .Naseema, A. and Anitha, N. 2003. Record of *Aspergillus candidus* Link on the fruit fly *B. dorsalis* (Hendel) infesting guava. *Insect Environ.* **9**(2):81.

Jiji, T.,Praveena, R., Anitha, N and Naseema, A. 2005.*Beauveria bassiana*(Bals) Vuill on *Sylepta derogata* Fab. and its cross infectivity on other major pests of vegetable. *Insect Environment,* **11**(1):22-23.

Klochko, M.D. 1969. The use of entomopathogenic fungi for controlling *Epilachna vigintioctopunctata* , *Inst. Zashch. Rast.* **1:** 26-31.

Milcami, Y., Yazawa, K. Fukushima, K., Arai, T., Udagawa, S. and Samson, S.A. 1989. Paecilotoxin production to suppress insect growth in clinical or terrestrial isolates of *Paecilomyces lilacinus* – an entomopathogenic fungi. *Mycopathologia* **108:**195-199.

Osborne, L.S. and Landa, Z. 1992. Biological control of whiteflies with entomopathogenic fungi. *Florida Entomol.* **75:**456-471.

Ramesh, K., Mohan, C.M., Lakshmi, K.A., Padmavathi, J. and Devi, K.U. 1999. *Beauveria bassiana* (Bals) Vuill. (Hyphomycetes, Moniliales) in cotton pest management: a field trial on the cotton leaf roller, *Sylepta derogata* Fabricius (Lepidoptera: Pyraustidae). *J. Entomol. Res.* **23(3):**267-271.

Samsinakova, A. and Kalalova, S. 2006. Artificial infection of scale insect with entomopathgenic fungi *Verticillium lecaniii* an *Aspergillus cadidus Biocontrol* **20(4):**1975.

Wraight, S.P., Carruthers, R.I., Bradley,C.A., Jaronski, S.T, Lacey, I. A., Wood, P. and Galainiwraight, S. 1998. Pathogenicity of the entomopathogenic fungi *Peacilomyces* spp. and *Beauveria bassiana* against the silver leaf white fly, *Bemisia argentifolii. J. Invert. Pathol.* **71:**217-226.

SECTION – VII
ISSUES

47

Organic Farming in Coconut: Developmental Issues

Thomas Mathew

INTRODUCTION

Coconut is an important crop of economic importance to many Asian and Pacific countries in the world. The crop provides livelihood securities and employment opportunities to major segments of the rural mass of these countries. Coconut is largely grown in 93 countries in the world. India ranks second in production of coconut holding a share of 26.65% of the world production followed by Indonesia which accounts for 26.95% share in the production. The other major coconut producing countries are Philippines, Vietnam, Thailand and Sri Lanka. In most of the producing countries about 50% of the production is consumed domestically and balance is made available for global trade. The bulk of the coconut produced in our country goes for the domestic consumption. However, till recently India has made some dent in export of coconut and coconut products especially to Middle East countries. The decelerating growth in production of coconut in the major exporting countries in the world like Philippines, Indonesia and Sri Lanka on account of displacement of coconut area for more profitable enterprises like oil palm cultivation, urbanization followed by real estate boom etc., have resulted deficit in global supply of coconut and its products. Indian coconut and coconut products are, therefore, now getting competitive advantage in occupying world market. However, in many of the EEC countries Indian coconut products are inaccessible due to the fact that the products do not satisfy the quality standards prescribed in the European market as well as for want of organic certifications. In view of the emerging economic scenario and also due to the tremendous growth in the production and productivity of coconut in the country it is inevitable to explore new markets for these products abroad for the stability of domestic market. Promotion of organic cultivation in coconut gardens is, therefore, considered inevitable.

Economic Importance

Coconut palm is considered as the benevolent and benign tree which provides food, drinks and shelter to mankind. Coconut kernel, the nucleus part of seed, forms an important ingredient in the diet of many people in India. Coconut milk, a liquid extracted from the grated wet kernel, is an indelible additive in culinary preparations to make the food more nutritious and tastier. Copra, the dried kernel, is one of the richest sources of vegetable oils. It is also used as dry fruits which command high demand in northern parts of India during winter. As edible oil its consumption is mostly confined to the state of Kerala and the boarder districts of Tamil Nadu and Karnataka. Coconut oil is an important raw material for many industrial products such as toilet soaps, liquid soaps, shaving cream and natural shampoos. Similarly coconut oil finds extensive use in the food industry due to its characteristics such as easy melting behaviour, resistance to oxidative rancidity, pleasing flavour and good digestibility besides using directly as hair oil and body oil. Coconut oil derivatives are used in most of the cosmetic products. The coconut oil cake, the by product of oil industry is in great demand and is a major ingredient in the manufacturing of cattle feed.

The inflorescence sap, traditionally called "neera", is produced from the unopened spathe/inflorescence of selected healthy palms, through a process known as tapping by skilled climbers. It has varied commercial application. Neera without fermentation is an ideal isotonic drink composed of vitamins, minerals and free nuclei. It is used for production of palm sugar, jaggery, vinegar, wine and arrack in many countries. Besides, the inflorescence is used to make ayurvedic medicines. Another important product that has got consumer demand, throughout the country, is tender coconut. It is a nutritious health drink which is diuretic and re-hydrating fluid. The tender coconut water is a base for many ayurvedic preparations. Of late tender coconut is also used for the manufacturing of wine. The water of mature nut yields various products such as vinegar, nata-de-coccoa, edible jelly and wine.

The shell is used as a fuel besides manufacturing various commercial products like shell powder, shell charcoal, shell based activated carbon, ice cream cups, buttons of garments, utility articles and show pieces. The soft bud known as cabbage or coconut "heart" is edible and nutritious. The matured coconut timber is a strong building material apart from making furniture, wall panels, show pieces and floor tiles. The plaited leaf of the palm is used for thatching, fencing, shading, etc. Dried leaves and other palm parts are used as fuel. The spindle leaf is used for decoration and costuming in folk dances. The midribs of leaves are used to make brooms, fish traps, and baskets and tongue cleaner. The husk yields fibre and pith. Coconut fibre is used for making an array of products ranging from coir mats to geo-textiles which command consumer demand through out the world. The coir and coir products form export commodities of the producing countries. Coconut pith, a by-product of coir industry is an ideal organic manure on composting, soil conditioner and rooting medium besides having many other end uses. The spathe and stipules are used as fuel and also for making handicrafts and other attractive articles. Coconut is an eco-friendly plant which helps to conserve soil, provides aesthetic beauty to the nature, perennial and less exhaustive and ideal tree for bio-hedging on the coastal ecosystem. Coconut farming, trade and processing activities sustain the livelihood of millions of people across the globe. It provides livelihood securities to more than 10 million people in 18 states and three Union Territories of India. It is a source of permanent income to five million farm families. Coconut contributes more than Rs. 8,900 crore to the country's GDP apart from an export earning of Rs. 695 crore. Coconut industrial sector alone generates more than 340 million man days of employment. Out of the total production of about 16,000 million coconuts, 8,000 million nuts is utilized for the production of coconut oil, 420 million nuts are converted into edible ball copra for using as dry fruit, 2,400 million nuts are consumed as health drink and the balance being used for edible and culinary purposes as well as for religious and social rituals. Coconut cultivation and industry in the country is, therefore, the back bone of all the producing states as it provides livelihood security, food security and employment guarantee to millions of people besides contributing considerably to States GDP. In short it is essential to guarantee the uninterrupted production of coconut in the country while considering any change in the cultivation practices, particularly when converting the gardens under modern agriculture to organic farming.

Coconut Palm: Its Distinct Features of Growth and Nature of the Industry

Coconut palm is a perennial horticulture crop, the economic life span of which lasts more than 50–60 years after planting. The cultivation of the palm involves heavy pre-bearing investments and continuous recurring expenditure for its maintenance. The coconut palm is, therefore, characterized by high rate of investments in the initial years and the realization of out put or the pay off in succeeding periods. Because of its unique nature of continuing growth and yielding phases, any change in the cultivation practices may upset the growth and the physiology of the crop. A vast stretch of area under coconut in the country is rain-fed. The success of coconut growing depends on the uniform distribution of an annual rainfall ranging between 1,000 mm and 3,000 mm. In states like Tamil Nadu, Andhara Pradesh, Karnataka, etc., the annual precipitation in most of the coconut growing districts varies from 700 mm to 1,500 mm with only 40–70 days rainy days in a year. Hence, a large number of coconut plantations in these states are maintained under irrigation. Since growth and reproductive phases of the plant proceed simultaneously, the palm requires nutrients and water regularly. The primordial formations of inflorescence, spike lets and buttons in coconut takes place 12–33 months before the appearance of the inflorescence. Any hindrance to successful growth of the tree during these productive phase results in reduction in yield and the cumulative effect of these adverse growths of the tree appears only after two-three years. Therefore, the crop production practices warrant the need for uninterrupted pattern of input application over longer periods to ensure continuous flow of returns. The Indian coconut industry is vulnerable to the influence of natural calamities. Serious incidence of pests and diseases, adverse climatic conditions, drought, cyclone, etc.

retards the productivity of the palms and often causes heavy loss to farmers. Various pathological agents are involved in different diseases, some of them are localized to specific part. A few of the diseases are lethal, while others reduce the vigor of the palms causing yield loss. The root (wilt) disease of Kerala, Tatipaka disease of Andhra Pradesh, Basal stem rot in Karnatak and Tamil Nadu, Crown chocking in Orrissa and North Eastern states, Leaf rot, Bud root and Stem bleeding are the major diseases causing damages to the crop and heavy loss to the farmers unless timely remedial measures are adopted. The recommended packages of practices for the control of most of these pests and diseases are resorted through only by chemical means. The epidemic and endemic outbreak of pests like, Eriophyid mite, Leaf eating cater pillar, etc., in recent years prompted the farmers and the policy makers to resort to massive application of chemicals which has caused considerable damages to environment. The morphological and physiological nature of the tree does not permit sudden structural changes in the maintenance and management of the crop.

Diversity in Coconut Cropping System: Impediment for Organic Farming

In many coconut growing states, coconut is largely cultivated as homestead gardens where the available space is fully exploited for the cultivation of other horticulture crops. Most of the palms are also used as standards for betel vine, pepper and other creepers. In states like Tamil Nadu, Karnataka and Andhra Pradesh, it is cultivated on a plantation scale where other commercial crops like banana, maize, ginger turmeric, etc., are raised as inter-crops. The success of cultivation of these crops depends on the timely application of inputs including irrigation. Planting of high value perennial horticulture crops like cocoa, black pepper, Nutmeg (*Myristica fragrans*), Clove(*Eugenia caryophyllus*), Cardamom (*Elettaria cardamomum),* Cinnamon(*Cinnamomum zeylanicum*), Vanilla (*Vanilla planifolia*) etc. are the common cultivation practices adopted by many innovative farmers and has become lucrative enterprises. The coconut based cropping system adopted by such farms render enhanced level of gross income and employment opportunities besides improving the productivity of the land. While the coconut under mono-cropping provides employment opportunities of only 150 man days per ha per year and net income of Rs10,400/ha, the coconut based cropping systems generate additional employment between 130 and 606 man days/ha/year. The rooting pattern of coconut is such that only 25% of the land area is utilized in the recommended spacing of 7.5 × 7.5 m, the remaining 75% space could effectively be tapped. As coconut canopy's space utilization is very low, plenty of sun light infiltrates and fall on the ground unutilized. The venation structure of the coconut crown and orientation of the leaves facilitates inter-cropping. As much as 56% of the sunlight is transmitted through the canopy during peak hours, the diffused sunlight facilitates growing of number of spice crops in the interspaces. Coconut based cropping system is important and most relevant activity in the present era for sustaining income from unit area. The cultural requirements of these crops under mixed cropping system with coconut warrants regular and balanced application of both organic and inorganic fertilizers and the proper adoption of plant protection measures apart from keeping the moisture level in the plantation at field capacity. Excess or deficits of any of these critical inputs will affect the health of plants which ultimately results in yield reductions among the main and complementary crops in the coconut based agro-eco-system. Even though organic farming in coconut plantations is advocated, proper restructuring and reengineering the existing gardens, which are practicing modern agronomic practices to traditional organic farming are the essential pre-requisite.

Organic Cultivation in Coconut: Current Status

Universally, coconut is cultivated under traditional agriculture except in countries like India, Sri Lanka and to some extent Philippines where coconut is maintained under modern agriculture. Coconut cultivation in India, Kerala in particular, has a recorded history of more than 3,000 years. It has been cultivated under traditional agriculture till 1960's. The aftermath of the green revolutions brought in indiscriminate use of chemical fertilizers and pesticides for the crop maintenance. Though the transforming traditional agriculture has resulted in enhanced production of coconut in the country, it has lead to the emergence of multiple problems in crop production. The emergence of new diseases and pests, exhaustive depletion of available nutrient and soil matter content and increasing phenomenon of soil salinity were the repercussions experienced in the field. In many places, the soil salinity and the depletion of soil have affected the plant health and its yield. The presence of pesticide residues in the harvested fruits as a consequence of this revolution has become an alarming issue.

The increase in purchasing capacity of a large segment of consumers, mainly the middle income groups and their desire for healthy and natural products, has lead to the genesis of a holistic approach in agriculture. A new school of thought, "Organic Agriculture" has emerged which has been reflected in all the sectors including coconut culture and industry. However, the prevailing cultivation practices adopted by coconut farmers in the country is inadequate to immediately go in for organic farming and hence, call for re-engineering and revitalization of coconut culture to avoid any ditch in current level of the production, instead to ensure an increasing trend in the production of coconut in the country at least to the level of India's demographic growth rate.

Coconut Production and Organic Farming: Current Situation an Analysis

India contributes about 16.51% in area and 26.65% in production of coconut in the world. The productivity of coconut in the country is the highest among the major coconut growing countries in the world. Unlike other world countries, India has the comparative advantage of having the crop grown under varied agro-climatic zones and hence, there is distinct difference in the pattern of distribution of coconut in the country. As a result of this unique distribution, a steady supply of coconut is ensured in the country through out the year. The crop is grown in 18 States and three Union Territories covering an area of 1.9468 million ha with a production of 14,811.1 million nuts (2005-06). The bulk of the coconut acreage is concentrated on the west Coast region of the country comprising the states of Kerala, Karnataka and Maharashtra followed by the East Coast areas of Tamil Nadu, Andhra Pradesh, Orissa, and Pondicherry. The Islands of Andaman and Nicobar, Lakshadweep and the coastal regions of Gujarat are the other traditional coconut areas. About 90% of area and 89% of production of coconut are emanated from the four southern states viz. Kerala, Karnataka, Tamil Nadu and Andhra Pradesh.

The biggest obstacle to the competitiveness of India's coconut sector is the low rate of returns from the holdings and the reduced input-output realization especially in the traditional coconut growing states due to the depletion of soil on account of continuous growing of coconut. The continuous growing of a single crop without enrichment of soil through organic recycling and crop rotation is considered the root cause for the poor performance of most of the coconut gardens in the country. Hence, enrichment of such soil is possible only through balanced application of organic and inorganic fertilizers along with the recycling of biomass. Unlike other commercial plantation crops like rubber, tea, etc., where stabilized nature of plantation managements are followed in most of the gardens, coconut gardens are characterized by varying nature of cropping system management resulting considerable fluctuations in the production from garden to garden and state to states. The coconuts cultivated in traditional areas are more vulnerable to this exceptional nature of management. The recurring nature of price fluctuations further aggravates the situations. Every crest in the price movement always offers better management of the gardens, whereas, a trough results in the negligence of the garden. Similarly, an unprecedented onset of drought results total negligence of garden sometimes for more than a year. Therefore, majority of the coconut gardens in the traditional coconut growing states are being maintained partially through scientific management and sometimes without application of any inputs. Comparing to traditional places, the cultivation of coconut in non-traditional places satisfy the scientific requirements and hence, its productivity in places of recent introduction particularly after 1980's is significantly high. Such gardens are laid out scientifically and maintained under proper care and managements with regular application of fertilizers and other plant protection chemicals. Hence, the perspective for practicing organic agriculture in coconut needs re-look and refinements. The coconut gardens in North-East, the Union Territory of Andaman & Nicobar Islands, Lakshadweep and Pondicherry and selected areas in the traditional states can be earmarked for practicing organic farming. In fact the continued coconut growing on the same land without recycling the residual bio-mass deplete the soil biodynamic properties and hence, considerably reduce the input use efficiency. Therefore, appropriate package of practices have to be evolved to convert initially small and uneconomic holdings into 100% organic to make them economically viable units through integrated cropping system managements. The scheme "Integrated farming in coconut holdings for productivity improvement" implemented by the Coconut Development Board in the country during the past couple of decades has helped in the creation of adequate infrastructure facilities at farm level for the continuous production of organic manure particularly vermicompost which has made some dent for transforming present cropping system management into organic agriculture.

Constraints

At present there is no reliable data available in the country on the extent of coconut cultivations which are fully maintained under organic farming. The resurgence of traditional organic farming, prompted many innovative coconut farmers in different parts of the country to go in for the new venture with the expectation of getting more leverage for the marketing of their products. However, their initiation of converting their coconut gardens which were hitherto nourished under modern agricultural practices, into total organic have bested with varied problems like deterioration of plant health, sudden decline in yield and income loss. A few farmers in Tamil Nadu and Andhra Pradesh who converted their coconut gardens into total organic with their acquired knowledge have been reported to fall into perils and dismays due to irreparable damages to plantations due to the sudden shock. It was revealed that the farmer in Tamil Nadu who has begun to change the modern farming to organic with a cut in inorganic fertilizer, worth Rs.2 lakh, had in turn resulted the decline in the income by four times which had prompted him to resort to the original modern farming immediately and thereby regained the income loss. Similarly a coconut grower in Andhra Pradesh substituted his entire area of coconut gardens for oil palm cultivation because his entire investment for converting his coconut garden was ruined on account of practicing organic farming. The multiplicity of emerging problems associated with the introduction of organic farming warrant the need for strategic reforms and development of appropriate and cost effective proven technologies for advocating the organic farming in coconut based farming systems. Contrary to the recommended dose of inorganic fertilizer the quantity of the nutrients replenished through organic sources was inadequate to retain the production and production capacity of the palms.

A healthy coconut palm, maintained under average management, annually requires 500 g of N, 320 g of P and 1200 g of K supplemented with 25–50 kg of Organic manure and 5 kg neem cake for the maintenance of plant health and its sustained production. With this nutrients recommendation, an area of 1 ha of coconut garden annually requires about 100 kg N, 65kg P, 2400kg K and 10mt of organic manure besides the requirements of 875 kg neem cake. The major requirements of these inputs are currently met through organic and inorganic bases. The organic requirements of palms have been mainly supplied through cow dung. The dwindling growth in the livestock population has now become a constraint for the easy availability of adequate quantity of quality organic manure. The absence of necessary enactments for the imposition quality control of organic manure encourages the distribution spurious materials to the farmers. Organic manures which are currently supplied through co-operatives and other organized delivery systems are bone meal, neem cake and bio-fertilizers. At present there is no mechanism to test these products and ensure the quality before it is applied to the plants. For example, the supply and distribution of quality neemcake which is composed of the minimum levels of N: P: K and oil content with the threshold level of moisture and silica is pausing problems. The neem cake mixed with coffee husks, coir pith, sand and common salt is seen available in plenty in the market with cheap price. Such materials, instead of giving expected result on the crop growth and yield, sometimes it damages the crops on account of high concentration of salt and heavy metals. Similarly in the case of bio-fertilizers, the absence of necessary mechanism to test the active micro-organisms at the desired level of population pause serious problems on the bio-efficacy of these products on improving the crop yield. Hence, inadequate facilities for quality assurance and supply of adequate quantity organic manures pause problems and concern for the introduction of 100 % organic cultivation in coconut and coconut based farming systems. The tiny and scattered nature of distribution of coconut gardens in the country is another major hurdle for the implementation of a certification programme in organic farming. The unorganized method of coconut cultivation is another serious problem pausing the introduction of organic farming as every garden is being maintained under varied cultivation practices which are different from garden to garden. For example, a coconut garden integrated with seasonal inter-crops or vegetables necessitate the application of inorganic fertilizers and pesticides at frequent intervals for getting more yield and reducing the crop loss due to pest and diseases.

CONCLUSION

The increasing concern on healthy living dovetailed with the consumer behaviour towards utilizing safety food created a new demand for organically grown products across the world. The change in consumer behavour coupled with the increase in demand for such products in turn offer premium prices for those products certified as organic products. The absence of an appropriate package of practices based on scientific studies is the major bottleneck for introduction of organic farming in coconut. Until a technology for 100% organic cultivation of coconut based farming system is standardized, it may be difficult to promote organic farming in coconut cultivation without affecting the current level of production. In the absence of a proven technology for organic coconut cultivation practices, probably a strategic plan need to be drawn by proper segmentations of existing coconut area and then introduce certifications. There are many areas where farmers are still practicing traditional farming without application of any inorganic fertilizers. Such areas need to be identified and earmarked as the benchmark for promoting organic farming. For example, whole coconut gardens in Lakshadweep can very well be declared as organically grown coconut as the application of inorganic fertilizers is banned through government orders. Similarly there are many states in North-Eastern regions which are also maintained under pure organic. The areas which are already maintained under traditional cultivations are, therefore, the ideal places for introducing the organic certifications. The Coconut Development Board (CDB), through its cluster approach, may assist farmers and act as catalytic agent for empowerment of them to jointly go in for certifications. Similarly the CDB can assist any entrepreneur to set up processing units for the production organic certified coconut products and their marketing, both in domestic and international markets.

SECTION – VIII
QUALITY CONTROL AND CERTIFICATION

48

Organic Farming: Quality Control Standards and Certification

Mathew Sebastian

INTRODUCTION

Organic farming can be defined as a system of managing agricultural holdings that implies major restrictions on chemo-synthetic fertilizers, pesticides, and pharmaceuticals. This method is based on varied crop farming practices and is concerned with protecting the environment and seeks to promote sustainable agricultural development. By respecting the natural capacity of plants, animals and the landscape, it aims to optimize quality in all aspects of agriculture and the environment. It allows powerful laws of nature to increase both agricultural yields and disease resistance. The products obtained from organic agriculture are called organic products.

An organic product has:

- No synthetic pesticides used in its production
- No synthetic herbicides used in its production
- No prohibited preservatives or additives used while processing
- No Genetically Modified Organisms (GMOs or GE)
- No Irradiation at any stage of processing, packing and transportation.

Organic agriculture encourages bio-diversity and the use of varieties resistant to pests and diseases. Restrictions on the use of fertilizers and plant protection chemicals ensure wholesome produce without residue of chemicals and eliminate the risk of contamination and pollution of natural environment.

According to the definition of the Codex Alimentarius, "organic agriculture is a holistic production management system which promotes and enhances agro-ecosystem health including biodiversity, biological cycles and soil biological activity". It emphasizes the use of management practices in preference to the use of off-farm inputs, taking into account the regional conditions that require locally adapted systems. This is accomplished by using, where possible, agronomic, biological and mechanical methods, as opposed to using synthetic materials, to fulfill any specific function within the system." Products labelled as "organic" are those certified as having been produced according to clearly defined organic production standards, for instance formulated in regulations.

THE WORLD OF ORGANIC AGRICULTURE

Organic agriculture is developing rapidly, and statistical information is now available from 141 countries of the world (data end 2007). Its share of agricultural land and farms continues to grow in many countries. The latest facts and figures related to organic agriculture world wide are furnished below.

32.2 million hectares are certified according to organic standards (data as at the end of 2007). "At the level of the geographical regions, growth was strongest in Latin America and Africa. With its vast grazing lands, Australia continues to account for the largest certified organic surface area, 12 million hectares, followed by Argentina (2.8 million hectares), and Brazil (1.8 million hectares). The greatest share of the global organic surface area is in Oceania (37.6%), followed by Europe (24.1%) and Latin America (19.9%). In terms of certified land under organic management as a proportion of national agricultural area, the Alpine countries, such as Austria (13.4%) and Switzerland (11%), top the statistics. The global market for organic products reached a value of over 46 billion US Dollars in 2007, with the vast majority of products being consumed in North America and Europe

Organic Quality Control

The organic quality control involves the following process.

A. Accreditation

The accreditation process ensures that the certification practices of certification bodies are acceptable, typically meaning that they are competent to inspect and certify third parties, behave ethically, and employ suitable quality assurance.

The accreditation systems prevalent internationally for offering organic certification services are:

Accreditation by National Accreditation Body(NAB), Govt. of India

The programme is implemented by the Government of India through the Ministry of Commerce and Industry as the apex body. The Ministry will constitute a National Steering Committee for NPOP which formulates the National Accreditation Policy and Programme to draw up standards and to regulate the use of the National Organic Certification Mark. The National Steering Committee also functions as the National Accreditation Body (NAB) which administers the National Accreditation Policy and Programme. The NAB is responsible for drawing up procedures for the evaluation and accreditation of certification programmes, as well formulating procedures to evaluate agencies implementing these programmes. The accreditation of inspection and certifying agencies is also the responsibility of the NAB.

ISO 65 (EN 45011) accreditation, which is the leading internationally accepted norm for certification bodies operating a product certification system.

B.Standards for Organic Agriculture

Minimum requirements for a farm or product to be certified as 'organic' are precisely defined by organic standards. There are organic standards on the national as well as international level. For certification, the standards of the target market or importing country are relevant. Certain private labels such as Naturland, Demeter or BIO SUISSE have additional requirements in addition to their national standards.

Organic standards have long been used to create an agreement within organic agriculture about what an "organic" claim on a product means, and to some extent, to inform consumers. Regional groups of organic farmers and their supporters began developing organic standards as early as in the 1940's. Currently there are hundreds of private organic standards worldwide; and in addition, organic standards have been codified in the technical regulations of more than 60 governments. Third-party organic certification was first instituted in the 1970's by the same regional organic farming groups that first developed organic standards. In the early years, the farmers inspected one another on a voluntary basis, according to quite a general set of standards. Today third-party certification is a much more complex and formal process. Although certification started as a voluntary activity, the market began to demand it for sales transactions, and now it is required by the regulations of many governments for any kind of an "organic" claim on a product label. There are sets of guidelines and norms to be complied with, followed by the inspection by the certification agency before certification. These certifications are basically carried out on the basis of statutory certification norms and the voluntary certification norms.

National Standards for Organic Production

In 2000, the Government of India released the National Standards for Organic Products (NSOP) under the National Programme for Organic Production (NPOP). Products sold or labeled as 'organic' thereafter need to be inspected and certified by a nationally accredited certification body

Major International Standards

UNITED STATES DEPARTMENT OF AGRICULTURE (USDA) NATIONAL ORGANIC PROGRAMME STANDARDS (USDA/NOP)

The production rules for organic production according to the U.S. Standards have been laid down in the National Organic Program (NOP) of the United States Department of Agriculture (USDA). This Regulation is in force within the U.S.A., but also sets criteria for importing organic products to the country. This program establishes national standards for the production and handling of organically produced products, including a National List of substances approved for and prohibited from use in organic production and handling. The Indian conformity assessment system for organic certification bodies used by National Accreditation Body, Govt. of India has been determined by USDA Agricultural Marketing Service (AMS) to be sufficient to ensure conformity with the technical standards of USDA's National Organic Programme.

EU REGULATION EC NO.834/2007

The regulations governs organic agriculture for European Union member states set out for a complete set of objectives, principles and basic rules for organic production. This regulation is most relevant for exports to Europe. Organic products imported into the European Community should be allowed to be placed on the Community market as organic, where they have been produced in accordance with production rules and subject to control arrangements that are in compliance with or equivalent to those laid down in Community legislation. In addition, the products imported under an equivalent system should be covered by a certificate issued by the competent authority, or recognized control authority or body of the third country concerned.

JAPANESE AGRICULTURAL STANDARDS (JAS)

Japanese Agricultural Standards are the standards for the agriculture industry maintained by the Japanese Government. They are comparable to Japanese Industrial Standards for food and agricultural products. The JAS rule admits the equivalence with the European organic certification for vegetable products (in European countries). For livestock products the equivalence with European certification is not admitted. The certification based on these standards is widely recognized across the three key organic markets of US, EU and Japan and is trusted by a large section of importers and consumers.

C. Inspection Process

Inspection is an on-site visit to verify that the performance of an operation is in accordance with specific standards. It involves evaluation and verification of agricultural production, processing and trading. The findings of the inspection are presented in a report which is submitted to certification committee for granting certification. A copy of the same is being given to the client.

The inspection includes the following:

- Interviews with persons responsible for production
- Physical inspection of fields, premises, processing equipments and machinery, storage area etc., Inspection of documentation and record keeping, etc.

- Testing for residue analysis is carried out if the inspector suspects the presence of chemicals or prohibited material in the farm/ processing.

D. Certification

Certified organic products are those that's production, processing, handling and trade have been verified by an accredited certification body as being in conformity with specified organic standards. Once a product is certified organic, the product can be labeled as organic. Certification is a defined procedure in which a certification body assesses a farm /processing/ trade and assures in writing that it meets the requirements of the organic standards. The inspector transmits his findings to the certification body as a written report. The certification body compares the results of the inspection with the requirements of the organic standards. A certification committee decides whether certification may be granted or not based on the inspection findings reported to them.

Certification decisions may be as follows:

- Approval
- Approval with conditions that must be fulfilled prior to certification
- Approval with conditions that must be fulfilled within a specified period of time.
- Denial: reasons for denial are clearly stated based on the standards
- Decertification: when the operator not giving free access, prohibited chemical use, etc.
- Deferred: More information is needed to make a decision

The establishment of an organic management system and building of soil fertility requires an interim period, the conversion period. The conversion period may not always be of sufficient duration to improve soil fertility and re-establish the balance of the ecosystem but it is the period in which all the actions required to reach these goals are started.

Plant products produced can be certified organic when the national standards requirements have been met during a conversion period of at least two years for annual crops before sowing. In the case of perennial crops other than grassland, the conversion period is at least three years before the first harvest of products. The accredited inspection and certification agency may decide in certain cases to extend or reduce the conversion period in the light of previous status of the land use but the period must equal or exceed twelve months. Crops harvested during conversion period are not considered fully organic; but products may be labeled as in conversion to organic.

Validity of Certification

Organic certificate issued according to NPOP remains valid for 12 months from the date of certification or until the next certification decision is made. Organic certificate issued according to USDA NOP is valid for 3 years. Organic certificate stands invalid in cases where you voluntarily surrender or your certification is suspended by the certification agency. Inspection is generally done once in a year. Additional inspections are conducted wherever found necessary.

Certification procedures for the certification of organic products should make it possible to track and control the flow of products from primary production at farm level through each stage of manufacturing right to the final consumer product. Producers and exporters will have to obtain certification against organic standards applicable in those markets, in which they intend to sell their products with an indication that they are organic. In other words, certification creates trust in organic labeling and promotes fair competition in the market place. The product from the organic certified farms is having higher demand and attracts a greater price in the domestic and international markets.

Inspection and Certification Process

The following steps are to be followed by a person for organic inspection and certification process:

- Select an accredited inspection and certification agency
- The selection of agency is based on the acceptability of the exporting country and the standards prescribed by that country
- File application form detailing preliminary information about the size of the farm/ units, location of units, activities, etc., to determine the cost of inspection and certification and based on application a cost of inspection and certification is determined and a contract regarding the same is signed between the producer and the certification body
- Based on the information provided by the producer, the agency carries out farm inspection on a mutually convenient day and time.
- After conducting the inspection, the agency provides the report of inspection, based on which, the agency decides whether or not to grant certification.
- If certification is not granted, the agency provides the reasons for rejection.
- Each year following the certification, the agency performs inspections to determine whether the requirements for certification are still met and issue certification decisions accordingly

Smallholder Group Certification with Internal Control System (ICS): An Approach to Overcome Resource Constraints in Certification of Organic Farms

A majority of agriculture practitioners worldwide are smallholders and are often located in remote areas with long travel times from one place to another. Furthermore, the overall revenue from their agricultural production is usually far too small to allow a viable farm inspection by some external inspection body for each farmer.

What is an ICS?

Smallholder farmers can be certified as a group with their own Internal Control System (ICS), which is a documented quality assurance system that allows the external certification body to delegate the annual inspection of individual group members to an identified body/ unit within the certified operator. A group of homogenous farmers, practicing organic agriculture come under one umbrella to form a small holder group, for the reason of getting organic certification, developing ICS for meeting the requirements of certification. The external inspection focuses on evaluation of the efficiency of the ICS and only a representative percentage of farmers must be inspected by the certification body directly, which greatly reduces the costs for certification of a large group of small farmers.

Smallholder farmers (farmers associations/ cooperatives or processors with contracted smallholder farmers) may be certified as a group under common marketing of the produce as a group, homogeneity of members in terms of location, production system, products. The minimum requirements for an ICS to be certified include written procedures and forms (ICS Manual), Contractual relationships, Internal Organic Standard, qualified personnel and a clear structure, training of farmers and ICS staff, internal inspections covering all the farmers at least twice annually and documented effective procedures for internal farm inspection, internal approval and dealing with non-compliances.

The inspection of a smallholder group focuses on the evaluation of the efficiency of the Internal Control System (ICS). In order to evaluate the internal Control System, a certain percentage of farmers are re-inspected by the inspector of Certification Body. This percentage strongly varies with the risk category of the operator and with the effectiveness of functioning of the ICS.

In ICS, there will be a single certificate for all individual production as well as processing and handling activities registered within the group. Individual operators within the group may not use the certification independently. The rationale for an ICS is to bring down the cost of organic certification to smallholders by establishing a group that can do much of the monitoring itself. Beside the reduction in cost of certification, it helps to have an effective quality management system in place and also create better marketing opportunities for organic products under certification.

Labeling of Certified Organic Products

The eligibility for labeling the organic products varies with countries as per the organic regulations prevalent in the respective countries.

The United States, the European Union and Japan have comprehensive organic legislation, and the term "organic" may be used only by certified producers. Being able to put the word "organic" on a food product is a valuable marketing advantage in today's consumer market, but does not guarantee the product is legitimately organic. Certification is intended to protect consumers from misuse of the term, and make buying organics easy.

However, the organic labeling made possible by certification itself usually requires explanation. In countries without organic laws, government guidelines may or may not exist, while certification is handled by non-profit organizations and private companies.

A trademark – "India Organic" will be granted on the basis of compliance with the National Standards for Organic Production (NSOP). Communicating the genuineness as well as the origin of the product, this trademark will be owned by the Government of India. Only such exporters, manufacturers and processors whose products are duly certified by the accredited inspection and certification agencies, will be granted the license to use of the logo which would be governed by a set of regulations.

Organic food products originating from non-European Union countries may be imported and marketed in the European Union with an organic label if it is accepted that the products are produced and certified according to procedures equivalent to those of the European Union. According to the Japan Agriculture Standard (JAS) regulations for product labeling, all products to be labeled as organic must be certified by a Registered Certification Organization (RCO) and must display the JAS logo, as well as the RCO's name. RCOs are required to be accredited by the Ministry of Agriculture, Fisheries and Forest (MAFF), Japan.

Other Sectors of Organic Certification

Organic Aquaculture

Most aquatic products are highly perishable and, therefore, very demanding with respect to organic quality assurance. To increase the customer trust in aquatic products, it is necessary to provide reliable information on their origin, the method and quality of production. This can be done by following an internationally accepted standard with easily recognizable labels. Credibility of standards and labels has to be built up and maintained by a sound quality system.

Currently many private and government bodies (including European Union and USDA NOP) are in the process of developing standards for sustainable aquaculture methods. Naturland-Germany, BioSuisse-Switzerland, and Soil Association-UK, etc., are some of the internationally accepted private labels for organic aquaculture. The entire chain of custody, hatcheries and feed mill has to be certified to get greater acceptance in the international market. The certifiers work in co-operation with independent inspection bodies experienced in quality assurance to guarantee the implementation and credibility of their label.

Organic Textiles

Around a quarter of the world's insecticides are used to grow conventional cotton, and ten percent of its pesticides, and at least 8,000 chemicals are used to turn raw material into clothes, towels, bedding and other items that we put next to our skin every day. The scientific community's jury is still out on the exact effects to human health of many of the chemicals used in cotton farming, with potential side effects including cancer and disruption to the hormonal and reproductive systems of the body.

Interestingly there was an increase of production of more than 61% with approximately 50,000 tons of cotton from organic cultivation–during the period 2006-07 which means there is a true organic textile boom.

The fashion industry has an enormous impact on the environment. Organic fashion means clothes which have been made with a minimum use of chemicals and with minimum damage to the environment. This includes chemicals used during every step in the process, from growing cotton, to the dying and finishing of the fabric. Organic textile products and garments are certified in accordance with the Global Organic Textile Standard (GOTS).

Organic Cosmetics

Today the organic philosophy has found a new application in the sector of cosmetics. Eco-Bio-Cosmetics are the substances used to enhance or protect the skin with the ingredients or use of raw materials from organic farming or from wild harvest. Through the practicing of organic life style we can attain a healthy and environmental friendly world. Today, adoption of organic style has been a trendsetter where respect for the environment, preservation of bio diversity and protection of natural resources are the key principles ensuring a better quality of life. Organic cosmetics are a great way to look beautiful and at the same time avoid the risks posed by synthetic cosmetics. Also as these cosmetic products are used daily organic products are the best option available.

Eco-Bio Hotels

Organic food is produced by farmers who avoid the use of chemical pesticides, fungicides and fertilizers. Farmers emphasize the use of renewable (or sustainable) resources and the conservation of soil and water to enhance environmental quality for future generations. Eco-Bio-Hotel means a process of developing and establishing an eco-bio friendly hotel through the practice of organically safe food chain. It should be considered with good organic practice and follow up. The conversion period will be three to six months for the certification. The main goal is to develop the environmental manage system through the procurement of environmentally friendly goods and services and to improve the energy and water conservation and waste management. The Eco-Bio-Hotel should be followed by good organic and hygienic practices to minimize the risk factors and for the better up-gradation of the organic system. It is also strictly recommended that the entire hotel should be converted to Eco-Bio-Hotel.

Key Issues to be Addressed for Certification

Organic Standards and Certification be Defended

The environmental damage and resource use associated with organic agriculture are much less than those of conventional agricultural systems. However, yields tend to be smaller, and the organically grown produce is often relatively expensive to the consumer. In some regions of the country, contamination of organic crops with genetically modified crops has become a problem. Genetically modified plants possess novel genes that impart beneficial characteristics such as herbicide resistance.

Development of Harmonized International Standards

A harmonized standard having international acceptance is yet to be developed.

Protection of Price Premium and Enhancement of Social Accountability

Organic producers face challenges in finding markets for their products and negotiating prices. However, organic agricultural systems will not become more widely adopted unless a number of socio-economic conditions change. First, the consumers will have to be willing to pay the somewhat higher costs for certified organic products.

Consumer Education

There should be a battling to gain consumer awareness of labels and the official marks of certification that differentiate their products from others.

More farmers has to be encouraged to convert to organics and capture greater margins, consumer's must perceive they can distinguish organic products and value them more highly than non-organic products

Linkage Between Organic and Other Labels

When consumers choose to buy organic, they need to know whether they are getting exactly what they pay for. The organic logo and labelling system is the mechanism that makes this possible and the consumer level awareness is needed.

Annexure – I

Certification and Inspection Procedure in Organic Agriculture

Certification is a system by which the conformity of products, services etc. as applicable to standards is determined and confirmed. Certification in organic agriculture generally refers to third-party certification. The independent body in this case is expected to be neutral or have a balance of interests.

In case of organic products, certification is primarily an acknowledgement that these products have been produced according to organic production standards. These standards may be standards of private associations or companies or certification bodies or of the State.

Certification of organic agriculture (OA) is primarily a certification of production system or product method. Certification of OA includes the certification of products and the certification of quality systems. It is primarily certification of a production system or a production method, including the product thereof. The term *organic*, as in organic *agriculture* or *organic products*, refers to certain farming and processing methods or products from such systems. Certification of OA uses concepts developed both in product certification and quality systems certification in addition to concepts developed from the field to meet the unique needs of the organic industry.

Third party certification is required mainly to create or enhance trust between farmers and consumers. The other advantages are:

- Facilitates production planning
- Facilitates marketing and extension
- Creates transparency
- Helps in authentication of the products
- Facilitates to introduce special support schemes for OA

- It strengthens the position of the primary producer
- The data collection in the process of certification is very useful for marketing and extension services.

Normally an organic certification body starts with the certification of the primary production and gradually expands its service offer upwards and downwards in the supply chain.

To be eligible for certification, all production sites (fields, stables, storage areas or processing units) must be identified and fulfil certain requirements including the ability to be inspected. This relates to conversion requirements, contamination risk, and separation from non-organic activities etc.

OA is a production system and this makes the production method the central object for certification. For farming systems, typical features include crop rotation, erosion control, and non-use of agrochemicals. Appropriate processing methods, ingredients and inputs must also be used in processing.

In order to certify the final product, the whole chain of custody (all steps in the handling of a product up to the point of final packaging) has to be certified, as the organic integrity can be compromised by switching, mixing or contamination at any stage.

Once certified, organic products are marketed carrying a certification mark indicating that the products are certified organic. The certification mark indicates conformity with certain standards and is in itself not a trademark. However, in most countries the certification mark is also registered as a trademark.

For products to enter a specific market for certified organic products, their products must be produced and certified according to the standards applicable in that market. If the markets have State regulations, these regulations must be complied with. In some cases, it may be an advantage to be certified by a certification body with a certification mark that is well known in the market.

For a product to be certified organic, all operations in the production chain, including farmers, processors, manufacturers, exporters, importers, wholesalers and retailers must be certified as acting in conformity with the standards and regulations of the certification programme concerned. Sometimes different certification bodies certify different operators in the product chain. Certification may be linked to the production of certain products, or it may relate to the operation such as, for instance the fields of a farm. Such differences depend on the certification scheme applied. Once certified, it is the operator who labels the product with the certification mark and it is the producer who is continuously responsible to the certification body and/or the owner of the mark for its correct use. It should be noted that certification of organic products is basically the certification of a production system, as opposed to the certification of a product. It is more complex than product certification because it cannot be based solely on measurable product characteristics.

To ensure objectivity, reliability and transparency, written policies and procedures are fundamental in a certification system. Policies and procedures need to consider and incorporate the basic principles of OA. All producers within a certification system should be bound by written agreement with clear conditions and consequences in case of non-conformity. These rules and requirements should be clearly documented.

Certification Procedures

Organic Certification is the first step towards organic farming practices. There are set guidelines and norms to be complied with, followed by the inspection by the certifying agency before certification. These certifications are basically carried out on the basis of statutory certification norms and the voluntary/civil certification norms. Typically the statutory certification norms are a subset of the voluntary certification norms.

The certification process may broadly be divided into two parts:

- Inspection (or control) to verify that production and handling are carried out in accordance with the standards against which certification is to be done
- Certification to confirm that production and handling conforms to standards.

A certification body may operate its own inspection activities; or inspection may be carried out on its behalf by external inspection bodies. The decision as to whether adequate confidence is provided that production and handling are in conformity with the standards rests with the certification committee. If a certification body has its own standards against which certification is carried out, it may also have a standards setting committee. Certification procedures for the certification of the organic products should make it possible to trade and control the flow of products from primary production at farm level through each stage of manufacturing right to the final consumer product.

Steps in Organic Certification

First step: The operator, i.e. the farmer, the processor, the trader or whoever is handling the product and needs certification and a certification body sign a contract, and then the certification body registers the operator. Farmers are required to provide basic information on the farm, such as size of the fields, crops, crop rotation, etc. For processors and packers, information is required on recipes, capacities, range of products, operations etc.

Second step: It is the inspection, which may be carried out by an inspection body on behalf of the certification body or by the body itself. On-site checks are carried out by the inspectors who follow a verification programme. At the level of farm or operators, inspections cover field and facilities, production practices, inputs and materials used, and records. The findings are presented in a report and submitted for evaluation, normally by a special certification committee representing various areas of competence and different sectors.

Third step: If the certification body is confident that the operation is in conformity with the standards against which the inspection and certification are carried out, a certificate will be forwarded to the producer, stating such conformity. The operator will be licensed to use such certificates or marks of certification on its products.

The inspection of a certification programme covers:

- agricultural production
- storage, processing
- transaction between parties
- labeling and certificates

The key elements of inspection are verification and evaluation. Certification is not a onetime procedure. It is carried out continuously on the basis of ongoing monitoring and inspection. The cost of certification varies. At the farm level it is generally a fixed amount, calculated from the number of days required for carrying out the inspection. For processors and traders there may be a fixed price, as well as a percentage of commercial value ranging between 0.3 and 1.0%.

Inspection and certification systems are used to verify the labelling of, and claim for, organically produced foods. Development of these systems should take into account the Principles for Food Import and Export Inspection and Certification, the Guideline for the Design, Operation, Assessment and Accreditation of Food Import and Export Inspection and Certification Systems.

The main steps in the certification process are:

- Operator requests information
- Certification body send application forms
- Operator submits application
- Application screened by Certification body
- Inspection visit
- Assessment of inspection visit
- Operator comment to report
- Certification Decision
- Operator commits to corrective actions
- Certificate sent to operator

Types of certification

1. Direct certification
2. Co-certification
3. Local certification

Certification Process

The main activities of certification programme are setting of standards, inspection and certification. The certification agency sets its own standards based on EC regulations/ IFOAM standards or country regulation. The producers make contact with certification agency for certification of farms. The certification agency provides information on the standards, fees, application, inspection, certification and appeal procedures to the producers before formal application is made. The producers have to furnish the following information along with the formal application.

- Information on products requested for certification
- Three year field history
- Acerage
- Farm maps
- Source of seed/planting material
- Fertility management
- Pest/Disease/Weed management
- Record keeping system
- Storage system
- Equipment
- Other areas

Once the formal application is made, the certification contract is executed regarding scope, obligations to be met, inspection and certification, sanction and appeals, duration of the contract, certification mark, fee structure etc. between the producer and the certification agency. Then the certification agency conducts the inspection for compliance with standards including risk assessment. The inspector after the inspection submits his report with documents to the certifying agency with recommendations. The certifying agency reviews the report and issues approval or denial for certification as the case may be. The certificate issued is valid for the current harvest. To renew the certificate the inspections are to be carried out on an annual basis. Organic certification programme is being further developed with the growth or organic market worldwide.

Inspection Programme

The normal frequency of inspection in organic certification is once in a year. For different types of production, different types of inspection regimes will be needed. It is recommended to develop an inspection programme based on risk analysis, instead of just relying on annual inspection. The risk analysis may lead to more inspection or longer time spent on operation with larger risks. During the inspection the following techniques are used.

a. Interview with for the production responsible persons

b. Inspection of fields, premises, processing equipment, storage etc

c. Inspection of paper work, book keeping, weighing tags etc

d. If judged necessary by the inspector, samples are taken for residue analysis

Inspection Procedures

Grower interview

Complete farm inspection paper work

Farm tour

TYPES OF INSPECTIONS

Initial Inspection

It will provide the basis for both initial certification decisions and for the future inspections. It is important that all facts are clearly reported and verified. A complete inspection is conducted, and all areas of potential non compliance are identified, along with the strengths and weakness of the operation. A comprehensive report is submitted to the certification programme.

Routine Inspection

Routine inspection, as the name implies, is the most common type of inspection. These inspections are normally scheduled in advance and are conducted in co-operation with the producer (operator). All aspects of (production) operation are examined, with special emphasis given to new crops, production areas and products or issues of special concern. This type of inspection can occur more frequently than annually in production with high risk.

Random inspection

These are inspections, which occur on a random basis, and are not triggered by complaints or suspicions. These inspections are more routine inspections which are conducted on an unscheduled spot-check basis. They can also be directed to certain area of concern for the certifier. They are part of an annual quality assurance programme implemented by the certifier as a check on the inspection programme. It is recommended that the certifier decide on a certain proportion of such random inspections to be made annually. They can also be directed to certain groups of products (with high risk). When operators apply for certification, they should agree to the possibility of "surprise" inspections.

Extra inspections

Extra inspections are scheduled for a variety of reasons, including:

- The inability of the inspector to cover the operation due to unforeseen circumstances,
- Additional land, crop, livestock, products, processes or facilities requested for

certification by a previously certified operator,

- Follow-up inspection of a facility which was not fully operated for certification during an initial or routine inspection, or where the inspector failed to do his job properly,
- Follow-up on issues which need verification of compliance (certification decisions)

Announced or Unannounced Inspections

Unannounced inspections are typically targeted for special situations where complaints have been filed or suspicions exist, and where there is concern that the operator may "cover up" if he is notified in advance.

Points of Attention During Inspection

Farm(s)

- Cropping plan and rotation, area under each crop
- History of past three years of each field where organic products are grown
- Manuring and storage of manure
- Origin of seed material and planting material
- Weed, pest and disease control, also fertilizers and sprayers used in conventional crops on the region are assessed
- Harvest estimates
- Storage possibilities
- Livestock holding on the farm, (number, housing/ feeding practices)
- Book keeping in the farm

Audit Trail

An Audit trail is a record keeping system by which a crop may be traced from field to storage to sale. It is a documentation of farm activities in a systematic way. It is the procedure of proof of using acceptable management practices and approved products. It is reviewed by the inspector to verify compliance with organic standards. All certification agencies require accurate audit trail. It is especially important for split operations where both conventional and organic crops are produced. The audit trail must clearly verify that no co-mingling of crop occurs. A good audit trail can also help the producer make sound management decisions.

Procedure for Certification for Growers Group

This system is based on the internal quality system and shall apply to producer groups, farmer's cooperatives, contract production and small scale processing units. The producers in the group will apply to similar production systems and the farms are to be in geographical proximity. Farms with land holding of 4 ha and above can also belong to a group but will have to be inspected annually. The total area of such farms shall be less than 50% of the total area of the group. Processors and exporters can be a part of the same group but will have to be inspected annually.

Constitution of the Group Organization

The group will have a legal status or constitution of the organization and shall be presented by an organizational chart. For implementation of the procedures to maintain the internal control system, responsibilities shall be delegated to individual members / committees for carrying out specific activities.

Certification Criteria for Growers Group

Internal Control System (ICS)

Group Certification is based on the concept of an Internal Control System comprising of the following:

- Implementation of the Internal Control System(ICS)
- Internal Standards
- Risk assessment

An Internal Control System (ICS) is the part of a documented quality assurance system that allows an external certification body to delegate the periodical inspection of individual group members to an identified body or unit within the certified operator. This means that the third party certification bodies only have to inspect the well-functioning of the system, as well as to perform a few spot-check re-inspection of individual smallholders.

The rationale behind ICSs for group certification is two-fold:

- to facilitate smallholder certification i.e. simplify certification and reduce its cost for smallholder through coordinated documentation
- to implement and maintain a high quality assurance system for organic standards in smallholder production

Group certification enables smallholders to access organic markets and enables developing countries to commercialize their products at the international level. ICSs also provide a good basis for sound quality systems to ensure organic quality and provide consumers with organic products at reasonable cost.

The external inspection agency shall evaluate by checking the ICS documentation, Staff qualifications and re-inspecting some farms as per the group certification norms.

How to Develop an ICS?

The following are the minimum requirements for setting up an ICS for grower groups:-

- Development of Internal Control System (ICS)

- Identification of producer groups
- Creation of awareness about group certification
- Identification of qualified personnel for maintaining the internal control system
- Give necessary training in production and ICS development
- Preparation of ICS manual containing policies and procedures
- Implementation of the policies and procedures
- Review and improvement of the ICS document for maintaining a harmonized ICS.

Annexure – II

Accreditation Agencies

The National Steering Committee has designated the following statutory boards/ authority as accreditation agencies:

1. **Agricultural and Processed Food Products**
   NCUI Building 3, Siri Institutional Area
   Export Development Authority (APEDA)
   August Kranti
   Marg, New Delhi – 110016
   Tel: 91-11-26513204, 26514572, 26534186
   Fax: 91-11-26526187
   E-Mail: eadq@apeda.com
   Website: www.apeda.com

2. **Coffee Board**
   The Chairman
   Coffee BoardGovernment of India
   Ministry of Commerce & Industry#1,
   Dr. B.R.Ambedkar Veedhi
   Bangalore – 560 001,
   Tel: +91-80- 22266991-994
   Fax:+91-80-22255557

3. **Spices Board**
   The Chairman
   Spices Board, Ministry of Commerce
   Government of India
   Sugandha Bhavan
   Cochin – 682 025
   Tel: 0484-2333610; Fax: 0484 233 1429
   E mail: mail@indianspices.com
   www.indianspices.com

4. **Tea Board**
   The Chairman, Tea Board 14
   B.T.M. Sarani
   Kolkata – 700 001
   Tel: +91 33 22351411 (11 lines)
   Fax . +91 33 22215715

5. **Coconut Development Board**
   The Chairman
   Coconut Development Board
   Kera Bhavan, SRVHS Road,
   Kochi – 682011
   Tel: 91-484-2376265, 2377267, 2376553
   Fax: 91-484-2377902
   Email: cdbkochi@dataone.in , cdbkochi@gmail.com

6 **Directorate of Cashew and Cocoa Development**
   The Director
   Directorate of Cashew nut & Cocoa Development
   M/o Agriculture, D/o Agri. & Co-operation
   8th & 9th Floor, Kera Bhavan
   SRV School Road
   Cochin – 682 011
   Phone: 0484-2377151 (O)
   Fax: 0484-2377239
   E-Mail: dccd@nic.in (O)

Annexure – III

List of Accredited Certification Bodies under NPOP

Fuction in India

1. **Bureau Veritas Certification India Pvt. Ltd.**
(Formerly known as BVQI (India) Pvt. Ltd.)
Mr. R. K. Sharma, (Director)
Marwah Centre, 6th Floor
Opp. Ansa Industrial Estate
Krishanlal Marwah Marg, Off Saki-Vihar Road
Andheri (East), Mumbai – 400 072
Tel: 022-66956300, 56956311
Fax No. 022-66956302/10
Email: scsinfo@in.bureauveritas.com

2. **Ecocert India Pvt. Ltd**
Dr. Selvam Daniel (Country Representative)
Sector-3, S-6/3 & 4
Gut No. 102, Hindustan Awas Ltd.
Walmi-Waluj Road, Nakshatrawadi
Aurangabad – 431 002
Tel. No: 0240-2377120, 2376949
Fax No.: 0240-2376866
Email: ecocert@sancharnet.in
Web: www.ecocert.in

3. **IMO Control Pvt. Ltd.**
Mr. Umesh Chandrasekhar (Director)
No. 3627, 1st Floor, 7th Cross
13th G, Main, H.A.L. 2nd Stage
Bangalore-560 008.
Tel. No: +91-80-25285883, 25201546, 25215780
Fax: 0091-80-25272185
Email: imoind@vsnl.com
Web: www.imo.ch

4. **Indian Organic Certification Agency (INDOCERT)**
Mr. Mathew Sebastian, (Executive Director)
Thottumugham
P.O. Aluva, Cochin – 683 105 (Kerala)
Telefax: 0484-2630908-09/2620943
Email: info@indocert.org
Web: indocert.org

5. **Lacon Quality Certification Pvt. Ltd.**
Mr. Bobby Issac,Director
Chenathra, Theepany
Thiruvalla – 689 101
Kerala
Tel: 0469 2606447 Fax: 0469 2631902
Email: info@laconindia.com
Web: www.laconindia.com

6 **Natural Organic Certification Association (NOCA)**
Mr. Sanjay Deshmukh
Chief Executive Officer
Row House Banglow No-2
E-10 Bldg. Sun Empire, Survey No. 7, 9 (Part)
Vadgaon- Budruk, Sinhgad Road
Pune – 411051
Mobile : 9822006586/9822148609
Tel: +91-20-65218063
Fax: +91-20-24105179
E-mail: nocaindia@gmail.com
Web site: www.nocaindia.com

7. **OneCert Asia Agri Certification Pvt. Ltd.**
Mr. Sandeep Bhargava (CEO)
Plot No. 8, Pratap Nagar Colony
(Near Glass Factory and Gopalpura bypass),
Tonk Road
Jaipur – 302017
Telefax No: 0141-2701882
Email: info@onecertasia.in
Web: www.onecertasia.in

8. **SGS India Pvt. Ltd.**
Dr. Manish Pande (Head – Food Services)
250 Udyog Vihar, Phase-IV
Gurgaon – 122015
Tel: +91-124-2399990
Mobile: +91-9871794640
Fax: +91-124-2399764
Email: manish.pande@sgs.com
Web: www.sgs.com

9. **Control Union Certifications**
(Formerly known as Skal International (India)
Mr. Dirk Teichert (Managing Director)
"Summer Ville", 8th Floor
33rd – 14th Road Junction
Off Linking Road, Khar (West)
Mumbai – 400052
Tel: 022-67255396/97/98/99
Fax: 022-67255394/95
Email: cuc@controlunion.in
cucindia@controlunion.com,controlunion@vsnl.com

10. **Uttarakhand State Organic Certification Agency (USOCA)**
Dr. B.S. Bisht (Director)
12/II Vasant Vihar
Dehradun – 248 006
Tel: 0135-2760861
Fax: 0135-2760734
Email: uss_opca@rediffmail.com
ua_usoca@yahoo.co.in

11. **APOF Organic Certification Agency (AOCA)**
Mr. K. Dorairaj (CEO)
26, 1st Floor, Govindappa Road
Off D.V.G. Road, Gandhi Bazar
Bangalore – 560004
Tel: +91-80-26677275, +91-80-41203848
Mobile: 09342349255/09886019021
Email: aocabangalore@yahoo.co.in
Website: www.aoca.in

12. **Rajasthan Organic Certification Agency (ROCA)**
Mr. Yashpal Mahawat (Director)
3rd Floor, Pant Krishi Bhawan
Janpath
Jaipur – 302005
Tel: 0141-2227104
Telefax: 0141-2227456
Email: rocajpr.cb@gmail.com

13. **Vedic Organic Certification Agency**
Dr. (Mrs.) M. Usha (Managing Director)
Plot No. 55, Ushodaya Enclave
Mythrinagar, Miyanagar
Hyderabad – 500050
Mobile: 09290450666
Tel: 040-65276784; Fax: 040-23045338
Email: voca_org@yahoo.com
usha_preetham@yahoo.co.in

14. **ISCOP**
Prof. Dr. K. K. Krishnamurthi (President)
Rasi building, 162/163, Ponnaiyarajapuram
Coimbatore – 641001
Mobile: 0-9443243119
Tel: 0422-2544199; 0422-6586060
E-mail: iscop_cbe@yahoo.in; profdrkkk@yahoo.com
Website: www.iscoporganiccertification.org

15. **Food Cert India Pvt. Ltd**
Mr. Srihari Kotela (Director)
Plot No: 452, H. No. 8-2-601/P/6
Road No: 10
Banjara Hills, Panchavathy Colony
Hyderabad – 500034
Tel: 91-40-23301618, 23301554, 23301582
Fax:91-40-23301583
Website: www.foodcert.in

16. **Aditi Organic Certifications Pvt. Ltd**
Mr. Narayana Upadhyaya (Director)
No. 531/A, Priya Chambers
Dr. Rajkumar Road
Rajajinagar, 1st Block
Bangalore – 560010
Tel: +91-80-32537879
Fax: +91-80-23373083
Mobile: +91-9845064286
Email: aditiorganic@gmail.com
Website: www.aditicert.net

Annexure – IV

Sample Form – 1

Application Form for Organic Certification

In order to help to estimate the cost of inspection and certification, please supply the following in-formation, by indicating as many details as possible and omitting items that do not apply.

1	**Company Name and legal form of the company:**		
2	**Responsible Person:**		
3	**Address** **Street and No:**		**P.O. Box:**
	Postal Code:	City	
	Department:	Country:	
	Phone:	Fax:	Email:
4	**Type of Operation to be Certified**: Farm (Crop Production) ☐ Farmers Group ☐ Wild Collection ☐ Beekeeping ☐ Group of Beekeepers ☐ Animal Husbandry ☐ Food Processing ☐ Export ☐		
4.1	**Crop Production**: Farm Location: Number of farms: Total hectares: Distance from farm(s) to address given under (3) above: km		
	Crop	Hectares	Approx. quantity harvested per year
4.2	**Animal Husbandry:** Farm location: No. of farmers Kind of animals: No. of animals: Approximate yearly production: No. of processing units (incl. packaging activities, etc.): Distance from farm(s) to address given under (3) above: km		
4.3	**Wild Collection:** Collection site: Total approximate collection area (km^2): No. of collectors: No. of local wholesalers: No. of processing units (e.g. freezing, drying): Distance from collection area to address given under (3) above: km		
	Wild species collected (please attach a list, if not enough space)		Approx. quantity harvested /year
4.4	**Beekeeping:** Location of hives: Total No. of beekeepers: No. of apiaries: No. of colonies: Approximate total production/year: No. of processing units (including packaging activities, etc.): Distance from apiaries to address given under (3) above: km		
4.5	**Food Processing:** Number of units: Distance from processing unit(s) to address given under (3) above: km **Type of processing:**		
	Raw material	Final Product	Annual capacity for organic
4.6	**Export:** Export office is same address as under (3): ☐ Other address: Exporter stores, packs or labels products: ☐ Exporter only handles paperwork: ☐		
5	Standard(s) for which you wish to become certified: ☐ **NPOP**, Government of India: http://www.apeda.com/organic ☐ **NOP** Final Rule, for the organic US-market: http://www.ams.usda.gov/nop/NOP/standards.html		
6	Do you have a copy of the standard(s) according to which you request certification? Hardcopy: ☐ Access through internet: ☐ No copy: ☐		
7	Have the **above mentioned units/products** ever been inspected and/or certified before? If so, Please enclose all information regarding the inspection(s) and/or certification(s), including reports of findings etc. What was the reason for termination of the contract with regard to the inspection and/or certification mentioned above?		

Date: Signature and Company stamp:

This part has to be filled in by ADITI! Examination of the application

1	The certification requirements have been defined clearly: Yes ☐ No ☐
2	Any differences between applicant and ADITI about cert. procedure have been cleared: Yes ☐ No ☐
3	ADITI is able to perform the cert. service (incl. aspects like domicile, language, and any other specific requirements) and application is accepted: Yes ☐ No ☐
4	Comments:

Date: Signature:

SAMPLE FORM – 2

APPLICATION FORM

Certification programmes Organic

(EU, USDA/NOP, JAS, COR*)* and Input

FOR CONTROL UNION CERTIFICATIONS (CU) INSPECTION & CERTIFICATION

NAME OF COMPANY APPLYING AND LEGAL STATUS (please use full name of company with e.g. Ltd., Inc, SAC, SA, SARL, BV)	***ADDRESS OF COMPANY*** (Street, post code, town, province, country, P.O. Box)
COMPANY'S LEGAL REPRESENTATIVE (name of person and function)	**COMPANY'S CONTACT PERSON** (complete if different from Legal Representative)
TELEPHONE / FAX NUMBER	***EMAIL*** *(and/or website)*

☒	**Applying for the first time**
☒	**Indicate changes** if you are already a CU client

Applying for organic certification program according to the requirements of:

☒	***European Union (EU 834/2007, 889/2008)***
☐	***United States (USDA/NOP)***
☒	***Japan (JAS)***
☒	***Canada (COR)***

Applying for certification program EU Input (based on EU 834/2007, 889/2008):

☒	**Crop protection material**
☒	**Fertilizers and Soil Improvement Materials**

Applying for certification program **Pet Food** (based on EU 834/2007, 889/2008):

☒	

1. Products

Please describe below the products you want to have certified. If it concerns changes mention all the products and indicate which one is **changed, added** or **withdrawn**.

Name Product	**Changed, added or withdrawn** (if applicable)

Additional for COR: in case you have products containing less than 70% organic ingredients, please indicate to CU.

2. Agricultural Units

Please describe below the activities of all agricultural units. If it concerns changes mention all the units and indicate which one is **changed, added** or **withdrawn**.

Name and address of unit	**Ha**	**Nr. of farmers***	**Changed, added or withdrawn**(if applicable)

* in case of Small Farmers Group

3. Processing units

Please describe below the activities of all processing units, including the central administration office. If it concerns changes mention all the units and indicate which one is **changed**, **added** or **withdrawn**.

Name and address of unit	**Process (-es)** (e.g. storage washing, cutting, selection, packing)	**Changed, added or withdrawn** (if applicable)

4. Please describe the location of the production/processing units, e.g. travel time between the different units, if applicable

Travel time between units, if applicable estimated time needed for travelling from nearest international airport etc.

5. Has the project and/or any farmers of it ever been registered, inspected or certified before by another Inspection/Certification Body?

If YES, please mention: the name of the inspection and/or certification body, year of application, the previous registration number, reason of changing inspection/certification body. Please enclose relevant documents concerning the previous inspection(s) / certification(s) (inspection report, certificate etc.)

Undersigned declares to have completed this Application Form truthfully

NAME COMPANY
LEGAL REPRESENTATIVE (The person of the company, accepted by the Chamber of Commerce registration as assignment authorized.)

DATE & SIGNATURE

Based on the above information, CU will draw up a no-obligation offer for a contract.

CU OFFICE USE ONLY! REVIEW of APPLICATION ***This part can be completed in CUSI as well***	Yes/No	*Comment*
Is the information in application form sufficient to make an offer (e.g. filled in completely, signature present and signed by the company requesting certification)?		(if no, give reasons)
Is it possible for CU to inspect and certify the project (e.g. sufficient inspection and expert inspector capacity, product within scope)?		(if no, give reasons)
For JAS: Applicant does not fall into any of the categories as described in the CU Business Rules to be prohibited to give to offer to?		
Conclusion: Can an offer be made to the Company? If one or more answered with NO, offer is not possible!		
Assessed by:		Date:

Sample Form – 3

APPLICATION FORM – CROP PRODUCTION

(All questions according to NSOP, USDA NOP and EU Reg. (EC) No 834/2007& 889/2008 except if marked)

(Kindly fill in English only)

Note: Please take note that validity period of this application expires after 3 months from the date of its issue. Kindly request for a new application from INDOCERT if the validity period has expired)

SECTION 1: General Information	
Name:	**For Office Use:** Issued on: _/_/_ Received on:_/_/_
Address: City: State:	Reviewed on: _/_/_ Initials: ______ Date of Inspection: _/_/_ INDOCERT No:
Phone/Fax:	Email:
Address for correspondence (if different from the above address): ☐ By courier ☐ By post ***Please note that the certificate will be sent to the correspondence address only.***	
Do you belong to any farmers organiza tion, ☐ Yes ☐ No If yes, name of the organization: Responsible person for the organization:	

Farm to be certified according to:

☐ Indian standards (NSOP) ☐ European standards (rules equivalent to Reg. EC No.834/2007 & 889/2008)
☐ American standards (USDA NOP) ☐ Others, specify………..

Have you applied for organic certification earlier: ☐ Yes ☐ No

If yes, please furnish the below mentioned details along with this application form:

1. Name of the Certification agency:
2. Year in which certification was applied for:
3. Result of certification (please enclose copy of the certificate).
4. Please enclose details on non compliances found, if any.
5. Please enclose details on the corrective actions taken by you to correct non compliances with evidence.

Indian Standards are available from **APEDA, Tea Board, Coffee Board, Spices Board, INDOCERT** office or can be downloaded from the following site www.apeda.com and the EU standards can be downloaded from the following link

http://eur-lex.europa.eu/en/index.htm

The latest version of NOP Program standards can be downloaded from:

http://www.ams.usda.gov/nop/NOP/standards.html

☐ Please send me a print copy of the Indian Standards (price: Rs.75/)

in English ☐ in Malayalam ☐

☐ Please send me a print copy of the European standards
in English (price: Rs.100/-) ☐ in Malayalam (price: Rs.200/-) ☐

☐ Please send me a print copy of the American standards (Rs.100/-)

SECTION 2: Farm Plan Information

How many farms do you have? ____________(Nos)

Total area of own farm: ____________(acres)

Area of organic farming section: ____________(acres)

Area of conventional farming: ____________(acres)

Do you manage any farm not owned by you? (Eg: of parents, sister, brother or any relative) ? Yes ? No

If Yes, Area in acres managed organically: ____________(acres)

If Yes, Area in acres managed conventionally: ____________(acres)

Total area to be brought under certification ____________(acres)

Please fill in Annex 1 of the application form:

Please list **all** crops grown by you, add separate sheet if needed

Name of the Crop	**Certification needed (yes/no)**	**Intercrop (if any)**	**Area under crop or No. of plants for tree crops (please specify No. of bearing and Non Bearing plants)**		**Expected crop yield (only organic crop)**
			Under organic cultivation	**Under conventional cultivation**	

Section: 3 Planting Material

What kind of planting materials did you use?

Certified Organic ☐ Conventional, untreated ☐ Conventional, treated ☐

Please explain in detail any seed/planting material treatment done:

Section 4: Soil fertility and crop management

Please tick the box against methods followed by you to maintain/improve the soil fertility:

☐ Crop rotation ☐ Intercropping ☐ Cover crops
☐ Incorporation of plant residue ☐ Green manuring ☐ Mulching
☐ Farm yard manure ☐ Compost ☐ Bio-dynamic preparations

Apart from the above I am also practicing the following methods:

____

____

____

Please list all manures, fertilizers, pesticides etc. brought from outside the farm.

Name	Purchased from	Crops on which it was used	Quantity (kgs) used & no of times it is used/year	Reason for its use
Manures and fertilizers				
Pesticides and bio-control				
Others				

Do you apply sewage sludge: ☐ Yes ☐ No

Is there soil erosion in your farm ☐ Yes ☐ No

The following measures are taken to control soil erosion:

Is your farm: Irrigated ☐ Rain fed ☐

Describe the **methods** followed by you to keep the **weeds** in your farm **under control**:

____

____

Describe the **methods** used by you for **pest/disease** management on your farm to keep the pest/disease **under control** **(if required use additional sheet)**

____

____

____

Section: 5 Processing

On-farm Processing:

Do you process any of your products on your farm: ☐Yes ☐No (eg. cleaning, cutting, drying, smoking etc.)

If yes, what is the method used and name the product of processing?

Off farm processing and Subcontracting:

Do you further process any of your farm produce outside your farm or by a contracted party?

☐ Yes ☐No

If yes, what are the methods of processing and what are the end products?

Name and address of the contracted party:

Are you aware that the off-farm processing units and storage facilities must be inspected?

☐Yes ☐No

Section 6: Harvest, Storage, Marketing and Transport

When do you harvest your main crops? ______________________________(Months)

How do you store your harvested produce?

☐ Wooden box ☐ Card board box ☐ Gunny bags ☐ others: __________

How many storage units/rooms do you have?

Where are they located? ______________________________

____

Do you have the same storage unit for organic and conventional produce? ☐ Yes ☐ No

Do you share your storage area with others? ☐ Yes ☐ No

How do you clean your storage rooms? ______________________________

What are the types of packaging materials used?

____

Where do you store the organic inputs for your farm?

Where do you store the conventional inputs for your farm?

Do you transport your products? ☐ Yes ☐ No

How do you transport your products? ☐ Own ☐ Contracted party ☐ Done by the buyer

Mode of transport?

How do you market your produce? To whom do you sell your products?

____

____

____

SECTION 6 Animal Husbandry					
Certification requested for animal or animal products: ☐ Yes ☐ No					
How many animals do you have:					
Cows:	___Nos.	Buffaloes:	___Nos.	Goats:	____Nos.
Cow calves:	___Nos.	Buff. calves:	___Nos.	Others: __________	____Nos.
What is the feed used for your animals?					
Do you feed maize, cotton or soya products ? If yes, please state origin and status.					
Do you buy animals from outside? If yes, please state the number, origin and status?					
Please describe the housing conditions for all mentioned animals?					
How is the dung/urine of the animals utilized? Do you sell it from your farm?					
Do you give your animals temporarily to other farmers for rearing? ☐ Yes ☐ No					
Mode of slaughtering the animals? None ☐ On-farm☐ Off-farm ? ☐					

Please describe in detail how you identify/differentiate the animals and the animal products during transportation and marketing?

Please list all kind of animal products that you want to market as organic.

The following documents may be kept ready at the time of inspection:

1. A detailed farm map showing the different plots, crops grown, house, cattle sheds, biogas plant, storage units, compost units etc.
2. All bills/invoices for purchase of all inputs to your farm
3. All bills/invoices for all sales of produce from your farm.

Declaration:

I declare that:

I assume the responsibility in understanding the contents of this application form provided in English and henceforth approve the certification body in adopting the above language in all of its correspondence and the inspection and certification related documents.

I was practicing organic farming since the last ____________________months/ years and the last date of chemical application was on ____________

I do hereby affirm that all information supplied to INDOCERT is true and accurate. I affirm my commitment and responsibility to know the respective Organic Standards as indicated in section 1 and to comply my crop production according to that standards and also keep updated records for all activities.

If the organic production rules are violated, I agree to be sanctioned according to the INDOCERT Scale of Sanctions.

If major changes in the organic farming system occur, e.g. purchase or disposal of land, I will inform INDOCERT immediately. I agree not to release any products resulting from these changes until the certification body has notified me accordingly.

I agree to keep a record of complaints about my organic farming activities, take appropriate action with respect to such complaints and document the action taken.

Name and Signature of the operator:........................

Place and Date:.....................

Annex:1

OPERATOR INFORMATION: LIST OF PLOTS (all plots must be listed corresponding to the farm map)				
No. of plot (according to survey no in map)	**Name of plot**	**Crops cultivated**	**Status (conventional, organic, in conversion)**	**Remarks**

SAMPLE FORM – 4

Part 1: Information format	
Name:	Enquire Received on:
Address: City: State:	Enquire Reviewed on: _________ Date of Inspection:_______________ Inquiry No:
Phone/Fax:	E-mail:

Do you belong to any farmers organization, ☐ Yes ☐ No

If yes, name of the organization:

Responsible person for the organization:

Farm to be certified according to:

Indian standards ☐NOP USDA ☐ Other

Have you applied for certification earlier: ☐ Yes ☐No

If yes, please attach copy of certificates/reports of the inspections of previous years.

Are you aware of the relevant Indian organic standards? ☐ Yes ☐ No

Indian Standards are available from NOCA office or can be downloaded from the following site www.apeda.com

☐ Please send me a print copy of the Indian Standards

in English ☐ in Regional Language. ☐

To visit your farm, please indicate distance and time required to reach from Pune:

Part 2: Farm Plan Information

How many farms do you have? ____________(Nos)

Area of own farm: ____________(acres)

Area of organic farming section: ____________(acres)

Area of conventional farming: ____________(acres)

Do you manage any farm not owned by you? (eg: of parents, sister, brother or any relative) ☐ Yes ☐ No

If Yes, Area in acres managed organically: ____________(acres)

If Yes, Area in acres managed conventionally: ____________(acres)

Are you interested in converting: ☐ whole farm ☐ part of your farm to organic farming:

Please fill in Annex 1 of the application form:

Please list **all** crops grown by you, add separate sheet if needed

Name of the Crop	Certification needed (yes/no)	Intercrop (if any)	Area under crop or No. of plants for tree crops (please specify No. of bearing and non-bearing plants)		Expected crop yield (only organic crop)
			Under organic cultivation	Under conventional cultivation	

Part : 3 Planting Material

What kind of planting materials did you use?

Certified Organic ☐ Conventional, untreated ☐ Conventional, treated ☐

Please explain in detail any seed/planting material treatment done:

If you are using conventional/conventional treated planting material please enclose a letter/note from shopkeepers regarding its non-availability in organic status.

Part 4: Soil fertility and crop management

Please tick the box against methods followed by you to maintain/improve the soil fertility:

☐ Crop rotation ☐ Intercropping ☐ Cover crops
☐ Incorporation of plant residue ☐ Green manuring ☐ Mulching
☐ Farm yard manure ☐ Compost ☐ Bio-dynamic preparations

Apart from the above I am also practicing the following methods:

____

____

____

Please list all manures, fertilizers, pesticides etc. brought from outside the farm.

Name	Purchased from	Crops on which it was used	Quantity (kgs) used & no of times it is used/year	Reason for its use
Manures and fertilizers:				
Pesticides and bio-control:				

Off farm processing:

Do you further process any of your farm produce outside your farm or by a contracted party?

☐ Yes ☐ No

If yes, what are the methods of processing and what are the end products?

Name and address of the contracted party:

Are you aware that the off-farm processing units and storage facilities must be inspected?

☐ Yes ☐ No

Part 5: Harvest, Storage and Marketing

When do you harvest your main crops?

_______________________________________(Months)

How do you store your harvested produce?

☐ Wooden box ☐ Card board box ☐ Gunny bags ☐ others: ____________

How many storage units/rooms do you have?

Where are they located?

Do you have the same storage unit for organic and conventional produce? ☐ Yes ☐ No

Do you share your storage area with others? ☐ Yes ☐ No

How do you clean your storage rooms?

____

What are the types of packaging materials used?

____

____

Where do you store the organic inputs for your farm?

Where do you store the conventional inputs for your farm?

How do you market your produce? To whom do you sell your products?

____

____

____

Part : 6:Animal Husbandry

Certification requested for animal or animal products: ☐Yes ☐No

How many animals do you have:

Cows: ___Nos. Buffaloes: ___Nos. Goats: ____Nos.

Cow calves: ___Nos. Buff. calves: ___Nos. Others: __________ ____Nos.

What is the feed used for your animals?

Do you feed maize, cotton or soya products? If yes, please state origin and status.

Do you buy animals from outside? If yes, please state the number, origin and status?

Please describe the housing conditions for all mentioned animals?

How is the dung/urine of the animals utilized?

Do you sell it from your farm?

Do you give your animals temporarily to other farmers for rearing? ☐ Yes ☐ No

Mode of slaughtering the animals? None On-farm off-farm?

Please describe in detail how you identify/differentiate the animals and the animal products during transportation and marketing?

Please list all kind of animal products that you want to market as organic.

The following documents may be kept ready at the time of inspection:

- A detailed farm map showing the different plots, crops grown, house, cattle sheds, biogas plant, storage units, compost units etc.
- All bills/invoices for purchase of all inputs to your farm
- All bills/invoices for all sales of produce from your farm.

Declaration:

I declare that:

I am practicing organic farming since the last ____________________months/ years and the last date of chemical application was on _____________

I do hereby affirm that all information supplied to NOCA is true and accurate. I affirm my commitment to meet the respective Organic Standards as indicated in section 1.

If the organic production rules are violated, I agree to be sanctioned according to the NOCA Scale of Sanctions.

If major changes in the organic farming system occur, e.g. purchase or disposal of land, I will inform INOCA immediately. I agree not to release any products resulting from these changes until the certification body has notified me accordingly.

I agree to keep a record of complaints about my organic farming activities, take appropriate action with respect to such complaints and document the action taken.

Signature of the farmer:........................

Place and Date:.....................

OPERATOR INFORMATION: LIST OF PLOTS (all plots must be listed corresponding to the farm map)				
No. of plot (According to survey no in map)	**Name of plot**	**Crops cultivated**	**Status (conventional, organic, in conversion)**	**Remarks**

DATA FOR ORGANIC PRODUCTS (2007-08)

Total production	9,76,646 M.T.
Total quantity exported	37,533 M.T
Value of total export	USD 100.4 million (Rs. 498 Crores)
Total area under Certification (including wild harvest)	2.8 million hectares (28 lakh Hec)
Total area under certified organic cultivation	0.45 million hectares (4.5 lakh Hec)
Share of Exports to total Production	4% approx.
Increase in Export Value over previous year	30% approx.

Commodity Wise Export Data

Commodity	Export Contribution (of volume)	Export Contribution (of value)	Export Contribution (in Value)
Cotton	43%	25%	USD 26.36 million (Rs.123.88 Cr)
Basmati Rice	15%	13%	USD 12.60 million (Rs.59.20 Cr)
Honey	11%	10%	USD 9.87 million (Rs.46.41 Cr)
Tea	8%	20%	USD 19.6 million (Rs.92.13 Cr)
Dry Fruits	7%	18%	USD 17.94 million (Rs.84.31 Cr)
Processed Food	5%	4%	USD 3.83 million (Rs.17.99 Cr)
Sesame	4%	2%	USD 1.94 million (Rs.9.13 Cr)
Spices	3%	4%	USD 4.27 million (Rs.20.09 Cr)
Medicinal & Herbal plants/products	2%	2%	USD 2.25 million (Rs.10.59 Cr)
Others	2%	2%	USD 1.75 million (Rs.5.05 Cr)